T0207429

Lecture Notes in Mathematics

Volume 2353

This series reports on new developments in all areas of mathematics and their applications - quickly, informally and at a high level. Mathematical texts analysing new developments in modelling and numerical simulation are welcome. The type of material considered for publication includes:

1. Research monographs
2. Lectures on a new field or presentations of a new angle in a classical field
3. Summer schools and intensive courses on topics of current research.

Texts which are out of print but still in demand may also be considered if they fall within these categories. The timeliness of a manuscript is sometimes more important than its form, which may be preliminary or tentative. Please visit the LNM Editorial Policy (https://drive.google.com/file/d/1MOg4TbwOSokRnFJ3ZR3ciEeKs9hOnNX_/view?usp=sharing)

Titles from this series are indexed by Scopus, Web of Science, Mathematical Reviews, and zbMATH.

Viorel Barbu • Michael Röckner

Nonlinear Fokker-Planck Flows and their Probabilistic Counterparts

Viorel Barbu
Alexandru Ioan Cuza University
Iasi, Romania

Octav Mayer Institute of Mathematics of
Romanian Academy
Iasi, Romania

Michael Röckner
Mathematics Faculty
University of Bielefeld
Bielefeld, Germany

Academy of Mathematics and Systems
Science Chinese Academy of Sciences
(CAS)
Beijing, China

ISSN 0075-8434 ISSN 1617-9692 (electronic)
Lecture Notes in Mathematics
ISBN 978-3-031-61733-1 ISBN 978-3-031-61734-8 (eBook)
https://doi.org/10.1007/978-3-031-61734-8

This Springer imprint is published by the registered company Springer Nature Switzerland AG
The registered company address is: Gewerbestrasse 11, 6330 Cham, Switzerland

Preface

This work is devoted to the existence, uniqueness and the long-time behaviour of solutions $u = u(t,x)$ to the Cauchy problem for the nonlinear Fokker–Planck equation

$$\frac{\partial}{\partial t} u - \sum_{i,j=1}^{d} D_{ij}^2(a_{ij}(x,u)u) + \mathrm{div}(b(x,u)u) = 0$$

with main emphasis on the special isotropic case

$$\frac{\partial u}{\partial t} - \Delta\beta(u) + \mathrm{div}(D(x)b(u)u) = 0.$$

The implications to the existence and uniqueness theory of probabilistically weak McKean–Vlasov stochastic differential equations are also treated in some detail. These equations are unanimously recognized nowadays as fundamental instruments to understanding the evolution of complex systems in disordered media as well as of mean-field processes. The key result for these equations is that, under appropriate assumptions on the nonlinear diffusion and drift terms, there is a selection $u = S(t)u_0$ in the class of weak (mild) solutions such that $S(t) : L^1(\mathbb{R}^d) \to L^1(\mathbb{R}^d)$ is a continuous semigroup of contractions in $L^1(\mathbb{R}^d)$ and this is by definition a *nonlinear Fokker–Planck flow*. In general, this semigroup is not unique but it captures, however, the essential features of the Fokker–Planck dynamics and will be referred to as the nonlinear Fokker–Planck flow. By the so-called *superposition principle*, this semigroup defines a flow of probabilistically weak solutions to corresponding McKean–Vlasov stochastic differential equations. The uniqueness of distributional solutions to the nonlinear Fokker–Planck equations in $L^1(\mathbb{R}^d)$ as well as the asymptotic behaviour of solutions and, in particular, the H-theorem, and the existence of invariant attractors are studied in some details.

This book is based on several papers the authors published in the last 5 years, and we would like to thank Elena Mocanu from the Octav Mayer Institute of Mathematics of Romanian Academy (Iaşi Branch) for her excellent type job.

Thanks are also due to anonymous referees for carefully checking this work and helping us to correct numerous inaccuracies and misprints.

Last but not least, we are grateful for the support from German Science Foundation (DFG) through CRC 1238, and to Romanian Ministry of Research, Innovation and Digitization through CNCS-UEFISCDI projects within PN-III-P4-PCE-2021-0006.

Iaşi, Romania
Bielefeld, Germany
March 2024

Viorel Barbu
Michael Röckner

Contents

Symbols and Notation

$\mathbb{R}^d$	the d-dimensional Euclidean space
$\mathbb{R}$	the real line $(-\infty, +\infty)$
$\mathbb{R}^+$	the half line $[0, +\infty)$
$x \cdot y$	the dot product of vectors $x, y \in \mathbb{R}^d$
$\lvert \cdot \rvert_X, \ \lVert \cdot \rVert_X$	the norm of a linear normed space X
∇f	the gradient of the map $f : X \to Y$
$\nabla \cdot f$	the divergence of vector field $f : \mathscr{O} \to \mathbb{R}^d \subset \mathbb{R}^d$
$L(X, Y)$	the space of linear continuous operators from X to Y
$\lVert \cdot \rVert_{L(X,Y)}$	the norm of $L(X, Y)$
$X^*, \ X'$	the dual of the space X
$(x, y), \ (x, y)_H$	the scalar product of the vectors $x, y \in H$ (a Hilbert space). If $x \in X, \ y \in X^*$, this is the value of y at x.
$D(A)$	the domain of the operator A
$A^*, \ A'$	the adjoint of the linear operator A
A^{-1}	the inverse of the operator A
$\{\Omega, \mathscr{F}, \mathbb{P}\}$	the probability space
$y'(t), \frac{dy}{dt}(t)$	the derivative of the function $y : [0, T] \to X$
$W^{1,p}([0, T]; X)$	the space $\{y \in AC([0, T]; \ y' \in L^p(0, T; X)\}$
$C([0, T]; X)$	the space of all continuous functions from $[0, T]$ to X
$C_w([0, T]; X)$	the space of all weakly continuous functions from $[0, T]$ to X
e^{At}	the C_0-semigroup generated by A
$H^k(\mathscr{O})$	Sobolev spaces of order k on $\mathscr{O} \subset \mathbb{R}^d$, $1 \leq p \leq \infty$

Chapter 1
Introduction

Nonlinear Fokker–Planck equations (NFPEs) describe in the *statistical mechanics* and, respectively, in the *mean field* theory [70], the dynamics of a set of interacting particles or of many body systems (See [53, 54]). These equations, which will be briefly described below, play a fundamental role in the study of nonlinear diffusion systems and in stochastic analysis as well. In particular, these equations are suitable to describe the so called *anomalous diffusion*, that is, the motion of particles in disordered media. Here, we shall study the following general form of NFPEs to describe the dynamics of the time-point probability density $u(t, x)$, which in physical models represents the spatial distributions of particles,

$$\begin{aligned}
&\frac{\partial u}{\partial t}(t, x) - \sum_{i,j=1}^{d} D_{ij}^2(a_{ij}(x, u(t, x))u(t, x)) + \operatorname{div}(b(x, u(t, x))u(t, x)) = 0, \\
&\qquad\qquad\qquad\qquad (t, x) \in (0, \infty) \times \mathbb{R}^d, \\
&u(0, x) = u_0(x), \; x \in \mathbb{R}^d, \; d \geq 1,
\end{aligned} \tag{1.1}$$

where $u_t = \frac{\partial u}{\partial t}$, $D_{ij}^2 = D_i D_j$, $D_i = \frac{\partial}{\partial x_i}$, $i = 1, 2, ..., d$, and $a_{ij} : \mathbb{R}^d \times \mathbb{R} \to \mathbb{R}$, $a_{ij} = a_{ji}$, $b = \{b_i\}_{i=1}^d : \mathbb{R}^d \times \mathbb{R} \to \mathbb{R}^d$ are given measurable functions. The initial data u_0 is either a bounded measure or an $L^1(\mathbb{R}^d)$ function.

We would like to emphasize here that a_{ij} and b in (1.1) can be also allowed to depend explicitly on t (see Chap. 3). But for simplicity we restrict to the time independent case in this introductory chapter.

Equation (1.1) has its origins in the *Liouville transport equation* from classical mechanics and we refer to the second term on the left-side of (1.1) as the *diffusion* term, while $\operatorname{div}(b(x, u)u)$ is the *drift* term. As a matter of fact, Eq. (1.1) can be

V. Barbu, M. Röckner, *Nonlinear Fokker-Planck Flows and their Probabilistic Counterparts*, Lecture Notes in Mathematics 2353,
https://doi.org/10.1007/978-3-031-61734-8_1

rewritten in the form of the classical continuity equation

$$\begin{aligned} &\frac{\partial u}{\partial t} + \operatorname{div} J(u) = 0 \text{ in } (0,\infty)\times\mathbb{R}^d,\\ &J_i(u) = -\sum_{j=1}^{d} D_i(a_{ij}(x,u)u) + b_i(x,u)u. \end{aligned} \tag{1.2}$$

The equation $J(u) = 0$ describes the stationary or equilibrium states of the system, while Eq. (1.1) (equivalently, (1.2)) operates far from equilibrium and describes the dynamic of the transient state.

The linear case, that is, $a_{ij} \equiv a_{ij}(x)$, $b(x,u) \equiv b(x)$, was originally introduced by Fokker and Planck [52, 85] and discovered independently by A. Kolmogorov in 1931.

A more general form of NFPE (1.1) is

$$\begin{aligned} &\frac{\partial}{\partial t}\mu_t - \sum_{i,j=1}^{d} D^2_{ij}(a_{ij}(x,\mu_t)\mu_t) + \operatorname{div}(b(x,\mu_t)\mu_t) = 0,\\ &\mu(0,\cdot) = \mu_0, \end{aligned} \tag{1.3}$$

where $\mu_t : [0,\infty) \to \mathcal{M}(\mathbb{R}^d)$ for every t is a bounded not necessarily nonnegative measure on $\mathbb{R}^d$. Equation (1.3) is considered in the distributional sense

$$\begin{aligned} &\int_0^\infty \Bigg(\int_{\mathbb{R}^d} \frac{\partial\varphi}{\partial t}(t,x)\mu_t(dx) + \sum_{i,j=1}^{d} a_{ij}(x,\mu_t)D^2_{ij}\varphi(t,x)\mu_t(dx)\\ &\quad + \sum_{i=j}^{d} b_i(x,\mu_t)D_i\varphi(t,x)\mu_t(dx)\Bigg) dt + \int_{\mathbb{R}^d} \varphi(0,x)\mu_0(dx) = 0,\\ &\qquad \forall \varphi \in C_0^\infty([0,\infty)\times\mathbb{R}^d). \end{aligned} \tag{1.4}$$

In the special case where the μ_t have densities with respect to Lebesgue measure dx, i.e., $\mu_t(dx) = u(t,x)dx$, $t \ge 0$, (1.3) reduces to

$$\begin{aligned} &\int_0^\infty \int_{\mathbb{R}^d} \Bigg(\frac{\partial\varphi}{\partial t}(t,x) + \sum_{i,j=1}^{d} a_{ij}(x,u(t,x))D^2_{ij}\varphi(t,x)\\ &\quad + \sum_{i=1}^{d} b_i(x,u(t,x))D_i\varphi(t,x)\Bigg) u(t,x)dxdt + \int_{\mathbb{R}^d} u_0(x)\varphi(0,x)dx = 0,\\ &\qquad \forall \varphi \in C_0^2([0,\infty)\times\mathbb{R}^d), \end{aligned} \tag{1.5}$$

that is, u is a *Schwartz distributional solution* to NFPE (1.1).

An important source of NFPEs of the form (1.1)–(1.3) is the description of Itô stochastic processes in terms of their probabilistic laws. Namely, if $\mu = (\mu_t)_{t\in(0,T)} \subset \mathscr{P}(\mathbb{R}^d)$ is a weakly continuous solution to (1.3), then under minimal integrability conditions on the coefficients there is a probabilistically weak solution X to the stochastic differential

$$dX(t) = b(X(t), \mu_t)dt + \sigma(X(t), \mu_t)dW(t),\ t \geq 0,$$
and, respectively, (1.6)
$$dX(t) = b(X(t), u(t, X(t)))dt + \sigma(X(t), u(t, X(t)))dW(t),\ t \geq 0,$$

on a probability space $(\Omega, \mathscr{F}, \mathbb{P})$ with filtration $(\mathscr{F}_t)_{t\geq 0}$ and $(\mathscr{F}_t)$-Brownian motion $W(t)$ with values in $\mathbb{R}^d$ such that $\mathscr{L}_{X(t)} = \mu_t$. Here, $\mathscr{L}_{X(t)}$ is the law of $X(t)$ under $\mathbb{P}$, and $\sigma = (2(a_{ij})_{i,j=1}^d)^{\frac{1}{2}}$. In fact, there is a close relationship between Eqs. (1.1) and (1.6). Indeed, by Itô's formula, under suitable conditions on b and σ, the time marginal laws $\mu_t = \mathscr{L}_{X(t)}$, $t \geq 0$, with $\mu_0 = \mathscr{L}_{X_0}$ of the solution $X(t)$, $t \geq 0$, to (1.6) satisfy a *nonlinear* Fokker-Planck equation. Indeed, for all $\varphi \in C_0^\infty([0, \infty) \times \mathbb{R}^d)$, we have

$$\int_{\mathbb{R}^d} \frac{\partial \varphi}{\partial t}(t, x)\mu_t(dx) + \int_{\mathbb{R}^d} \varphi(0, x)\mu_0(dx) + \int_0^t \int_{\mathbb{R}^d} L_{\mu_s}\varphi(s, x)\mu_s(dx)ds = 0, \tag{1.7}$$

where

$$L_\mu\varphi(x) := \sum_{i,j=1}^d a_{ij}(x, \mu)D_{ij}^2\varphi(x) + \sum_{i=1}^d b_i(x, \mu)D_i\varphi(x), \tag{1.8}$$

is the corresponding Kolmogorov operator. Equivalently,

$$\frac{\partial}{\partial t}\mu_t = L_{\mu_t}^*\mu_t \text{ with } \mu_0 \text{ given.} \tag{1.9}$$

In this work, we shall go in the opposite direction, that is, the construction of microscopic dynamics in (1.6) from NFPEs. Namely, we first solve (1.7) and, using the obtained $\mu_t, t \geq 0$, one obtains, via the superposition principle, briefly described in Chap. 5, a (probabilistically) weak solution to (1.6) with the time marginal laws of $X(t), t \geq 0$, given by these $\mu_t,\ t \geq 0$.

Equation (1.6) is called in the literature the *McKean–Vlasov equation* (or the *mean-field SDE*). In the special case $a_{ij} = \delta_{ij},\ i, j = 1, ..., d,\ b(x, u) \equiv D(x)$, Eq. (1.1) reduces to the Smoluchowski equation

$$u_t - \Delta u + \operatorname{div}(Du) = 0 \text{ in } (0, \infty) \times \mathbb{R}^d, \tag{1.10}$$

while (1.6) is just the Langevin equation

$$dX = D(X)dt + \sqrt{2}\, dW(t). \tag{1.11}$$

These equations describe the diffusion of a field generated by an external potential and it is associated with the standard Boltzmann–Gibbs theoretical model in statistical mechanics. Previously, the diffusion equation $u_t - \Delta u = 0$, which is the simplest Fokker–Planck equation, was used by A. Einstein to describe the Brownian motion and, more precisely, the dynamics of the probability density of Brownian particles. Later on, A. Fokker in 1914 and M. Planck in 1917 have proposed Eq. (1.1) to describe the evolution of Brownian particles in a fluid under the influence of friction.

In the special case $b \equiv 0$, Eq. (1.1) reduces to the *porous media equation*

$$\begin{aligned} &u_t - \operatorname{div} q(u) = 0 \text{ in } (0, \infty) \times \mathbb{R}^d, \\ &q(u) = \left\{ \sum_{j=1}^{d} D_j(a_{ij}(x, u)u) \right\}_{i=1}^{d}, \end{aligned} \tag{1.12}$$

which was extensively studied in the literature in connection with mathematical models of nonlinear diffusion processes and also of diffusion processes with phase transition (see, e.g., [56, 99]).

The McKean–Vlasov equation is also closely linked to the phenomenon of the *propagation of chaos*, which describes the limit behaviour of interacting mean-field particles.

In the following, we shall focus and treat in more details the special (isotropic) case

$$a_{ij}(x, y) \equiv \delta_{ij} \frac{\beta(y)}{y}, \quad b(x, y) \equiv D(x)b(y), \quad x \in \mathbb{R}^d, \quad y \in \mathbb{R}, \tag{1.13}$$

where $\beta : \mathbb{R} \to \mathbb{R}$ is a monotonically nondecreasing smooth function and $D : \mathbb{R}^d \to \mathbb{R}^d$ is a measurable, locally Lebesgue integrable function. Then (1.1) reduces to a nonlinear Fokker–Planck equation with the isotropic diffusion $\Delta\beta(u)$ and the drift term $D(x)b(y)$. Namely,

$$\begin{aligned} &u_t - \Delta\beta(u) + \operatorname{div}(D(x)b(u)u) = 0 \text{ in } (0, \infty) \times \mathbb{R}^d, \\ &u(0, x) = u_0(x), \quad x \in \mathbb{R}^d. \end{aligned} \tag{1.14}$$

In statistical mechanics such an equation describes the dynamics of probability densities far from equilibrium and are associated with a generalized entropy

$$S[u] = \int_{\mathbb{R}^d} F(u(x))dx,$$

where

$$F''(u) \equiv -\frac{\beta'(u)}{|b(u)u|} \leq 0, \ \forall u \in \mathbb{R}.$$

The corresponding McKean–Vlasov equation (1.6) reads in this case as

$$\begin{aligned} dX(t) &= D(X(t))b\left(\frac{d\mathscr{L}_{X(t)}}{dx}(X(t))\right)dt + \sqrt{\frac{2\beta\left(\frac{d\mathscr{L}_{X(t)}}{dx}(X(t))\right)}{\frac{d\mathscr{L}_{X(t)}}{dx}(X(t))}}\, dW(t), \\ &\qquad t \geq 0, \\ X(0) &= X_0. \end{aligned} \tag{1.15}$$

A special case of (1.14) is the classical *conservation law equation*

$$\begin{aligned} &u_t + \operatorname{div}(a(u)) = 0 \text{ in } (0,\infty) \times \mathbb{R}^d, \\ &u(0,x) = u_0(x), \quad x \in \mathbb{R}^d, \end{aligned} \tag{1.16}$$

$a \equiv a(x,u)$, which, in general, is well posed in the class of entropy solutions and its analysis plays an important role in identifying the correct notion of generalized solutions for NFPE (1.14).

As mentioned earlier, the statistical interpretation of NFPE (1.1) requires that the solution u should be nonnegative for nonnegative initial data u_0 and conserves the normalization, that is,

$$\begin{aligned} &u(t,x) \geq 0, \ \forall (t,x) \in (0,\infty) \times \mathbb{R}^d, \\ &\int_{\mathbb{R}^d} u(t,x)dx = \int_{\mathbb{R}^d} u_0(x)dx = 1, \ \forall t \geq 0, \end{aligned} \tag{1.17}$$

and so, this is the basic class where the solutions u to (1.1) will be chosen though the existence theory for problem (1.1) is applicable to a more general space of solutions.

As a matter of fact, formally, NFPE (1.1) can be written in the Banach space $\mathscr{X} = L^1(\mathbb{R}^d)$ as the Cauchy problem

$$\begin{aligned} &\frac{d}{dt}u(t) + Au(t) = 0, \ t \geq 0, \\ &u(0) = u_0, \end{aligned} \tag{1.18}$$

where $u : [0,\infty) \to L^1(\mathbb{R}^d)$, while the operator $A : D(A) \subset L^1(\mathbb{R}^d) \to L^1(\mathbb{R}^d)$ is defined as an extension in $L^1(\mathbb{R}^d)$ of the differential operator

$$A_0 y = -\sum_{k,j=1}^{d} D_{ij}^2(a_{ij}(x,y)y) + \operatorname{div}(b(x,y)y) \text{ in } \mathscr{D}'(\mathbb{R}^d). \tag{1.19}$$

(Here, $D^2_{ij} = D_i D_j$ and div are taken in the sense of the Schwartz distributions space $\mathscr{D}'(\mathbb{R}^d)$ on $\mathbb{R}^d$.) The main idea behind the existence and well-posedness of NFPE (1.1) pursued in this work is to construct such an operator A as an m-accretive operator in the space $L^1(\mathbb{R}^d)$ and apply the general existence theory for the mild (generalized) solutions to the Cauchy problem in Banach spaces (see Sect. 6.1). More precisely, under suitable hypotheses on a_{ij} and b, the operator A generates a continuous semigroup of contractions $S(t) : L^1(\mathbb{R}^d) \to L^1(\mathbb{R}^d)$ and $u(t) = S(t)u_0$ is a *mild solution* to (1.18). This means that $u(t) = \lim_{h\to 0} u_h(t)$ in $L^1(\mathbb{R}^d)$, $\forall t \geq 0$, where u_h is for each $h > 0$ the solution to the equation

$$\begin{aligned} &\frac{1}{h}(u_h(t) - u_h(t-h)) + Au_h(t) = 0, \ t \geq 0, \ x \in \mathbb{R}^d, \\ &u_h(t) = u_0, \ \forall t < 0. \end{aligned} \tag{1.20}$$

Such a function, which is also a *distributional solution* in the sense of (1.4), will be called in the following a *mild solution* to the NFPE (1.1) and, in general, it is not unique because it is dependent on the realization A of A_0 in the class of m-accretive operators on $L^1(\mathbb{R}^d)$. Since A is obtained in the following as a limit in the sense of graphs, i.e., $(I + \lambda A)^{-1}u_0 = \lim_{\varepsilon\to 0}(I + \lambda(A_0)_\varepsilon)^{-1}u_0$ in L^1, where $(A_0)_\varepsilon$ is a smooth approximation of A_0 and $S(t)u_0 = \lim_{\varepsilon\to 0} \exp(-tA_\varepsilon)$, $t \geq 0$, the solution $u = S(t)u_0$ can be viewed as a *viscosity-weak* (mild) solution to (1.1). Moreover, in many cases, which will be made precise later on, this is the unique mild solution to (1.1). Otherwise, $u(t) = S(t)u_0$ can be viewed as an autonomous flow (*Fokker–Planck flow*) in the class of mild solutions. The Fokker–Planck flow $S(t)$ describes the evolution of open systems far from equilibrium and the eventual transition to equilibrium. This is the *semigroup approach* to NFPEs to be developed in Chap. 2. Section 2.1 is devoted to the existence, via nonlinear semigroup theory, of mild solutions to NFPE (1.1) and it completes the authors' work [11] in several directions. Related results were previously established in special cases in [10, 19]. In Sect. 2.2, one proves the existence and the uniqueness of mild solutions to NFPE (1.14) under appropriate assumptions on β, D and b, while in Sect. 2.3 the uniqueness of distributional solutions to (1.14) is proved. These results were established in the authors' work [21], but previous versions are in [13, 15] (see, also, [24]). In Sect. 2.4 one proves the existence of mild solutions to NFPE (1.14) under more general assumptions on the diffusion term (β monotonically increasing and possibly degenerate at the origin). It is also proved the smoothing effect on initial data of a nonlinear Fokker–Planck flow and the existence for a Radon measure as initial data. Earlier results for nonlinear porous media equations were established by H. Brezis and A. Friedman [32] (see also [83, 84, 100]). The main reference here is the authors' work [12], but, however, there is no overlap. The case of NFPEs with fractional Laplacian is also briefly discussed in Sect. 2.6 following the authors' work [17]. In Sect. 2.9, we have used the work [8].

An important feature of the mild solution $u(t) = S(t)u_0$ is that it is structurally stable, that is, stable with respect to perturbation of coefficients of Fokker–Planck equations. This property follows from the Trotter–Kato theorem for nonlinear semigroups of contractions (Theorem 6.9 in Chap. 6) and will be frequently invoked in the following.

It should be emphasized that in order to simplify the treatment we did not put the existence theory as well as longtime behaviour of solutions into their most general setting. Many results given herein could be extended in several directions and some important problems remain to be studied. This is the case with the uniqueness of mild and distributional solutions for the nonlinear Fokker–Planck equation (2.1) and the treatment of existence and asymptotic behaviour of solutions u to Eq. (2.80) in the degenerate case.

In Chap. 3 it is studied, following [18], the existence and uniqueness of strong solutions in Sobolev spaces for nonlinear Fokker–Planck equations with time-dependent coefficients.

Chapter 4 is devoted to the convergence of transient solutions $u(t) = S(t)u_0$ of NFPE (1.14) to stationary solutions to NFPEs. In Sect. 4.1, one proves the H-theorem for the NFPE (1.14) for strictly monotone diffusion functions β and drift terms $D = -\nabla\Phi$. The approach follows the standard Lyapunov function method, but the main difficulty here is due to the fact that the mild solutions to NFPEs are not differentiable and so the standard manipulations with Lyapunov functions, which are described in the literature on the H-theorem, are formal. In Sect. 4.2, it is studied via classical method based on LaSalle–Dafermos invariance principle, the longtime behaviour of solutions in case where β is degenerate. The main result here is that the ω-limit set $\omega(u_0)$ corresponding to a mild solution, is relatively compact and the nonlinear Fokker–Planck flow $S(t)$ is an isometry on $\omega(u_0)$. The main references here are the authors works [16, 19]. As regards the literature on the convergence to equilibrium for linear Fokker–Planck equations, the works [2, 27, 28, 38, 39, 74, 92] should be primarily cited. Finally, an ergodic result for the Fokker–Planck flow is briefly discussed in Sect. 4.3 following the work [14].

It should be emphasized that, for asymptotic results established here as well as for the ergodic properties of the Fokker–Planck flow $S(t)$, the fact that $S(t)$ is a C_0-semigroup of contractions on the set $\mathscr{P}$ of all probability densities u plays a crucial role.

The *mild solution* u defined above is a weak solution to (1.1), which is appropriate though not identical with other concepts of a generalized solution for (1.1). We shall briefly describe below one of them.

Entropy Solutions to NFPE

The entropy solution was first defined by S. Kružkov [69] for the conservation law equation (1.16).

A function $u \in C([0,\infty); L^1(\mathbb{R}^d))$ is said to be an *entropy solution* to (1.16) if

$$\frac{\partial}{\partial t}|u-k| + \operatorname{div}(\operatorname{sign}(u-k)(a(u)-a(k))) \le 0 \text{ in } \mathscr{D}'(0,\infty)\times\mathbb{R}^d, \tag{1.21}$$

for all $k \in \mathbb{R}$ and $u(0,x) = u_0(x)$, $\forall x \in \mathbb{R}^d$. More generally, (1.21) can be replaced by the entropy inequality

$$\frac{\partial}{\partial t} S(u) + \operatorname{div} \eta'(u) \le 0 \text{ in } \mathscr{D}'(0,\infty) \times \mathbb{R}^d, \tag{1.22}$$

where (S, η) is an entropy pair for Eq. (1.16), that is, $S : \mathbb{R} \to \mathbb{R}$ is convex and

$$\eta(r) = \left\{ \int_0^r b_i(s) S'(s) ds \right\}_{i=1}^d, \ b_i(s) = a_i'(s), \ a = \{a_i\}_{i=1}^d.$$

The existence and uniqueness of an entropy solution was proved in [69] for $u_0 \in L^1 \cap L^\infty$ and $a = a(t,x,u)$ where a is either in $C^3([0,\infty) \times \mathbb{R}^d \times \mathbb{R})$, a_u bounded on compacts, or $a = a(u)$ with $\frac{a(u)}{u} \in L^\infty(\mathbb{R}^d)$. Later on, M.G. Crandall [42] has proven the existence of an entropy solution to (1.16) by rewriting this equation in the form (1.18) with a suitable m-accretive operator in $L^1(\mathbb{R}^d)$.

This concept was extended to NFPEs of the form (1.7) by J.A. Carillo [37] and G.Q. Chen, B. Perthame [41] (see, also, [1, 58, 59]). For the sake of simplicity, we shall recall this definition (see [41]) in the special case of Eq. (1.14) where $Db(u)u \equiv f(u)$, $f : \mathbb{R} \to \mathbb{R}^d$ and $f'(r) = a(r)$, $r \in \mathbb{R}$.

The function $u \in L^\infty((0,\infty) \times \mathbb{R}^d)$ is called an *entropy* to (1.14) if the following conditions hold

$$D_{x_i}(\beta'(u))^{\frac{1}{2}} \in L^2((0,\infty) \times \mathbb{R}^d), \ i = 1, \dots, d, \tag{1.23}$$

and for any convex function $S \in C^2(\mathbb{R})$,

$$\frac{\partial}{\partial t} S(u) + f(u) \cdot \nabla_x S(u) - \operatorname{div}(\beta'(u) \nabla_x S(u)) = -m_S - n_S \text{ in } \mathscr{D}'((0,\infty) \times \mathbb{R}^d), \tag{1.24}$$

where

$$m_S(t,x) = \int_{\mathbb{R}^d} S''(\xi) m(t,x,\xi) d\xi, \ n_S(t,x) = \int_{\mathbb{R}^d} S''(\xi) n(t,x,\xi) d\xi$$

and m, n are bounded measures on $(0,\infty) \times \mathbb{R}^d \times \mathbb{R}$.

In this formulation, the entropy solution can be seen as a renormalized solution to (1.14) in terms of DiPerna and P.L. Lions [47].

We may, equivalently, write (1.24) as

$$\begin{aligned} \frac{\partial}{\partial t} \mathscr{X}(\xi,u) + a(\xi) \cdot \nabla_x \mathscr{X}(\xi,u) - \beta(\xi) \Delta_x \mathscr{X}(\xi,u) &= \partial_\xi m(t,x,\xi) \\ &\text{in } \mathscr{D}'((0,\infty) \times \mathbb{R}^d), \end{aligned} \tag{1.25}$$

where, $\mathscr{X}$ is the so called *kinetic function*

$$\mathscr{X}(\xi, u) = \begin{cases} 1 & \text{if } 0 < \xi < u, \\ -1 & \text{if } u < \xi < 0, \\ 0 & \text{otherwise.} \end{cases}$$

In the special case of Eq. (1.16), (1.25) is the kinetic formulation of the entropy inequality (1.22) due to P.L. Lions et al. [71].

By definition, a *kinetic solution* to (1.14) is a function $u \in L^1((0,\infty); L^1(\mathbb{R}^d))$ which satisfies (1.23), (1.25) and

$$\int_{\mathbb{R}^d} \psi(\xi) n(t, x, \xi) = d\psi(u)\beta'(u) \left| \sum_{i=1}^d D_i u \right|^2, \ \forall (t, x) \in (0, \infty) \times \mathbb{R}^d,$$

for all $\psi \in C_0^\infty(\mathbb{R}),\ \psi \geq 0$ and

$$\int_0^\infty \int_{\mathbb{R}^d} (m+n)(t, x, \xi) dt\, dx \in L^\infty(\mathbb{R}).$$

It turns out that any entropy solution is a kinetic solution and, as shown in [41], if $a' \in L^\infty_{loc}(\mathbb{R}; \mathbb{R}^d)$ and $\beta' \geq 0$, $\beta' \in L^\infty_{loc}(\mathbb{R})$, then for each $u_0 \in L^1(\mathbb{R}^d) \cap L^\infty(\mathbb{R}^d)$ there is a unique entropy solution u and this result extends to all $u_0 \in L^1(\mathbb{R}^d)$ in the case of the kinetic solution.

The notion of entropy solution for Eq. (1.14) (and, more generally, for NFPE (1.1)) is derived from the classical vanishing viscosity approach of (1.16) via the regularized equation

$$\begin{aligned} &\frac{\partial u}{\partial t} - \varepsilon \Delta\beta(u) + \operatorname{div} a(u) = 0 \text{ in } (0, \infty) \times \mathbb{R}^d, \\ &u(0, x) = u_0(x), \end{aligned} \tag{1.26}$$

which has a unique strong solution u_ε.

If we take a smooth entropy function S and compute $\frac{\partial}{\partial t} S(u)$, we get

$$\frac{\partial}{\partial t} S(u_\varepsilon) - \varepsilon \Delta S(u_\varepsilon) + \operatorname{div}(S'(u_\varepsilon)\nabla\beta(u_\varepsilon)) + a(u_\varepsilon) \cdot \nabla S(u_\varepsilon) = -m_\varepsilon(t, x) - n_\varepsilon(t, x),$$

where

$$\begin{aligned} m_\varepsilon(t, x) &= \varepsilon S'''(u_\varepsilon(t, x))|\nabla u_\varepsilon(t, x)|^2 \\ n_\varepsilon(t, x) &= d S''(u_\varepsilon(t, x))\beta'(u_\varepsilon(t, x))|\nabla u_\varepsilon(t, x)|^2. \end{aligned}$$

Letting (formally) $\varepsilon \to 0$, we get for $u = \lim_{\varepsilon \to 0} u_\varepsilon$ Eq. (1.24), which implies in particular the entropy inequality

$$\frac{\partial}{\partial t} S(u) + \operatorname{div} G(u) \geq 0, \tag{1.27}$$

where

$$G(u) = F(u) - \beta'(u)\nabla_x S(u), \; F = (F_i)_{i=1}^d,$$
$$F_i'(r) \equiv a_i(r)S'(r),$$

This means (see, e.g., [50], p. 124) that u is an entropy solution to (1.14) with the entropy flux $(S(u), G(u))$. It should be emphasized that the mild solution u to (1.1) obtained here (see Sect. 2.2 below) via nonlinear semigroup theory is not necessarily an entropy solution and, also, it is not clear whether an entropy or kinetic solution is a mild solution.

As a matter of fact, though the flow $\{t \to u(t, u_0)\}$ of kinetic solutions is a nonexpansive continuous semigroup in $L^1(\mathbb{R}^d)$, it is not necessarily generated by an m-accretive operator A. When this happens, its generator A is not simply the closure of A_0 in $L^1(\mathbb{R}^d)$, but should be defined by $Au = f$ if u is an entropy (or kinetic) solution to $A_0(u) = f$. This means that the class of mild solutions constructed here is larger than that of entropy solutions and the procedure developed here allows to construct a mild solution flow for (1.1) for situations where entropy solutions cannot be found. Of course, in statistical physics the entropy solution is a convenient concept of solution, but for stochastic analysis the mild solutions are significant as well.

Notation $L^p(\mathbb{R}^d)$, $1 \leq p \leq \infty$ (denoted L^p) is the space of all Lebesgue measurable and p-integrable functions on $\mathbb{R}^d$, with the standard norm $|\cdot|_p$. $(\cdot,\cdot)_2$ denotes the inner product in L^2. By L^p_{loc} we denote the corresponding local space. Let $C^k(\mathbb{R})$ denote the space of continuously differentiable functions up to order k and $C_b(\mathbb{R})$ the space of continuous and bounded functions on $\mathbb{R}$. For any open set $\mathscr{O} \subset \mathbb{R}^d$, let $W^{k,p}(\mathscr{O})$, $k \geq 1$, denote the standard Sobolev space on $\mathscr{O}$ and by $W^{k,p}_{\mathrm{loc}}(\mathscr{O})$ the corresponding local space. We set $W^{1,2}(\mathscr{O}) = H^1(\mathscr{O})$, $W^{2,2}(\mathscr{O}) = H^2(\mathscr{O})$, $H_0^1(\mathscr{O}) = \{u \in H^1(\mathscr{O}),\ u = 0 \text{ on } \partial\mathscr{O}\}$ where $\partial\mathscr{O}$ is the boundary of $\mathscr{O}$. By $H^{-1}(\mathscr{O})$ we denote the dual space of $H_0^1(\mathscr{O})$ (of $H^1(\mathbb{R}^d)$, respectively, if $\mathscr{O} = \mathbb{R}^d$). We shall also set $H^k = H^k(\mathbb{R}^d)$, $k = 1, 2$, and $H^{-1} = H^{-1}(\mathbb{R}^d)$. For each $k \geq 1$, $C^k(\mathscr{O})$ is the space of continuously differentiable functions up to order k, and by $C_b^k(\mathscr{O})$ the space $\{u \in C^k(\mathscr{O});\ D^\alpha u \in L^\infty(\mathscr{O});\ |\alpha| \leq k\}$, where $D^\alpha = D_1^{\alpha_1} \ldots D_m^{\alpha_m}$, $|\alpha| = \alpha_1 + \alpha_2 + \cdots + \alpha_m$, $D_i = \frac{\partial}{\partial x_i}$, $i = 1, \ldots, d$. $C_0^\infty(\mathscr{O})$ is the space of infinitely differentiable real-valued functions with compact support in $\mathscr{O}$ and $\mathscr{D}'(\mathscr{O})$ is the dual of $C_0^\infty(\mathscr{O})$, that is, the space of Schwartz distributions on $\mathscr{O}$. Lip($\mathbb{R}$) is the space of real-valued Lipschitz functions on $\mathbb{R}$ with the norm denoted by $|\cdot|_{\mathrm{Lip}}$. $C([0,\infty); L^1(\mathbb{R}^d))$ is the space of continuous functions $y : [0,\infty) \to L^1(\mathbb{R}^d)$.

Let $C_0^\infty([0,\infty) \times \mathbb{R}^d)$ denote the space of all $\varphi \in C^\infty([0,\infty) \times \mathbb{R}^d)$ such that *support* $\varphi \subset K$, where K is compact in $[0,\infty) \times \mathbb{R}^d$. We shall also use the following notations:

$$\beta'(r) \equiv \frac{d}{dr}\beta(r),\ b'(r) \equiv \frac{d}{dr}b(r),\ \beta_r(x,r) = \frac{\partial}{\partial r}\beta(x,u),$$

$$y_t = \frac{\partial}{\partial t}y,\ \Delta y = \sum_{i=1}^d \frac{\partial^2}{\partial x_i^2}y,\ \nabla y = \left\{\frac{\partial y}{\partial x_i}\right\}_{i=1}^d,\ \operatorname{div} u = \sum_{i=1}^d \frac{\partial u_i}{\partial x_i},\ u = \{u_i\}_{i=1}^d.$$

In the following, we shall denote the Euclidean norm in all spaces $\mathbb{R}^d$, $d \geq 1$, by the same symbol $|\cdot|$.

By $\mathcal{P} = \mathcal{P}(\mathbb{R}^d)$ we denote the set of all probability densities u on $\mathbb{R}^d$, that is,

$$\mathcal{P} = \left\{y \in L^1(\mathbb{R}^d);\ y \geq 0,\ \text{a.e. in } \mathbb{R}^d,\ \int_{\mathbb{R}^d} y(x)dx = 1\right\}. \tag{1.28}$$

Let $\mathcal{M}(\mathbb{R}^d)$ denote the space of all signed Radon measures on $\mathbb{R}^d$ of bounded variation. A sequence $\{\mu_n\} \subset \mathcal{M}(\mathbb{R}^d)$ is said to be converging to μ in the $\sigma(\mathcal{M}(\mathbb{R}^d), C_b(\mathbb{R}^d))$ topology if

$$\lim_{n\to\infty} \int_{\mathbb{R}^d} \psi\, d\mu_n = \int_{\mathbb{R}^d} \psi\, d\mu,\ \forall \psi \in C_b(\mathbb{R}^d). \tag{1.29}$$

The function $t \to \mu(t) \in \mathcal{M}(\mathbb{R}^d)$ is said to be *weakly continuous* or *narrowly continuous* on $[0,\infty)$ if, for every $\psi \in C_b(\mathbb{R}^d)$, the function $t \to \int_{\mathbb{R}^d} \psi\, d\mu(t)$ is continuous on $[0,\infty)$.

Chapter 2
Existence of Nonlinear Fokker–Planck Flows

We shall discuss in this chapter the existence and uniqueness of mild solutions to NFPE (1.1), that is,

$$\begin{aligned} & u_t - \sum_{i,j=1}^{d} D_{ij}^2(a_{ij}(x,u)u) + \sum_{i=1}^{d} D_i(b_i(x,u)u) = 0 \text{ in } (0,\infty)\times\mathbb{R}^d, \\ & u(0,x) = u_0(x),\ x \in \mathbb{R}^d, \end{aligned} \tag{2.1}$$

with emphasis on the isotropic diffusion case (1.14), that is,

$$\begin{aligned} & u_t - \Delta\beta(x,u) + \operatorname{div}(D(x)b(u)u) = 0 \text{ in } (0,\infty)\times\mathbb{R}^d, \\ & u(0,x) = u_0(x),\ x \in \mathbb{R}^d, \end{aligned} \tag{2.2}$$

where $\beta : \mathbb{R} \to \mathbb{R}$, $D : \mathbb{R}^d \to \mathbb{R}^d$ and $b : \mathbb{R} \to \mathbb{R}$ are to be made precise below.

2.1 Existence for the NFPE (2.1)

We shall study herein the existence for (2.1), under two different sets of hypotheses on the coefficients a_{ij} and b, as follows.

(H1) $a_{ij} \in C^2(\mathbb{R}^d\times\mathbb{R}) \cap C_b(\mathbb{R}^d\times\mathbb{R})$, $(a_{ij})_x \in C_b(\mathbb{R}^d\times\mathbb{R};\mathbb{R}^d)$, $a_{ij} = a_{ji}$, $a_{ij}(x,u) = a_{ij}(x,|u|)$, $x\in\mathbb{R}^d$, $u\in\mathbb{R}$, $i,j=1,\dots,d$.

(H2) $\sum_{i,j=1}^{d} (a_{ij}(x,u) + (a_{ij})_u(x,u)u)\xi_i\xi_j \ge \gamma|\xi|^2$, $\forall \xi\in\mathbb{R}^d$, $x\in\mathbb{R}^d$, $u\in\mathbb{R}$, $\gamma > 0$.

(H3) $b_i \in C_b(\mathbb{R}^d\times\mathbb{R}) \cap C^1(\mathbb{R}^d\times\mathbb{R})$, $i=1,\dots,d$.

V. Barbu, M. Röckner, *Nonlinear Fokker-Planck Flows and their Probabilistic Counterparts*, Lecture Notes in Mathematics 2353,
https://doi.org/10.1007/978-3-031-61734-8_2

(H1)′ $a_{ij}(x,u) \equiv a_{ij}(u)$, $a_{ij} \in C^2(\mathbb{R}) \cap C_b(\mathbb{R})$, $a_{ij} = a_{ji}$, $a_{ij}(u) = a_{ij}(|u|)$, $\forall i,j = 1,\dots,d$.

(H2)′ $\sum_{i,j=1}^{d} (a_{ij}(u) + (a_{ij})_u(u)u)\xi_i\xi_j \geq 0$, $\forall \xi \in \mathbb{R}^d$, $u \in \mathbb{R}$.

(H3)′ $b_i(x,u) \equiv b_i(u)$, $b_i \in C_b(\mathbb{R}) \cap C^1(\mathbb{R})$, $b_i(0) = 0$, $i = 1,\dots,d$.

Here $(a_{ij})_u = \frac{\partial}{\partial u} a_{ij}$ and $(a_{ij})_x = \nabla_x a_{ij}$.

Remark 2.1 We note that the condition

$$a_{ij}(x,u) = a_{ij}(x,|u|),\ x \in \mathbb{R}^d,\ u \in \mathbb{R},$$

i.e., axial symmetry in u is only used in the proof of Lemma 2.3, and as can be seen from that proof can be relaxed to assuming that, for some $c \in (0,\infty)$,

$$|a_{ij}(x,u) - a_{ij}(x,-u)| \leq c|u| \text{ for all } x \in \mathbb{R}^d,\ u \in \mathbb{R},\ 1 \leq i,j \leq d.$$

On the other hand, since in applications to SDEs the solution u to (2.1) is nonnegative, the axial symmetry condition is automatically satisfied by redefining $a_{ij}(u)$ for $u < 0$.

We also note that in the case where the mild solution to (2.1) is to be found in $L^\infty((0,T)\times\mathbb{R}^d)$, then in Hypotheses (H1)–(H3) and (H1)′–(H3)′, $C_b(\mathbb{R}^d \times \mathbb{R})$ and $C_b(\mathbb{R})$ can be replaced by $C(\mathbb{R}^d \times \mathbb{R})$ and $C(\mathbb{R})$, respectively.

The first set of hypotheses, that is (H1)–(H3), allows for nondegenerate FPEs with x-dependent coefficients only, while the second set (H1)′–(H3)′ allows for degenerate nonlinear FPEs, however, with x-independent coefficients.

Now we define in the space $L^1 = L^1(\mathbb{R}^d)$ the operator $A_0 : D(A_0) \subset L^1 \to L^1$,

$$\begin{aligned} A_0(u) &= -\sum_{i,j=1}^{d} D_{ij}^2(a_{ij}(x,u)u) + \operatorname{div}(b(x,u)u),\ \forall u \in D(A_0), \\ D(A_0) &= \{u \in L^1;\ A_0(u) \in L^1\}, \end{aligned} \tag{2.3}$$

where $D_{ij}^2 = D_iD_j$, $D_i = \partial/\partial x_i$, and div are taken in the sense of Schwartz distributions on the space $\mathscr{D}'(\mathbb{R}^d)$. We note that, since $a_{ij}(x,u)u, b_i(x,u)u \in L^1$, for $u \in L^1$, the distribution $A_0(u) \in \mathscr{D}'(\mathbb{R}^d)$ is well defined. Clearly, $(A_0, D(A_0))$ is a closed operator on L^1.

We define now the *mild solution* which is the basic notion of a weak solution to NFPE (2.1) to be used in this work. (See Definition 2.4.)

Definition 2.1 The continuous function $u : [0,\infty) \to L^1$ is called a *mild solution* to NFPE (2.1) if, for each $T > 0$,

$$u(t) = \lim_{h\to 0} u_h(t) \text{ in } L^1,\ \forall t \in [0,T], \tag{2.4}$$

where $u_h : [0, T] \to L^1$ is the step function

$$u_h(t) = u_h^k,\ \forall t \in [kh, (k+1)h),\ k = 0, 1, \dots, N_h = \left[\tfrac{T}{h}\right], \tag{2.5}$$

$$u_h^{k+1} + hA_0(u_h^{k+1}) = u_h^k,\ \forall k = 0, 1, \dots, N_h;\ u_h^0 = u_0. \tag{2.6}$$

Equivalently (see (1.20)),

$$\begin{aligned}\frac{1}{h}\int_{\mathbb{R}^d}(u_h^{k+1}(x) - u_h^k(x))\psi(x)dx - \sum_{i,j=1}^d \int_{\mathbb{R}^d}(a_{ij}(x, u_h^{k+1}(x))D_{ij}^2\psi(x)\\ - b(x, u_h^{k+1}(x))u_h^{k+1}(x)\cdot\nabla\psi(x))dx = 0,\ \forall\psi \in C_0^\infty(\mathbb{R}^d),\end{aligned} \tag{2.7}$$

where $u_h : [0, \infty) \to L^1$ is given by

$$\begin{aligned}&\frac{1}{h}(u_h(t) - u_h(t-h)) + A_0u_h(t) = 0,\ \forall t > 0,\\ &u_h(t) = u_0, \qquad\qquad\qquad\qquad\qquad \forall t < 0.\end{aligned} \tag{2.8}$$

Roughly speaking, a mild solution to (2.1) is a t-continuous L^1-valued function u which is the limit of the finite difference scheme for (2.1). Of course, in order that (2.4)–(2.6) be well defined, it is necessary to have that $R(I + \lambda A_0) = L^1$, $\forall\lambda > 0$, and this problem will be treated later on.

We also note that a mild solution u to (2.1) is a distributional solution to (2.1), that is (1.5) holds. Indeed, by (2.4)–(2.6) we have, for all $\varphi \in C_0^\infty([0, T) \times \mathbb{R}^d)$,

$$\begin{aligned}&\int_0^\infty\int_{\mathbb{R}^d}\left(\sum_{i,j=1}^d a_{ij}(x, u)uD_{ij}^2\varphi + b(x, u)\cdot\nabla\varphi\, u\right)dxdt\\
&\quad = \lim_{h\to 0}\int_0^T\int_{\mathbb{R}^d}\left(\sum_{i,j=1}^d a_{ij}(x, u_h(t, x))u_h(t, x)D_{ij}^2\varphi(t, x)\right.\\
&\qquad\left. + b(x, u_h(t, x))u_h(t, x)\cdot\nabla\varphi(t, x)\right)dxdt\\
&\quad = \lim_{h\to 0}\sum_{k=1}^{N_h}\int_{kh}^{(k+1)h}\int_{\mathbb{R}^d}\left(\sum_{i,j=1}^d a_{ij}(x, u_h^k(x))u_h^k(x)D_{ij}^2\varphi(t, x)\right.\\
&\qquad\left. + b(x, u_h^k(x))u_h^k(x)\cdot\nabla\varphi(t, x)\right)dxdt\end{aligned}$$

$$= \lim_{h\to 0} \sum_{k=1}^{N_h} \int_{kh}^{(k+1)h} \int_{\mathbb{R}^d} \frac{1}{h}(u_h^k(x) - u_h^{k-1}(x))\varphi(t,x)dxdt$$

$$= \lim_{h\to 0} \frac{1}{h} \int_h^\infty \int_{\mathbb{R}^d} (u_h(t,x) - u_h(t-h,x))\varphi(t,x)dxdt$$

$$= \lim_{h\to 0} \left(\int_0^\infty u_h(t,x) \int_{\mathbb{R}^d} \frac{1}{h}(\varphi(t,x) - \varphi(t+h,x))dxdt \right.$$

$$\left. - \frac{1}{h} \int_0^h \int_{\mathbb{R}^d} u_h(t,x)\varphi(t+h,x)dxdt \right)$$

$$= - \int_0^\infty \int_{\mathbb{R}^d} u \frac{\partial\varphi}{\partial t} dxdt - \int_{\mathbb{R}^d} u(0,x)\varphi(0,x)dx,$$

and so (1.5) follows.

We shall first study equation the NFPE (2.1) under Hypotheses (H1)–(H3). For this purpose, we consider the equation

$$u + \lambda A_0(u) = f, \ f \in L^1, \ \lambda > 0. \tag{2.9}$$

We have

Proposition 2.1 *Under Hypotheses* (H1)–(H3), *there is* $\lambda_0 > 0$ *such that, for each* $f \in L^1$ *and* $\lambda \in (0, \lambda_0)$, *Eq.* (2.9) *has at least one solution,* $u \in L^1$, *that is,*

$$R(I + \lambda A_0) = L^1, \ \forall \lambda \in (0, \lambda_0). \tag{2.10}$$

More precisely, there is a family of operators $\{J_\lambda\}_{\lambda\in(0,\lambda_0)} : L^1 \to D(A_0) \subset L^1$ *such that for all* $\lambda, \lambda_1, \lambda_2 \in (0, \lambda_0)$; $f, f_1, f_2 \in L^1$,

$$J_\lambda(0) = 0, \ (I + \lambda A_0)(J_\lambda(f)) = f. \tag{2.11}$$

$$|J_\lambda(f_1) - J_\lambda(f_2)|_1 \le |f_1 - f_2|_1, \tag{2.12}$$

$$J_{\lambda_2}(f) = J_{\lambda_1}\left(\frac{\lambda_1}{\lambda_2} f + \left(1 - \frac{\lambda_1}{\lambda_2}\right) J_{\lambda_2}(f) \right), \tag{2.13}$$

$$\int_{\mathbb{R}^d} J_\lambda(f)dx = \int_{\mathbb{R}^d} f\,dx, \tag{2.14}$$

$$J_\lambda(f) \ge 0, \ \textit{a.e. in } \mathbb{R}^d \textit{ if } f \ge 0, \textit{ a.e. in } \mathbb{R}^d, \tag{2.15}$$

$$J_\lambda(L^1) \textit{ is dense in } L^1, \tag{2.16}$$

$$J_\lambda(L^1 \cap L^2) \subset H^1. \tag{2.17}$$

We postpone for the time being the proof of Proposition 2.1 and note that a mild solution u to (2.1) is just the mild solution to the Cauchy problem (see Sect. 6.2)

$$\frac{du}{dt} + A_0(u) = 0, \quad t \geq 0, \; u(0) = u_0.$$

However, the convergence of the approximating scheme (2.4)–(2.6) for $h \to 0$ requires the m-accretivity or, eventually, the quasi-m-accretivity of the operator A_0 in L^1 (see Theorem 6.5) which, in general, is not the case here. Indeed, by Proposition 2.1 it does not follow that $(I + \lambda A_0)^{-1}$ is single-valued and nonexpansive, but only that it has a section J_λ with this property. However, one can construct an m-accretive section of A_0 as follows.

Define the operator $A : D(A) \subset L^1 \to L^1$ by

$$\begin{aligned} A(u) &= A_0(u), \; \forall u \in J_\nu(f), \; f \in L^1, \\ D(A) &= \{u \in J_\nu(f), \; f \in L^1\}, \end{aligned} \tag{2.18}$$

where $\nu \in (0, \lambda_0)$ is arbitrary but fixed. By (2.13), we see that the domain $D(A)$ of A and the operator A itself are independent of ν, and therefore

$$A(u) = A_0(J_\lambda(f)), \; u = J_\lambda(f), \; f \in L^1, \; \lambda \in (0, \lambda_0). \tag{2.19}$$

Moreover, we see that $u + \lambda A(u) = u + \lambda A_0(u) = f$ if $u = J_\lambda(f)$ and, therefore,

$$(I + \lambda A)^{-1} f = J_\lambda(f), \; \forall \lambda \in (0, \lambda_0), \; f \in L^1. \tag{2.20}$$

Finally, it follows by (2.12) that A is accretive, that is (see Sect. 6.1),

$$|u - v + \lambda(Au - Av)|_1 \geq |u - v|_1, \; \forall u, v \in D(A), \; \lambda \in (0, \lambda_0),$$

and the latter extends to all $\lambda > 0$. Hence, the operator $A \subset A_0$ is m-accretive in L^1 and by (2.16) it follows also that $D(A)$ is dense in L^1, that is, $\overline{D(A)} = L^1$.

Then, by Theorem 6.5, the operator A generates a C_0-semigroup $S(t)$ of contractions on L^1, that is, for each $u_0 \in L^1$,

$$S(t)u_0 = \lim_{n \to \infty} \left(I + \frac{t}{n} A\right)^{-n} u_0 \text{ strongly in } L^1, \; \forall t \geq 0, \tag{2.21}$$

uniformly in t on compact intervals. We also have

$$\begin{aligned} &S(t+s)u_0 = S(t)S(s)u_0, \; \forall t \geq s \geq 0, \; \forall u_0 \in L^1, \\ &|S(t)u_0^1 - S(t)u_0^2|_1 \leq |u_0^1 - u_0^2|_1, \; \forall u_0^1, u_0^2 \in L^1, \; t \geq 0. \end{aligned}$$

Moreover, by (2.10)–(2.18) it follows also that

$$S(t)(\mathscr{P}) \subset \mathscr{P}, \ \forall t \geq 0.$$

The exponential formula (2.21) means that $u(t) = S(t)u_0$ is a mild solution to the Cauchy problem

$$\begin{aligned} &\frac{du}{dt} + Au = 0 \ \text{ on } (0, \infty), \\ &u(0) = u_0, \end{aligned} \tag{2.22}$$

that is (see Definition 6.2), $u \in C([0, \infty); L^1)$, $u = \lim_{h \to 0} u_h(t)$ in L^1 for each $t \in [0, T]$, $0 < T < \infty$, where

$$\begin{aligned} &u_h(t) = u_h^j, \ \forall t \in [jh, (j+1)h), \ j = 0, 1, \ldots, N = \left[\tfrac{T}{h}\right], \\ &u_h^{j+1} + hA(u_h^{j+1}) = u_h^j, \ j = 0, 1, \ldots, N, \\ &u_h^0 = u_0. \end{aligned} \tag{2.23}$$

Taking into account that $(I + hA)^{-1}u_h^j \in (I + hA_0)^{-1}u_h^j$, $\forall j$, it follows that *the mild solution* $u(t) = S(t)u_0$ *to the Cauchy problem* (2.22) *is a mild solution to NFPE* (2.1). (It should be mentioned however that, for each m-accretive operator A, the corresponding solution to (2.22) is unique (in the sense of mild solutions), while that of NFPE (2.1) given in Definition 2.1, in general, not.)

We have, therefore, the following existence theorem for NFPE (2.1).

Theorem 2.1 *Assume that Hypotheses* (H1)–(H3) *hold. Then, there is a continuous semigroup of contractions* $S(t) : [0, \infty) \to L^1$ *such that, for each* $u_0 \in L^1(\mathbb{R}^d)$, $u = u(t, u_0) = S(t)u_0$ *is a mild solution to NFPE* (2.1) *and*

$$|u(t, u_0^1) - u(t, u_0^2)|_1 \leq |u_0^1 - u_0^2|_1, \ \forall u_0^1, u_0^2 \in L^1, \ t \geq 0, \tag{2.24}$$

$$u \geq 0 \text{ a.e. in } (0, \infty) \times \mathbb{R}^d \text{ if } u_0 \geq 0 \text{ a.e. in } \mathbb{R}^d, \tag{2.25}$$

$$\int_{\mathbb{R}^d} u(t, x)dx = \int_{\mathbb{R}^d} u_0(x)dx, \forall u_0 \in L^1, \ t \geq 0. \tag{2.26}$$

Moreover, u *is a solution to Eq.* (2.1) *in the sense of Schwartz distributions on* $(0, \infty) \times \mathbb{R}^d$, *that is,*

$$\begin{aligned} &\int_0^\infty \int_{\mathbb{R}^d} (u(t, x)\varphi_t(t, x) + \sum_{i,j=1}^d a_{ij}(x, u(t, x))u(t, x)D_{ij}^2\varphi(t, x) \\ &+b(x, u)u(t, x) \cdot \nabla_x\varphi(t, x))dt\,dx + \int_{\mathbb{R}^d} u_0(x)\varphi(0, x)dx = 0, \\ &\forall \varphi \in C_0^\infty([0, \infty) \times \mathbb{R}^d). \end{aligned} \tag{2.27}$$

We shall call such a semigroup $S(t)$ a *Fokker–Planck flow generated by NFPE* (2.1) and, as mentioned above, in general it is not the unique mild solution to NFPE (2.1) because it is dependent on the operator A, which by construction is given by a family of the resolvent operators $\{J_\lambda\}$ satisfying (2.11)–(2.15). However, for reasons which will be clear later on, we also can call such a solution $u(t) = S(t)u_0$ a *viscosity solution* to NFPE (2.1). We shall also see later on that for Eq. (2.2), under certain conditions on β, D and b_0, the mild solution to (2.1) is unique.

Proof of Proposition 2.1 In the following, we shall simply write

$$a_{ij}(u) = a_{ij}(x, u),\ x \in \mathbb{R}^d,\ u \in \mathbb{R},\ i, j = 1, \dots, d.$$

We set

$$\begin{aligned} a^*_{ij}(u) &\equiv a_{ij}(x, u)u,\ x \in \mathbb{R}^d,\ u \in \mathbb{R},\ \forall i, j = 1, \dots, d,\\ b(x, u) = \{b_i(x, u)\}_{i=1}^d,\ & b^*(x, u) = b(x, u)u,\ x \in \mathbb{R}^d,\ u \in \mathbb{R}. \end{aligned}$$

We note that, by (H2), we have

$$\sum_{i,j=1}^d (a^*_{ij})_u(x, u)\xi_i\xi_j \ge \gamma|\xi|^2,\ \forall \xi \in \mathbb{R}^d,\ x \in \mathbb{R}^d,\ u \in \mathbb{R}, \tag{2.28}$$

where $\gamma > 0$. We shall first prove Proposition 2.1 under the following additional hypothesis

(K) *$(a^*_{ij})_u \in C_b(\mathbb{R}^d \times \mathbb{R})$ and, for some $C \in (0, \infty)$, $g \in L^2$ and $1 \le i, j \le d$,*

$$|(a^*_{ij})_u(x, u) - (a^*_{ij})_u(x, \bar{u})| \le C|u - \bar{u}|, \tag{2.29}$$

$$|\widetilde{b}^*(x, u) - \widetilde{b}^*(x, \bar{u})| \le g(x)|u - \bar{u}|,\ \forall u, \bar{u} \in \mathbb{R},\ x \in \mathbb{R}^d, \tag{2.30}$$

where $\widetilde{b}^*_i := b^*_i - \sum_{j=1}^d (a^*_{ij})_{x_j}$ and $\widetilde{b}^* := (\widetilde{b}^*_1, \dots, \widetilde{b}^*_d)$. □

We rewrite Eq. (2.9) as

$$u - \lambda \sum_{i,j=1}^d D^2_{ij}(a^*_{ij}(u)) + \lambda\, \mathrm{div}(b^*(x, u)) = f \text{ in } \mathscr{D}'(\mathbb{R}^d).$$

If $D_j u \in L^1_{\mathrm{loc}}$, $j = 1, \dots, d$, then this equation can be equivalently written as

$$u - \lambda \sum_{i,j=1}^d D_i((a^*_{ij})_u(u)D_j u + (a^*_{ij})_{x_j}(x, u)) + \lambda\, \mathrm{div}(b^*(x, u)) = f \text{ in } \mathscr{D}'(\mathbb{R}^d). \tag{2.31}$$

We also set

$$\begin{aligned}b_\infty &= \sup\{|b_i(x,u)|;\ (x,u)\in\mathbb{R}^d\times\mathbb{R},\ i=1,\dots,d\},\\ c_\infty &= \sup\{|(a_{ij})_{x_j}(x,u)|;\ (x,u)\in\mathbb{R}^d\times\mathbb{R},\ i,j=1,\dots,d\}.\end{aligned}$$

For each $N>0$, we set $B_N=\{\xi\in\mathbb{R}^d;\ |\xi|<N\}$. We have

Lemma 2.1 *Let $f\in L^2$ and $\lambda_0\in\left(0,\frac{1}{2}\gamma(b_\infty+c_\infty)^{-2}\right)$. Then, for all $N\in\mathbb{N}$, $\lambda\in(0,\lambda_0]$ there is at least one solution $u_N\in H_0^1(B_N)$ to the equation*

$$\begin{aligned}&u-\lambda\sum_{i,j=1}^{d} D_{ij}^2(a_{ij}^*(u))+\lambda\,\mathrm{div}(b^*(x,u))=f \text{ in } B_N,\\ &u=0 \text{ on } \partial B_N,\end{aligned}\tag{2.32}$$

and such that

$$\|u_N\|^2_{L^2(B_N)}+\lambda\gamma\|\nabla u_N\|^2_{L^2(B_N)}\le 2\|f\|^2_{L^2(B_N)}.\tag{2.33}$$

Proof We consider the operator $F:L^2(B_N)\to L^2(B_N)$ defined by $F(v)=u\in H_0^1(B_N)$, where u is the solution to the linear elliptic problem

$$\begin{aligned}&u-\lambda\sum_{i,j=1}^{d} D_i((a_{ij}^*)_v(x,v)D_ju+(a_{ij})_{x_j}(x,v)u)+\lambda\,\mathrm{div}(b(x,v)u)=f\\ &\qquad\qquad\text{in } \mathscr{D}'(B_N),\\ &u=0 \text{ on } \partial B_N.\end{aligned}\tag{2.34}$$

By (2.28) and, since all coefficients in (2.34) are bounded by (H1), (H3) and (K), it follows via the Lax-Milgram lemma that, for each $v\in L^2(B_N)$ and $\lambda\in(0,\lambda_0]$, (2.34) has a unique solution $u=F(v)\in H_0^1(B_N)$. Moreover, by (2.34), (2.28) and (H1)–(H3), we see that

$$\begin{aligned}&\|u\|^2_{L^2(B_N)}+\gamma\lambda\|\nabla u\|^2_{L^2(B_N)}\le\lambda b_\infty\|\nabla u\|_{L^2(B_N)}\|u\|_{L^2(B_N)}\\ &+c_\infty\lambda\|u\|_{L^2(B_N)}\|\nabla u\|_{L^2(B_N)}+\|f\|_{L^2(B_N)}\|u\|_{L^2(B_N)}\\ &\le\frac{\gamma\lambda}{2}\|\nabla u\|^2_{L^2(B_N)}+\frac{\lambda}{2\gamma}(b_\infty+c_\infty)^2\|u\|^2_{L^2(B_N)}+\frac{1}{2}(\|u\|^2_{L^2(B_N)}+\|f\|^2_{L^2(B_N)}).\end{aligned}\tag{2.35}$$

Hence, for $\lambda\in(0,\lambda_0]$, (2.33) holds for u replacing u_N.

Furthermore, if $v_n \to v$ in $L^2(B_N)$ and $u_n = Fv_n$, since $(a_{ij})_{x_j}$, b and $(a^*_{ij})_v$ are continuous and bounded by (H1), (H3) and (K), by Lebesgue's dominated convergence theorem we have

$$b(x, v_n) \to b(x, v),\ (a^*_{ij})_v(x, v_n) \to (a^*_{ij})_v(x, v),\ (a_{ij})_{x_j}(x, v_n) \to (a_{ij})_{x_j}(x, v)$$

strongly in $L^2(B_N)$. Then, along a subsequence we have, by (2.33),

$$u_n \to u \text{ weakly in } H^1_0(B_N), \text{ strongly in } L^2(B_N).$$

Now, letting $n \to \infty$ in (2.34), where $v = v_n$ and $u = u_n$, that is,

$$\int_{B_N} \Big(u_n\psi + \lambda \sum_{i,j=1}^{d} ((a^*_{ij})_v(x, v_n)D_j u_n + (a_{ij})_{x_j}(x, v_n)u_n)\Big) D_i\psi dx$$

$$-\lambda \int_{B_N} u_n b(x, v_n) \cdot \nabla\psi dx = \int_{B_N} f\psi dx,\ \forall\psi \in C^\infty_0(B_N),$$

we see that $u = Fv$ and, therefore, F is continuous on $L^2(B_N)$.

Moreover, since the Sobolev space $H^1_0(B_N)$ is compactly embedded in $L^2(B_N)$, by formula (2.33) we see that $F(L^2(B_N))$ is relatively compact in $L^2(B_N)$. We set $K = \overline{\text{Conv}}\, F(L^2(B_N))$ and note that K is closed, convex, compact in $L^2(B_N)$ and $F(K) \subset K$. Then, by the Schauder fixed point theorem, F has a fixed point $u_N \in K$ which, clearly, is a solution to (2.32). Also, by (2.35), it follows that estimate (2.33) holds.

Lemma 2.2 *Let $f \in (L^1 \cap L^2)(\mathbb{R}^d)$ and $0 < \lambda \le \lambda_0$. Then Eq. (2.9) has at least one solution $u \in H^1(\mathbb{R}^d)$ which satisfies the estimate*

$$|u|^2_2 + \gamma\lambda|\nabla u|^2_2 \le 2|f|^2_2. \tag{2.36}$$

Proof Consider a sequence $\{N\} \to \infty$ and let $u_N \in H^1_0(B_N)$ be a solution to (2.32) given by Lemma 2.1. By (2.33), we have

$$\|u_N\|_{H^1_0(B_N)} \le C,\ \forall N,$$

and so, along a subsequence, again denoted $\{N\}$, we have

$$u_N \to u \text{ weakly in } H^1(\mathbb{R}^d),\ \text{ strongly in } L^2_{\text{loc}}(\mathbb{R}^d). \tag{2.37}$$

Then, letting $N \to \infty$ in the equation

$$u_N - \lambda \sum_{i,j=1}^{d} D_i((a^*_{ij})_u u_N D_j u_N + (a_{ij})_{x_j}(x, u_N)u_N) + \lambda\, \text{div}(b(x, u_N)u_N) = f \text{ in } B_N,$$

or, more precisely, in its weak form

$$\int_{\mathbb{R}^d} u_N\psi\,dx+\lambda\sum_{i,j=1}^{d}\int_{\mathbb{R}^d}((a_{ij}^*)_u(u_N)D_ju_N+(a_{ij})_{x_j}(x,u_N)u_ND_i\psi)dx$$
$$-\lambda\sum_{i=1}^{d}\int_{\mathbb{R}^d} b(x,u_N)u_N\cdot\nabla\psi\,dx=\int_{\mathbb{R}^d} f\psi\,dx,\ \forall\psi\in C_0^\infty(B_N),$$

as in the proof of Lemma 2.1, by (H1), (H3), (K) and (2.37) we infer that $u\in H^1(\mathbb{R}^d)$ is a solution to (2.9). Also, estimate (2.36) follows by (2.33). This completes the proof of Lemma 2.2.

Now, we come back to the proof of Proposition 2.1. We first prove that, for each $f\in L^1\cap L^2$ and $\lambda\in(0,\lambda_0]$, the solution $u=u(\lambda,f)\in H^1$ to Eq. (2.9) from Lemma 2.2 satisfies

$$|u(\lambda,f_1)-u(\lambda,f_2)|_1\le|f_1-f_2|_1,\ \forall f_1,f_2\in L^1\cap L^2. \tag{2.38}$$

Here is the argument. We set $u_i=u(\lambda,f_i)$, $i=1,2$, and $f=f_1-f_2$, $u=u_1-u_2$. Then, we have

$$\begin{aligned}u-\lambda\sum_{i,j=1}^{d}D_{ij}^2(a_{ij}^*(x,u_1)-a_{ij}^*(x,u_2))&\\+\lambda\,\mathrm{div}(b^*(x,u_1)-b^*(x,u_2))=f\text{ in }\mathcal{D}'(\mathbb{R}^d).&\end{aligned} \tag{2.39}$$

More precisely, since $u_i\in H^1(\mathbb{R}^d)$, equation (2.39) can be taken in the variational form

$$\begin{aligned}\int_{\mathbb{R}^d}\Big(u\psi+\lambda\sum_{i,j=1}^{d}D_i(a_{ij}^*(x,u_1)-a_{ij}^*(x,u_2))D_j\psi&\\-\lambda(b^*(x,u_1)-b^*(x,u_2))\cdot\nabla\psi\Big)dx=\int_{\mathbb{R}^d} f\psi\,dx,\ \forall\psi\in H^1.&\end{aligned} \tag{2.40}$$

We set

$$\mathcal{X}_\delta(r)=\begin{cases}1 & \text{for } r\ge\delta,\\ \dfrac{r}{\delta} & \text{for } |r|<\delta,\\ -1 & \text{for } r\le-\delta,\end{cases} \tag{2.41}$$

where $\delta>0$. The function $\mathcal{X}_\delta$ is a Lipschitzian approximation of the multivalued signum function $\mathrm{sign}(r)=r/|r|$ for $r\ne0$; $\mathrm{sign}(0)=\{r\in\mathbb{R};\ |r|\le1\}$ and it will intervene frequently in the following.

We note that, since $u \in H^1$, it follows that $\mathscr{X}_\delta(u) \in H^1$ and

$$\sum_{i,j=1}^{d} D_i D_j (a^*_{ij}(x,u_1) - a^*_{ij}(x,u_2)) + \operatorname{div}(b^*(x,u_1) - b^*(x,u_2)) \in H^{-1}.$$

Furthermore, the integral

$$\int_{\mathbb{R}^d} \left(\sum_{i,j=1}^{d} D_j (a^*_{ij}(x,u_1) - a^*_{ij}(x,u_2)) D_i \mathscr{X}_\delta(u) \right. \\ \left. + (b^*(x,u_1) - b^*(x,u_2)) \cdot \nabla \mathscr{X}_\delta(u) \right) dx$$

is well defined.

Since (H1), (H3) imply that $b(\cdot, u_k)$, $(a^*_{ij})_{x_j}(\cdot, u_k) \in L^2$ and (K) implies that $(a^*_{ij})_u(\cdot, u_k) D_j u_k \in L^2$ for $k = 1$ or 2, taking in (2.40) $\psi = \mathscr{X}_\delta(u)$ yields

$$\begin{aligned} &\int_{\mathbb{R}^d} u \mathscr{X}_\delta(u) dx + \lambda \sum_{i,j=1}^{d} \int_{\mathbb{R}^d} ((a^*_{ij})_u(x,u_1) D_j u_1 - (a^*_{ij})_u(x,u_2) D_j u_2 \\ &\quad + (a^*_{ij})_{x_j}(x,u_1) - (a^*_{ij})_{x_j}(x,u_2)) D_i \mathscr{X}_\delta(u) dx \\ &= \lambda \int_{\mathbb{R}^d} (b^*(x,u_1) - b^*(x,u_2)) \cdot \nabla(\mathscr{X}_\delta(u)) dx + \int_{\mathbb{R}^d} f \mathscr{X}_\delta(u) dx. \end{aligned} \tag{2.42}$$

For $\widetilde{b}^*$ as in (2.30), we set

$$\begin{aligned} I^1_\delta &= \int_{\mathbb{R}^d} (\widetilde{b}^*(x,u_1) - \widetilde{b}^*(x,u_2)) \cdot \nabla(\mathscr{X}_\delta(u)) dx \\ &= \int_{\mathbb{R}^d} (\widetilde{b}^*(x,u_1) - \widetilde{b}^*(x,u_2)) \cdot \nabla u \mathscr{X}'_\delta(u) dx \\ &= \frac{1}{\delta} \int_{[|u| \le \delta]} (\widetilde{b}^*(x,u_1) - \widetilde{b}^*(x,u_2)) \cdot \nabla u \, dx. \end{aligned}$$

Let $E_\delta = \{x \in \mathbb{R}^d;\ |u(x)| \le \delta\}$. By (2.30) and $u_1, u_2 \in L^2$, it follows that

$$\lim_{\delta \to 0} \frac{1}{\delta} \int_{E_\delta} |(\widetilde{b}^*(x,u_1) - \widetilde{b}^*(x,u_2)) \cdot \nabla u| dx \le |g|_2 \lim_{\delta \to 0} \left(\int_{E_\delta} |\nabla u|^2 dx \right)^{\frac{1}{2}} = 0,$$

because $u \in H^1(\mathbb{R}^d)$ and $\nabla u = 0$, a.e. on $\{x;\ u(x) = 0\}$. This yields

$$\lim_{\delta \to 0} I^1_\delta = 0. \tag{2.43}$$

On the other hand, we have

$$\begin{aligned}
I_\delta^2 &= \int_{\mathbb{R}^d} \sum_{i,j=1}^d ((a_{ij}^*)_u(x,u_1)D_j u_1 - (a_{ij}^*)_u(x,u_2)D_j u_2) D_i(\mathscr{X}_\delta(u))dx \\
&= \frac{1}{\delta}\int_{E_\delta} \sum_{i,j=1}^d ((a_{ij}^*)_u(x,u_1)D_j u_1 - (a_{ij}^*)_u(x,u_2)D_j u_2) D_i u\, dx \\
&= \frac{1}{\delta}\int_{E_\delta} \sum_{i,j=1}^d (a_{ij}^*)_u(x,u_1)D_j u D_i u\, dx \\
&\quad + \frac{1}{\delta}\int_{E_\delta} \sum_{i,j=1}^d ((a_{ij}^*)_u(x,u_1) - (a_{ij}^*)_u(x,u_2))D_j u_2 D_i u\, dx \\
&= K_1^\delta + K_2^\delta.
\end{aligned} \tag{2.44}$$

By (2.28), it follows that $K_1^\delta \geq 0$ and, by (2.29) we obtain that $\lim\limits_{\delta\to 0} K_2^\delta = 0$, so, by (2.44) it follows that

$$\liminf_{\delta\to 0} I_\delta^2 \geq 0.$$

By (2.42) and (2.43), this yields

$$|u|_1 \leq |f|_1,\ \forall \lambda \in (0,\lambda_0], \tag{2.45}$$

as claimed. We note that in the above proof we may replace our specially constructed solutions $u(\lambda, f_i)$, $i = 1, 2$, by any two solutions u_1, u_2 of (2.9) such that $u_1, u_2 \in D(A_0) \cap H^1$. In particular, we have then $u_1 = u_2$ if $f_1 = f_2 \in L^1 \cap L^2$.

To resume, we have shown so far that under assumptions (H1)–(H3) and (K), for each $f \in L^1 \cap L^2$, Eq. (2.9) has, for $\lambda \in (0, \lambda_0]$, a unique solution $u(\lambda, f) \in H^1(\mathbb{R}^d) \cap D(A_0)$, which satisfies (2.36) and (2.38).

Now, we give up Hypothesis (K) and assume that a_{ij}, b_i satisfy Hypotheses (H1)–(H3) only. Consider, for $\varepsilon > 0$, the regularized functions

$$a_{ij}^\varepsilon(x,u) := \psi_\varepsilon(x,u)a_{ij}(x,u) + (1-\psi_\varepsilon(x,u))\max(\gamma,\widetilde{\gamma})\delta_{ij}, \tag{2.46}$$

$$b_i^\varepsilon(x,u) := \psi_\varepsilon(x,u)b_i(x,u), \tag{2.47}$$

for $1 \leq i,j \leq d$, $x \in \mathbb{R}^d, u \in \mathbb{R}, \widetilde{\gamma} := d \sup\limits_{1\leq i,j\leq d} |a_{ij}|_\infty$ such that $\sum\limits_{i,j=1}^d a_{ij}(x,u)\xi_i\xi_j \leq \widetilde{\gamma}|\xi|^2$, $\forall u \in \mathbb{R}$, $\xi \in \mathbb{R}^d$ and $\psi_\varepsilon(x,u) := \varphi(\varepsilon|u|)\varphi(\varepsilon|x|^2)$ with $\varphi \in C^\infty(\mathbb{R})$, $1_{(-\infty,1]} \leq \varphi \leq 1 - 1_{[2,\infty)}$, $-2 \leq \varphi' \leq 0$. Clearly, all $a_{ij}^\varepsilon, b_i^\varepsilon$

satisfy (H1), (H3) and (K). It only remains to prove that a^ε_{ij} satisfies (H2). To this end, define for $x \in \mathbb{R}^d$, $u \in \mathbb{R}$, $1 \le i, j \le d$,

$$a^{\varepsilon,*}_{ij}(x,u) := a^\varepsilon_{ij}(x,u)u$$

and let $\xi = (\xi_1, \dots, \xi_d) \in \mathbb{R}^d$. Then, by an elementary computation,

$$\begin{aligned}
&\sum_{i,j=1}^d (a^{\varepsilon,*}_{ij})_u(x,u)\xi_i\xi_j \\
&\quad = \psi_\varepsilon(x,u)\sum_{i,j=1}^d (a^*_{ij})_u(x,u)\xi_i\xi_j + (1-\psi_\varepsilon(x,u))\max(\gamma,\widetilde{\gamma})|\xi|^2 \\
&\quad\quad + \varepsilon u 1_{\{u>0\}}\varphi'(\varepsilon u)\varphi(\varepsilon|x|^2)\left(\sum_{i,j=1}^d a_{ij}(x,u)\xi_i\xi_j - \max(\gamma,\widetilde{\gamma})|\xi|^2\right) \\
&\quad \ge \psi_\varepsilon(x,u)\gamma + (1-\psi_\varepsilon(x,u))\max(\gamma,\widetilde{\gamma}) \ge \gamma,
\end{aligned} \tag{2.48}$$

where in the second to last step we used (H2), the assumption that $\varphi' \le 0$ and the definition of $\widetilde{\gamma}$.

We set $b^\varepsilon = \{b^\varepsilon_i\}^d_{i=1}$ and note that

$$|b^\varepsilon|_\infty \le |b|_\infty \quad \text{and} \quad |(a^\varepsilon_{ij})_{x_j}|_\infty \le 8\left(\max_{i,j}|a_{ij}|_\infty + \max(\gamma,\widetilde{\gamma})\right) + C_\infty =: \widetilde{C}_\infty.$$

Hence, by Lemmas 2.1 and 2.2, for $\lambda_0 \in \left(0, \frac{1}{2}\gamma(b_\infty + \widetilde{C}_\infty)^{-2}\right)$, as shown above, the equation

$$u_\varepsilon - \lambda\sum_{i,j=1}^d D^2_{ij}a^{\varepsilon,*}_{ij}(x,u_\varepsilon) + \lambda\,\mathrm{div}(b^\varepsilon(x,u_\varepsilon)u_\varepsilon) = f \tag{2.49}$$

has, for each $\lambda \in (0, \lambda_0]$ and $f \in L^2 \cap L^1$, a unique solution $u_\varepsilon = u_\varepsilon(\lambda, f) \in H^1(\mathbb{R}^d) \cap D(A^\varepsilon_0)$ satisfying (2.36) and (2.38), where $(A^\varepsilon_0, D(A^\varepsilon_0))$ is defined as $(A_0, D(A_0))$ with $u^\varepsilon_{ij}, b^\varepsilon_i$ replacing a_{ij}, b_i, respectively. Hence

$$|u_\varepsilon(\lambda, f_1) - u_\varepsilon(\lambda, f_2)| \le |f_1 - f_2|_1,\ \forall\varepsilon > 0,\ f_1, f_2 \in L^1 \cap L^2. \tag{2.50}$$

$$|u_\varepsilon(\lambda, f)|^2_2 + \gamma\lambda|\nabla u_\varepsilon(\lambda, f)|^2_2 \le 2|f|^2_2,\ \forall\varepsilon > 0,\ f \in L^1 \cap L^2. \tag{2.51}$$

For $\varepsilon \in (0,1)$, let us define operators $J^\varepsilon_\lambda : L^1 \to D(A_0) \subset L^1$, $\lambda \in (0, \lambda_0]$, by

$$J^\varepsilon_\lambda(f) := u_\varepsilon(\lambda, f),\ f \in L^1.$$

Then, J_λ^ε, $\lambda \in (0, \lambda_0]$, satisfy the nonlinear resolvent equation, i.e., for $\lambda_1, \lambda_2 \in (0, \lambda_0]$, $f \in L^1$,

$$J_{\lambda_2}^\varepsilon(f) = J_{\lambda_1}^\varepsilon\left(\frac{\lambda_1}{\lambda_2} f + \left(1 - \frac{\lambda_1}{\lambda_2}\right) J_{\lambda_2}^\varepsilon(f)\right). \tag{2.52}$$

By (2.45) it suffices to prove (2.52) for $f \in L^1 \cap L^2$. But then $J_\lambda^\varepsilon(f) \in D(A_0) \cap H^1$, $\lambda \in (0, \lambda_0]$. So, by the uniqueness result above, we only have to prove that

$$(I + \lambda_1 A_0^\varepsilon)(J_{\lambda_2}^\varepsilon(f)) = \frac{\lambda_1}{\lambda_2} f + \left(1 - \frac{\lambda_1}{\lambda_2}\right) J_{\lambda_2}^\varepsilon(f).$$

But this equality is obvious.

Now, for $\varepsilon \to 0$, it follows by the compactness of $H^1(\mathbb{R}^d)$ in $L^2_{\rm loc}$ that along a subsequence (depending on λ and f), again denoted ε, we have

$$u_\varepsilon(\lambda, f) \longrightarrow u = u(\lambda, f) \text{ strongly in } L^2_{\rm loc}, \text{ weakly in } H^1, \tag{2.53}$$

and so, by (2.46) and (2.47) we have, eventually along a subsequence,

$$\begin{aligned} a_{ij}^{\varepsilon,*}(x, u_\varepsilon(x)) &\longrightarrow a_{ij}^*(x, u(x)), \text{ a.e. } x \in \mathbb{R}^d \text{ and in } L^2_{\rm loc}, \\ b_i^{\varepsilon,*}(x, u_\varepsilon(x)) &\longrightarrow b_i^*(x, u(x)), \text{ a.e. } x \in \mathbb{R}^d \text{ and in } L^2_{\rm loc}. \end{aligned}$$

Hence, for $\varepsilon \to 0$,

$$\begin{aligned} D_{ij}^2(a_{ij}^{\varepsilon,*}(x, u_\varepsilon)) &\longrightarrow D_{ij}^2(a_{ij}^*(x, u)) \text{ in } \mathscr{D}'(\mathbb{R}^d), \\ \operatorname{div}(b^\varepsilon(x, u_\varepsilon)u_\varepsilon) &\longrightarrow \operatorname{div}(b(x, u)u) \text{ in } \mathscr{D}'(\mathbb{R}^d). \end{aligned}$$

and so it follows that $u = u(\lambda, f)$ is a solution to (2.9), as an equation in $\mathscr{D}'(\mathbb{R}^d)$. Moreover, by (2.50) it follows that

$$|u(\lambda, f_1) - u(\lambda, f_2)|_1 \le |f_1 - f_2|_1, \ \forall f_1, f_2 \in L^1 \cap L^2, \tag{2.54}$$

and by (2.53) that $u(\lambda, f) \in H^1$, $\forall f \in L^1 \cap L^2$.

Noting that by construction $u(\lambda, 0) = 0$, (2.54) implies that $u(\lambda, f) \in L^1$ for all $f \in L^1 \cap L^2$. So, by (2.9), $u(\lambda, f) \in D(A_0)$.

Now, we fix $f \in L^1$ and consider a sequence $\{f_n\} \subset L^1 \cap L^2$ such that $f_n \to f$ in L^1 and consider the corresponding solutions $u_n = u(\lambda, f_n)$ to (2.9). By (2.54), it follows that

$$|u_n - u_m|_1 \le |f_n - f_m|_1, \ \forall n, m \in \mathbb{N}.$$

Hence, there is $u^* = \lim_{n\to\infty} u_n$ in L^1.

By (2.9) it follows that also $A_0(u_n) \to y$ in L^1. Since $(A_0, D(A_0))$ is closed on L^1, we conclude that $u^* \in D(A_0)$, $y = A(u^*)$ and that thus u^* solves (2.9) as an equation in L^1.

Moreover, (2.54) follows for all $\lambda \in (0, \lambda_0)$; $f_1, f_2 \in L^1$.

We set $J_\lambda(f) = u(\lambda, f) := u^*$. Then

$$J_\lambda(f) + \lambda A_0(J_\lambda(f)) = f. \tag{2.55}$$

Now, let us check that (2.12)–(2.16) hold. We already know that

$$|J_\lambda(f_1) - J_\lambda(f_2)|_1 \le |f_1 - f_2|_1, \ \forall f_1, f_2 \in L^1, \text{ for all } \lambda \in (0, \lambda_0).$$

We shall deduce (2.13) from (2.52) using the following lemma, for which we introduce the following function

$$\Phi(x) := (1 + |x|^2)^{\frac{1}{2}}, \ x \in \mathbb{R}^d. \tag{2.56}$$

Lemma 2.3 *Let $f \in C_0(\mathbb{R}^d)$, $\lambda \in (0, \lambda_0]$. Then there exists a function $C_\Phi : [0, \infty)^3 \to [0, \infty)$ increasing in each of its variables and only depending on Φ and the dimension d such that*

$$\begin{aligned}&\sup_{\varepsilon>0} \int_{\mathbb{R}^d} |J_\lambda^\varepsilon(f)|\Phi\, dx \\ &\le \lambda C_\Phi \left(\max(\gamma, \tilde{\gamma}), \sup_{1<i,j<d} |a_{ij}|_\infty, \sup_{1\le i,j\le d} |(a_{ij})_{x_j}| \right) |f|_1 + \int_{\mathbb{R}^d} |f|\Phi\, dx.\end{aligned} \tag{2.57}$$

Furthermore,

$$J_\lambda^\varepsilon(f) \to J_\lambda(f) \text{ in } L^1,$$

as $\varepsilon \to 0$ along a subsequence (depending on λ and f). In particular, (2.57) *holds with $J_\lambda(f)$ replacing $J_\lambda^\varepsilon(f)$.*

Proof By (2.53), $J_\lambda^\varepsilon(f) \to J_\lambda(f)$ in L^1_{loc}, as $\varepsilon \to 0$ along a subsequence. Hence, it suffices to prove (2.57).

To see this, we note that for every sequence $f_n \in L^1$, $n \in \mathbb{N}$, which convergess in L^1_{loc} and $c \in (0, \infty)$

$$\begin{aligned}|f_n - f_m|_1 &= \int_{\{\Phi>c\}} |f_n - f_m| dx + \int_{\{\Phi\le c\}} |f_n - f_m| dx \\ &\le \frac{2}{c} \sup_{k\in\mathbb{N}} \int_{\mathbb{R}^d} |f_k|\Phi\, dx + \int_{\{\Phi\le c\}} |f_n - f_m| dx, \ m, n \in \mathbb{N}.\end{aligned}$$

Since $\{\Phi \le c\}$ is compact, we deduce

$$\lim_{m\to\infty} \sup_{n\ge m} |f_n - f_m|_1 \le \frac{2}{c} \sup_{k\in\mathbb{N}} \int_{\mathbb{R}^d} |f_k|\Phi\, dx,\ \forall c \in (0,\infty).$$

To prove (2.57), we set $u_\lambda^\varepsilon := J_\lambda^\varepsilon(f)$ and multiply Eq. (2.55) by $\Phi_\nu \mathcal{X}_\delta(u_\lambda^\varepsilon)$, where $\Phi_\nu := \Phi \exp(-\nu\Phi)$ and $\delta, \nu \in (0,\infty)$, and integrate over $\mathbb{R}^d$ to get

$$\begin{aligned}
\int_{\mathbb{R}^d} u_\lambda^\varepsilon \Phi_\nu \mathcal{X}_\delta(u_\lambda^\varepsilon)dx =& -\lambda \sum_{i,j=1}^d \int_{\mathbb{R}^d} D_j(a_{i,j}^{\varepsilon,*}(x,u_\lambda^\varepsilon))D_i(\Phi_\nu \mathcal{X}_\delta(u_\lambda^\varepsilon))dx \\
&+\lambda \int_{\mathbb{R}^d} b^{\varepsilon,*}(x,u_\lambda^\varepsilon)\cdot \nabla(\Phi_\nu \mathcal{X}_\delta(u_\lambda^\varepsilon))dx \\
&+ \int_{\mathbb{R}^d} f\,\Phi_\nu \mathcal{X}_\delta(u_\lambda^\varepsilon)dx \\
=& -\lambda \sum_{i,j=1}^d \int_{\mathbb{R}^d} D_j(a_{i,j}^{\varepsilon,*}(x,u_\lambda^\varepsilon))\mathcal{X}_\delta(u_\lambda^\varepsilon)D_i\,\Phi_\nu\, dx \\
&-\lambda \sum_{i,j=1}^d \int_{\mathbb{R}^d} (a_{i,j}^{\varepsilon,*})_u(x,u_\lambda^\varepsilon)D_j u_\lambda^\varepsilon \mathcal{X}_\delta'(u_\lambda^\varepsilon)D_i u_\lambda^\varepsilon \Phi_\nu\, dx \\
&+\lambda \int_{\mathbb{R}^d} \widetilde{b}^{\varepsilon,*}(x,u_\lambda^\varepsilon)\cdot \nabla u_\lambda^\varepsilon \mathcal{X}_\delta'(u_\lambda^\varepsilon)\Phi_\nu\, dx \\
&+\lambda \int_{\mathbb{R}^d} b^{\varepsilon,*}(x,u_\lambda^\varepsilon)\cdot \nabla\Phi_\nu \mathcal{X}_\delta(u_\lambda^\varepsilon)dx \\
&+ \int_{\mathbb{R}^d} f\,\Phi_\nu \mathcal{X}_\delta(u_\lambda^\varepsilon)dx,
\end{aligned}$$

where $\widetilde{b}_i^{\varepsilon,*} := b_i^{\varepsilon,*} - \sum_{j=1}^d (a_{ij}^{\varepsilon,*})_{x_j}$ and $b^{\varepsilon,*} = (b_1^{\varepsilon,*},\dots,b_d^{\varepsilon,*})$. Hence, first using (2.48) and that $\mathcal{X}_\delta'(u_\lambda^\varepsilon)\Phi_\nu \ge 0$ and then letting $\delta \to 0$, as in the proof of (2.43), using that $u_\lambda^\varepsilon, \nabla u_\lambda^\varepsilon \in L^2$, we find

$$\begin{aligned}
\int_{\mathbb{R}^d} |u_\lambda^\varepsilon|\Phi_\nu\, dx \le& -\lambda \sum_{i,j=1}^d \int_{\mathbb{R}^d} D_j(a_{i,j}^{\varepsilon,*}(x,u_\lambda^\varepsilon))\mathrm{sign}\, u_\lambda^\varepsilon\, D_i\,\Phi_\nu\, dx \\
&+\lambda\|b\|_\infty \int_{\mathbb{R}^d} |u_\lambda^\varepsilon|\,|\nabla\Phi_\nu|dx + \int_{\mathbb{R}^d} |f|\Phi_\nu\, dx.
\end{aligned} \tag{2.58}$$

But the integrand of the first integral in the right hand side of (2.58), by (2.46) and the axial symmetry assumption in (H1) can be written as

$$D_j(a_{i,j}^{\varepsilon,*}(x,|u_\lambda^\varepsilon|)D_i\Phi_\nu - ((a_{ij}^{\varepsilon,*})_{x_j}(x,|u_\lambda^\varepsilon|) - \mathrm{sign}\, u_\lambda^\varepsilon (a_{i,j}^{\varepsilon,*})_{x_j}(x,u_\lambda^\varepsilon))D_i\Phi_\nu.$$

Taking into account that

$$\nabla \Phi_\nu = (1-\nu\Phi)\nabla\Phi e^{-\nu\Phi},\ \ D_jD_i\Phi_\nu = (1-\nu\Phi)(D_jD_i\Phi + \nu D_i\Phi D_j\Phi)e^{-\nu\Phi}$$

and letting $\nu \to 0$ in (2.58), we hence find

$$\begin{aligned}\int_{\mathbb{R}^d} |u_\lambda^\varepsilon|\Phi\,dx \leq &\lambda \sum_{i,j=1}^d (|a_{i,j}|_\infty + \max(\gamma,\widetilde{\gamma}))|D_jD_i\Phi|_\infty |u_\lambda^\varepsilon|_1 \\ &+\lambda \sum_{i,j=1}^d 2|(a_{i,j}^\varepsilon)_{x_j}|_\infty |\nabla\Phi|_\infty |u_\lambda^\varepsilon|_1 + \int_{\mathbb{R}^d} |f|\Phi\,dx.\end{aligned} \tag{2.59}$$

Since, as an easily calculation shows, we have

$$|(a_{ij}^\varepsilon)_{x_j}|_\infty \leq 4\sqrt{2}(|a_{ij}|_\infty + \max(\gamma,\widetilde{\gamma}) + |(a_{ij})_{x_j}|_\infty), \tag{2.60}$$

and by (2.45)

$$|u_\lambda^\varepsilon|_1 \leq |f|_1 \text{ for all } \varepsilon > 0,\ \lambda \in (0,\lambda_0],$$

we obtain from (H1), (H2) and (2.59) that (2.57) holds and thus the assertion of the lemma follows. □

Now, we can prove (2.13). Let $\mathscr{G} \subset C_0(\mathbb{R}^d)$, $\mathscr{G}$ countable and dense in L^1. Then, by Lemma 2.3 and a diagonal argument there exists $\varepsilon_n \to 0$ such that

$$J_\lambda^{\varepsilon_n} f \to J_\lambda f \text{ in } L^1,\ \forall f \in \mathscr{G},\ \lambda \in (0,\lambda_0] \cap \mathbb{Q}.$$

By (2.45) and (2.52), it follows that, for all $\lambda_1, \lambda_2 \in (0,\lambda_0] \cap \mathbb{Q}$,

$$J_{\lambda_2}(f) = J_{\lambda_1}\left(\frac{\lambda_1}{\lambda_2} f + \left(1 - \frac{\lambda_1}{\lambda_2}\right) J_{\lambda_2}(f)\right),\ \forall f \in \mathscr{G},$$

which by (2.12) then extends to all $f \in L^1$ and hence, again by (2.12),

$$|J_{\lambda_2}(f) - J_{\lambda_1}(f)|_1 \leq 2\left(1 - \frac{\lambda_1}{\lambda_2}\right)|f|_1,\ \forall f \in L^1.$$

So, by the local Lipschitz continuity of $\lambda \mapsto J_\lambda(f)$ on $(0,\lambda_0] \cap \mathbb{Q}$ this map extends to all $\lambda \in (0,\lambda_0]$. It is not clear whether for $\lambda \in (0,\lambda_0] \setminus Q$ the resulting extension $\widetilde{J}_\lambda(f)$ coincides with $J_\lambda(f)$ as defined above. But from now on we replace the latter by the first and use the same symbol, so redefine $J_\lambda(f) := \widetilde{J}_\lambda(f)$. Then, it is easy to see that $J_\lambda(f)$ indeed also solves (2.55) for all $f \in L^1$, $\lambda \in (0, \lambda_0]$ and retains all properties proved so far for J_λ. Furthermore, (2.13) follows by density and thus

$$D(A) := J_\lambda(L^1)\ (\subset D(A_0))$$

is independent of $\lambda \in (0, \lambda_0]$ and we can define

$$A : D(A) \to L^1, \quad A := A_{0\uparrow D(A)}.$$

Then, for all $\lambda \in (0, \lambda_0]$,

$$I + \lambda A : D(A) \to L^1$$

is a bijection and J_λ is its inverse.

As regards (2.15), it follows by multiplying (2.9) with $\mathrm{sign}(u^-)$ (or, more exactly, by $\mathscr{X}_\delta(u^-)$ and letting $\delta \to 0$) and integrating over $\mathbb{R}^d$. Finally, (2.14) follows by integrating (2.9) over $\mathbb{R}^d$. It remains to prove (2.16). To this purpose, we shall prove that $C_0^\infty(\mathbb{R}^d) \subset \overline{D(A)}$, $\forall \lambda \in (0, \lambda_0)$. If $f \in C_0^\infty(\mathbb{R}^d)$ and $u_\lambda = J_\lambda(f)$, that is,

$$u_\lambda + \lambda A_0(u_\lambda) = f, \tag{2.61}$$

we know that $|u_\lambda|_1 \le |f|_1$, $\forall \lambda > 0$, and since $u_\lambda \in H^1$, we also have by (2.51)

$$|u_\lambda|_2^2 + \lambda |\nabla u_\lambda|_2^2 \le C(|f|_2^2 + 1), \ \forall \lambda > 0. \tag{2.62}$$

It follows also by (2.61) that

$$(u_\lambda, A_0(u_\lambda))_2 + \lambda |A_0(u_\lambda)|_2^2 = (f, A_0(u_\lambda))_2.$$

Taking into account that, by Hypothesis (H2) we have, for some $\alpha_1 > 0$ and $\alpha_2 \ge 0$,

$$\begin{aligned}(u_\lambda, A_0(u_\lambda))_2 &= \sum_{i,j=1}^d \int_{\mathbb{R}^d} D_j(a_{ij}^*(x, u_\lambda)) D_i u_\lambda dx - \int_{\mathbb{R}^d} b^*(x, u_\lambda) \cdot \nabla u_\lambda dx \\ &\ge \alpha_1 |\nabla u_\lambda|_2^2 - \alpha_2 |u_\lambda|_2^2,\end{aligned}$$

while

$$(f, A_0(u_\lambda)) = \sum_{i,j=1}^d \int_{\mathbb{R}^d} D_j(a_{ij}^*(x, u_\lambda)) D_i f\, dx - \int_{\mathbb{R}^d} \nabla f \cdot b^*(x, u_\lambda) dx.$$

Hence, we get that $\lambda|A_0(u_\lambda)|_2^2 + \frac{\gamma}{4}|\nabla u_\lambda|_2^2 \le C_f < \infty$, $\forall \lambda \in (0, \lambda_0)$, and, therefore, $\lambda A_0(u_\lambda) \to 0$ in L^2 as $\lambda \to 0$. Then, by (2.61) we see that, for $\lambda \to 0$, $u_\lambda \to f$ in L^2, and, therefore, since $L^2 \subset L^1_{\rm loc}$, we have

$$u_\lambda \to f \text{ in } L^1_{\rm loc} \text{ as } \lambda \to 0. \tag{2.63}$$

Hence, by (2.57) it follows that $u_\lambda \to f$ in L^1, as $\lambda \to 0$, and so $f \in \overline{D(A)}$. □

Existence for the Degenerate NFPE (2.1)

We consider here Eq. (2.1), that is,

$$\begin{aligned} &u_t - \sum_{i,j=1}^d D_{ij}^2(a_{ij}(u)u) + \sum_{i=1}^d D_i(b_i(u)u) = 0 \text{ in } \mathscr{D}'((0,\infty) \times \mathbb{R}^d), \\ &u(0,x) = u_0(x), x \in \mathbb{R}^d, \end{aligned} \tag{2.64}$$

in the special case where the coefficients a_{ij} and b_i are independent of x and satisfy Hypotheses (H1)′–(H3)′.

Proceeding as in the previous case, consider the operator $A_1 : D(A_1) \subset L^1 \to L^1$ defined by

$$\begin{aligned} A_1 u &= -\sum_{i,j=1}^d D_{ij}^2(a_{ij}(u)u) + \sum_{i=1}^d D_i(b_i(u)u) \text{ in } \mathscr{D}'(\mathbb{R}^d), \\ D(A_1) &= \left\{ u \in L^1; -\sum_{i,j=1}^d D_{ij}^2(a_{ij}(u)u) + \sum_{i=1}^d D_i(b_i(u)u) \in L^1 \right\}. \end{aligned} \tag{2.65}$$

We have

Lemma 2.4 *Assume that* (H1)′–(H3)′ *hold. Then, there is* $\lambda_0 > 0$ *such that for each* $\lambda \in (0, \lambda_0)$ *there is* $J_\lambda : L^1 \to L^1$ *such that* $(I + \lambda A_1)J_\lambda(f) = f$, $\forall f \in L^1$, *and conditions* (2.12)–(2.16) *hold.*

Proof We shall prove that, for each $\lambda \in (0, \lambda_0)$ and $f \in L^1$, the equation

$$u - \lambda \sum_{i,j=1}^d D_{ij}^2(a_{ij}(u)u) + \lambda \sum_{i=1}^d D_i(b_i(u)u) = f \text{ in } \mathscr{D}'(\mathbb{R}^d) \tag{2.66}$$

has a solution $u = u(\lambda, f)$ which satisfies the estimate

$$|u(\lambda, f_1) - u(\lambda, f_2)|_1 \le |f_1 - f_2|_1, \ \forall f_1, f_2 \in L^1. \tag{2.67}$$

To this end, we set, for each $\varepsilon > 0$,

$$a_{ij}^{\varepsilon}(r) = a_{ij}(r) + \varepsilon\delta_{ij},\ i, j = 1, \dots, d,\ r \in \mathbb{R}, \tag{2.68}$$

where δ_{ij} is the Kronecker symbol and we approximate (2.66) by

$$u - \lambda \sum_{i,j=1}^{d} D_{ij}^2(a_{ij}^{\varepsilon}(u)u) + \lambda \sum_{i=1}^{d} D_i(b_i(u)u) = f \text{ in } \mathscr{D}'(\mathbb{R}^d). \tag{2.69}$$

Equivalently,

$$u + \lambda A_1^{\varepsilon}(u) = f, \tag{2.70}$$

where $A_1^{\varepsilon} : L^1 \to L^1$ is the operator

$$A_1^{\varepsilon}(u) = -\sum_{i,j=1}^{d} D_{ij}^2(a_{ij}^{\varepsilon}(u)u) + \sum_{i=1}^{d} D_i(b_i(u)u),\ \forall u \in D(A_1^{\varepsilon}),$$

$$D(A_1^{\varepsilon}) = \left\{ u \in L^1;\ -\sum_{i,j=1}^{d} D_{ij}^2(a_{ij}^{\varepsilon}(u)u) + \sum_{i=1}^{d} D_i(b_i(u)u) \in L^1 \right\}.$$

We shall prove first that, for each $f \in L^1$, there is a solution $u = u_{\varepsilon}(\lambda, f)$ satisfying (2.67) for $0 < \lambda < \lambda_0$.

Since a_{ij}^{ε} and b_i satisfy, for each $\varepsilon > 0$, Hypotheses (H1)–(H3), Proposition 2.1 implies the existence of a solution $u_{\varepsilon} = u_{\varepsilon}(\lambda, f)$ to (2.69) in $L^1(\mathbb{R}^d)$ for each $f \in L^2$ if $0 < \lambda \le \lambda_0^{\varepsilon}$, where, according to Lemma 2.1, $\lambda_0^{\varepsilon} < \frac{1}{2}\varepsilon(b_{\infty} + c_{\infty})^{-2}$.
Moreover, one has

$$|u_{\varepsilon}(\lambda, f_1) - u_{\varepsilon}(\lambda, f_2)|_1 \le |f_1 - f_2|_1,\ \forall f_1, f_2 \in L^1,\ \lambda \in (0, \lambda_0^{\varepsilon}). \tag{2.71}$$

Now, fix $f \in L^1$. Note that we have, for all $\varepsilon > 0$ and $\lambda \in (0, \lambda_0^{\varepsilon})$,

$$\int_{\mathbb{R}^d} (I + \lambda A_1^{\varepsilon})^{-1} f\, dx = \int_{\mathbb{R}^d} f\, dx,\ \ \forall f \in L^1, \tag{2.72}$$

$$(I + \lambda A_1^{\varepsilon})^{-1} f \ge 0,\ \text{a.e. in } \mathbb{R}^d \text{ if } f \in L^1,\ f \ge 0,\ \text{a.e. in } \mathbb{R}^d, \tag{2.73}$$

while (2.71) yields

$$|I + \lambda A_1^{\varepsilon})^{-1} f_1 - (I + \lambda A_1^{\varepsilon})^{-1} f_2|_1 \le |f_1 - f_2|_1,\ \forall f_1, f_2 \in L^1,\ \varepsilon > 0. \tag{2.74}$$

It can be shown, however, as mentioned earlier, that $(I+\lambda A_1^\varepsilon)^{-1}$ extends to all $\lambda > 0$ by a well known argument based on the resolvent equation

$$(I+\lambda A_1^\varepsilon)^{-1}f = (I+\lambda_0^\varepsilon A_1^\varepsilon)^{-1}\left(\frac{\lambda_0^\varepsilon}{\lambda}f + \left(1-\frac{\lambda_0^\varepsilon}{\lambda}\right)(I+\lambda A_1^\varepsilon)^{-1}f\right),\ \lambda > \lambda_0^\varepsilon.$$

Now, we are going to let $\varepsilon \to 0$ in (2.69). To this purpose, we set for $f \in L^1$ and the solution u_ε to (2.69)

$$u_h^\varepsilon(x) = u_\varepsilon(x+h) - u_\varepsilon(x),\ \ f_h(x) = f(x+h) - f(x),\ x, h \in \mathbb{R}^d.$$

Since a_{ij}^ε and b_i^ε are independent of x, we see that the function $x \to u^\varepsilon(x+h)$ is the solution to (2.69) for $f(x) \equiv f(x+h)$. Then, by (2.71) it follows that

$$|u_h^\varepsilon|_1 \le |f_h|_1,\ \forall h \in \mathbb{R}^d,\ \varepsilon > 0,$$

and so, by the Riesz–Kolmogorov compactness theorem ([103], p. 275), it follows that $\{u^\varepsilon\}$ is compact in $L^1_{loc}(\mathbb{R}^d)$. Hence, along a subsequence $\{\varepsilon\} \to 0$ we have

$$u_\varepsilon \to u \text{ strongly in } L^1_{loc}(\mathbb{R}^d).$$

Since $|u_\varepsilon|_1 \le |f|_1$, $\forall \varepsilon > 0$, by (2.57) it follows that $u \in L^1$ and $u_\varepsilon \to u$ in L^1. Letting $\varepsilon \to 0$ in (2.69), where $u = u_\varepsilon$, and taking into account that

$$a_{ij}^\varepsilon(u_\varepsilon)u_\varepsilon \to a_{ij}(u)u,\ \ b_i(u_\varepsilon)u_\varepsilon \to b_i(u)u,\ \text{ a.e. in } \mathbb{R}^d,$$

while by (H1)′, (H3)′,

$$|a_{ij}^\varepsilon(u_\varepsilon)| + |b_i(u_\varepsilon)| \le C,\ \text{ a.e. in } \mathbb{R}^d,$$

where C is independent of ε, we see that $u = u(\lambda, f)$ is a solution to (2.69). We set $J_\lambda^1(f) = u(\lambda, f)$. Letting $\varepsilon \to 0$ in (2.72)–(2.74), we see that

$$|J_\lambda^1(f_1) - J_\lambda^1(f_2)|_1 \le |f_1 - f_2|_1,\ \forall \lambda \in (0, \lambda_0),\ f_1, f_2 \in L^1,$$

$$\int_{\mathbb{R}^d} J_\lambda^1(f)dx = \int_{\mathbb{R}^d} f\,dx,\ \forall f \in L^1,$$

$$J_\lambda^1(f) \ge 0,\ \text{ a.e. in } \mathbb{R}^d \text{ if } f \ge 0,\ \text{ a.e. in } \mathbb{R}^d.$$

Then, we define as in Proposition 2.1 (see (2.19)) the m-accretive operator

$$A^* : D(A)^*) \subset L^1 \to L^1,$$
$$A^*(u) = A_1(J_\lambda^1(u)),\ \forall u \in D(A^*) = J_\lambda^1(L^1),\ \lambda \in (0, \lambda_0).$$

We note that, as in the previous case, $J^1_\lambda(f) = (I + \lambda A^*)^{-1} f, \forall f \in L^1$ and $\lambda > 0$, and Proposition 2.1 remains true in this case too.

We also have $\overline{D(A^*)} = L^1$. Here is the argument. If $f \in C_0^\infty(\mathbb{R}^d)$ and $u_\lambda = J^1_\lambda(f)$, we have by the equation $u_\lambda + \lambda A_1(u_\lambda) = f$ that (2.61) holds and this implies that, for $\lambda \to 0$,

$$u_\lambda \to f \text{ strongly in } H^{-1} \text{ and weakly in } L^2. \tag{2.75}$$

On the other hand, if $u^h_\lambda(x) \equiv u_\lambda(x+h) - u_\lambda(x)$, we see, as above, by the equation $u^h_\lambda + \lambda(A_1(u_\lambda)) = f^h$ that

$$|u^h_\lambda|_1 \leq |f|_1, \forall \lambda > 0,\ h \in \mathbb{R}^d.$$

Hence, again by the Riesz–Komogorov compacity theorem, $\{u_\lambda\}$ is compact in $L^1_{\rm loc}$ and, therefore, by (2.75) we have, for $\lambda \to 0$, $u_\lambda \to f$ in $L^1_{\rm loc}$. Next, by Lemma 2.3 it follows that $u_\lambda \to f$ in L^1 for $\lambda \to 0$ and so $f \in \overline{D(A^*)}$, as claimed.

Then, once again by the Crandall–Liggett theorem (Theorem 6.5), for each $u_0 \in \overline{D(A_1)} = L^1$, the Cauchy problem

$$\begin{aligned} &\frac{du}{dt} + A^* u = 0,\ t > 0, \\ &u(0) = u_0, \end{aligned} \tag{2.76}$$

has a unique mild solution $u \in C([0, \infty); L^1)$, $u(t) = S_1(t)u_0$, where S_1 is a continuous semigroup of contractions on L^1 generated by the m-accretive operator A^*. We get

Theorem 2.2 *Under Hypotheses* (H1)′–(H3)′, *there is a continuous semigroup of contractions* $S_1(t)$ *on* L^1 *such that, for each* $u_0 \in L^1$, $u(t) = S_1(t)u_0$ *is a mild solution to Eq.* (2.64). *Moreover, this solution satisfies* (2.24)–(2.26) *and is a solution to* (2.64) *in the sense of Schwartz distributions on* $(0, \infty) \times \mathbb{R}^d$, *that is,* (2.27) *holds.*

Remark 2.2 One might suspect that under Hypotheses (H1)–(H3) ((H1)′–(H3)′, respectively) the mild solution u to (2.1) is unique, but this is an open problem. We shall see, however, in the next section that this is true in the isotropic case (2.2) under suitable assumptions on β, b and D.

Consider now the McKean-Vlasov SDE (1.6) on $(0, T)$. Namely,

$$\begin{aligned} dX(t) &= b\left(X(t), \frac{d\mathcal{L}_{X(t)}}{dx}(X(t))\right)dt + \sigma\left(X(t), \frac{d\mathcal{L}_{X(t)}}{dx}(X(t))\right)dW(t), \\ &\qquad 0 \leq t \leq T, \\ X(0) &= \xi_0, \end{aligned} \tag{2.77}$$

on $\mathbb{R}^d$, where $W(t)$, $t \geq 0$, is an $(\mathcal{F}_t)_{t\geq 0}$-Brownian motion on a probability space $(\Omega, \mathcal{F}, P)$ with normal filtration $(\mathcal{F}_t)_{t\geq 0}$ and $\xi_0 : \Omega \to \mathbb{R}^d$ is $\mathcal{F}_0$-measurable such that

$$P \circ \xi_0^{-1}(dx) = u_0(x)dx, \ u_0 \in \mathcal{P}. \tag{2.78}$$

Furthermore, the functions $b = (b_1, \ldots, b_d) : \mathbb{R}^d \times \mathbb{R} \to \mathbb{R}^d$ and $\sigma : \mathbb{R}^d \times \mathbb{R} \to L(\mathbb{R}^d; \mathbb{R}^d)$ are measurable. Let $a_{ij} := (\sigma\sigma^T)_{ij}$, $1 \leq i, j \leq d$. We know, by the transposition principle (see Theorem 5.1 in Chap. 5), that the existence of a distributional solution $u = u(t, x)$, which is weakly continuous in t from $[0, \infty)$ to L^1 (or narrowly continuous), implies the existence of a probabilistically weak solution X to (2.77). Then, by Theorems 2.1 and 2.2, from Theorem 5.1 invoked above we get

Theorem 2.3 *Suppose that* a_{ij}, b_i, $1 \leq i, j \leq d$, *satisfy either* (H1)–(H3) *or* (H1)′–(H3)′. *Then, under assusmption* (2.78) *there exists a weak solution in the probabilistic sense to SDE* (2.77). *Furthermore, for the solution u in Theorems* 2.1 *and* 2.2, *respectively, with* $u(0, \cdot) = u_0$, *we have the "probabilistic representation"*

$$u(t, x)dx = P \circ X(t)^{-1}(dx), \ t \geq 0. \tag{2.79}$$

As seen in the following section, in the special case of the isotropic NFPE (2.2), that is,

$$a_{ij}(x, u) \equiv \delta_{ij}\beta_0(x, u), \ b(x, u) \equiv D(x)b_0(u),$$

where δ_{ij} is the Kroneckder symbol, $\beta_0 : \mathbb{R}^d \times \mathbb{R} \to \mathbb{R}$, $D : \mathbb{R}^d \to \mathbb{R}^d$, $b_0 : \mathbb{R} \to \mathbb{R}$, Theorem 2.2 follows under weaker hypotheses on a_{ij} and b.

2.2 Existence and Uniqueness of Mild Solutions to Isotropic NFPE

We shall study here Eq. (2.2), that is,

$$\begin{aligned} &u_t(t, x) - \Delta\beta(x, u(t, x)) + \mathrm{div}(D(x)b(u(t, x))u(t, x)) = 0, \\ &\qquad\qquad\qquad (t, x) \in (0, \infty) \times \mathbb{R}^d, \\ &u(0, x) = u_0(x), \ x \in \mathbb{R}^d, \end{aligned} \tag{2.80}$$

where $\beta : \mathbb{R}^d \times \mathbb{R} \to \mathbb{R}$ is monotonically nondecreasing and $D : \mathbb{R}^d \to \mathbb{R}^d$, $b : \mathbb{R} \to \mathbb{R}$ are given functions to be made precise below.

As mentioned earlier, the Cauchy problem (2.80) with the conditions

$$u(t,x) \geq 0,\ \forall t \in [0,\infty) \text{ and a.e. } x \in \mathbb{R}^d, \tag{2.81}$$

$$\int_{\mathbb{R}^d} u(t,x)dx = \int_{\mathbb{R}^d} u(0,x)dx = 1,\ \forall t \geq 0, \tag{2.82}$$

is relevant in statistical mechanics to describe the anomalous diffusion in a disordered environment, mean field game theory, as well as in stochastic analysis, where it is used to reproduce the microscopic dynamics of the solution $X(t)$ to the corresponding McKean–Vlasov SDE, that is,

$$\begin{aligned} &dX(t) = D(X(t))b(u(t,X(t))dt + \sqrt{\frac{2\beta(X(t),u(t,X(t)))}{u(t,X(t))}}\,dW(t),\\ &X(0) = X_0,\\ &\mathcal{L}_{X(t)}(dx) = u(t,x)dx,\ t \geq 0, \end{aligned} \tag{2.83}$$

by the macroscopic evolution of its time marginal law denoted by $\mathcal{L}_{X(t)}$.

In order to represent (2.80) as a nonlinear Cauchy problem (1.18) with A m-accretive in L^1, we shall use herein various hypotheses on coefficients β, D and b. We shall consider first the following hypotheses.

(i) $\beta \in C^2(\mathbb{R}^d \times \mathbb{R})$, $\beta_r(x,r) > 0$, $\forall r \neq 0$, $x \in \mathbb{R}^d$; $\beta(x,0) \equiv 0$ *and*

$$\sup\{|\Delta_x \beta(x,r)|;\ r \in \mathbb{R}\} \in L^1, \tag{2.84}$$

$$|\beta(x,r)| \leq \alpha_1 |r|,\ \forall r \in \mathbb{R},\ x \in \mathbb{R}^d, \tag{2.85}$$

where $\alpha_1 > 0$.

(ii) $D \in L^\infty(\mathbb{R}^d;\mathbb{R}^d)$, div $D \in L^m_{\rm loc}(\mathbb{R}^d)$, $(\text{div } D)^- \in L^\infty$, it where $m > \frac{d}{2}$ if $d \geq 2$, $m = 1$ if $d = 1$.

(iii) $b \in C^1(\mathbb{R}) \cap C_b(\mathbb{R})$, $b(r) \geq 0$, $\forall r \in \mathbb{R}$.

(iv) $|b^*(r) - b^*(\bar{r})| \leq \alpha_2 |\beta(x,r) - \beta(x,\bar{r})|$, $\forall r, \bar{r} \in \mathbb{R}$, $x \in \mathbb{R}^d$, *where* $b^*(r) \equiv b(r)r$ *and* $\alpha_2 > 0$.

We note that (iv) is equivalent to

(iv)′ $|b'(r)r + b(r)| \leq \alpha_2 \beta_r(x,r)$, $\forall r \in \mathbb{R}$, $x \in \mathbb{R}^d$,

and, in particular, it holds if $b^* \in C^1_b(\mathbb{R})$ and $\beta_r > \alpha_2 > 0$ on $\mathbb{R}^d \times \mathbb{R}$, that is, if $r \to \beta(x,r)$ is nondegenerate.

The mild solution to (2.80) is defined as in the previous case. Namely,

Definition 2.2 A function $u : [0, \infty) \to L^1$ is said to be a *mild solution* to (2.80) if $u \in C([0, \infty); L^1)$ and we have

$$u(t) = \lim_{h\to 0} u_h(t) \text{ in } L^1, \ \ \forall t \in [0, \infty), \tag{2.86}$$

uniformly on compacts in $[0, \infty)$, where, for each $T > 0$, $u_h : [0, T] \to L^1$ is the step function,

$$u_h(t) = u_h^j, \ \forall t \in [jh, (j+1)h), \ j = 0, 1, \dots, N = \left[\tfrac{T}{h}\right], \tag{2.87}$$

$$u_h^{j+1} - h\Delta\beta(x, u_h^{j+1}) + h \operatorname{div}(Db(u_h^{j+1})u_h^{j+1}) = u_h^j \text{ in } \mathscr{D}'(\mathbb{R}^d), \tag{2.88}$$

$$u_h^j \in L^1, \ \forall j = 0, \dots, N; \ u_h^0 = u_0. \tag{2.89}$$

If we denote by $A_0 : L^1 \to L^1$ the operator

$$\begin{aligned} A_0(u) &= -\Delta\beta(x, u) + \operatorname{div}(Db(u)u), \ \forall u \in D(A_0), \\ D(A_0) &= \left\{u \in L^1; -\Delta\beta(x, u) + \operatorname{div}(Db(u)u) \in L^1\right\}, \end{aligned} \tag{2.90}$$

where Δ and div are taken in the sense of Schwartz distributions on $\mathbb{R}^d$, then the system (2.87)–(2.89) can be written as

$$\begin{aligned} &u_h^{j+1} + hA_0(u_h^{j+1}) = u_h^j, \ j = 0, 1, \dots, N, \\ &u_h^0 = u_0. \end{aligned}$$

As mentioned earlier, this means that u is a *mild solution* to the Cauchy problem

$$\begin{aligned} &\frac{du}{dt} + A_0(u) = 0, \ \forall t \geq 0, \\ &u(0) = u_0, \end{aligned} \tag{2.91}$$

in the sense of Definition 6.2, and so, for each $u_0 \in \overline{D(A_0)}$ problem (2.91) has a unique mild solution $u \in C([0, \infty); L^1)$, if A_0 is m-accretive in L^1, that is,

$$R(I + \lambda A_0) = L^1, \ \forall \lambda > 0, \tag{2.92}$$

$$|(I + \lambda A_0)^{-1}u - (I + \lambda A_0)^{-1}v|_1 \leq |u - v|_1, \ \ \forall u, v \in L^1, \tag{2.93}$$

where $|\cdot|_1$ is as usually the norm of $L^1 = L^1(\mathbb{R}^d)$.

Moreover, the solution $u = u(t, u_0) \equiv S(t)u_0 = e^{-tA_0}u_0$ defines a semigroup of contractions on $\overline{D(A_0)}$, that is, $u(t, u_0) \in \overline{D(A_0)}, \forall t \geq 0$, and

$$u(t+s, u_0) = u(t, u(s, u_0)), \ \forall t, s, \geq 0, \ u_0 \in \overline{D(A_0)}, \tag{2.94}$$

$$|u(t, u_0) - u(t, \bar{u}_0)|_1 \leq |u_0 - \bar{u}_0|_1, \ \forall t \geq 0, \ u_0, \bar{u}_0 \in \overline{D(A_0)}, \tag{2.95}$$

and, as seen by (2.86)–(2.89), $u = u(t, u_0)$ is given by the exponential formula

$$u(t, u_0) = \lim_{n\to\infty} \left(I + \frac{t}{n} A_0\right)^{-n} u_0 \ \text{ in } L^1, \ \forall t \geq 0, \tag{2.96}$$

uniformly on compacts in $[0, \infty)$. In the previous case treated in Sect. 2.1, we have constructed from the operator A_0 (see (2.18)) an m-accretive section $A \subset A_0$ such that $(I + \lambda A)^{-1} f \subset (I + \lambda A_0)^{-1} f$, $f \in L^1$, $\lambda > 0$. Then, for each $u_0 \in \overline{D(A)}$, the Cauchy problem

$$\begin{aligned} &\frac{du}{dt} + A(u) = 0, \ \ t \geq 0, \\ &u(0) = u_0, \end{aligned} \tag{2.97}$$

has a unique mild solution u, which is just a mild solution to (2.80) in the sense of Definition 2.2. However, as mentioned earlier, this does not imply the uniqueness of the mild solution to (2.80) because, in general, the operator A defined as in (2.18) is not unique. In fact, in this case it depends on the approximating equation used to solve the equation $u + \lambda A_0(u) = f$ and so, Theorems 2.1 and 2.2 do not provide the uniquness of mild solutions to (2.80), but only the existence of a semigroup-flow $S(t)$ of solutions. For the uniqueness one should request further conditions. Here, we shall prove that, *under Hypotheses* (i)–(iv), *the operator A_0 is itself m-accretive* (that is, $I+\lambda A_0$ is invertible for all $\lambda > 0$, and also (2.93) holds) and so *the Crandall & Liggett existence theorem is applicable to the Cauchy problem* (2.91) *to derive not only the existence, but also the uniqueness of a mild solution u to* (2.80).

The main existence and uniqueness result is Theorem 2.4.

Theorem 2.4 *Under Hypotheses* (i)–(iv), *for each $u_0 \in L^1$ there is a unique mild solution $u = u(t, u_0)$ to Eq.* (2.80) *which is also a distributional solution. Moreover,* (2.94)–(2.96) *hold and, if $u_0 \in \mathcal{P}$, then $u(t) \in \mathcal{P}, \forall t \geq 0$, that is,* (2.81) *and* (2.82) *hold. Finally, if $u_0 \in L^1 \cap L^\infty$, then $u \in L^\infty((0, T) \times \mathbb{R}^d)$ for all $T > 0$.*

Since Hypothesis 5.2 in Theorem 5.1 in Chap. 5 is fulfilled by assumptions (ii), (iii) and (2.85), Theorem 2.4 implies that the corresponding McKean–Vlasov SDE (2.83) has a probabilistically weak solution and the probabilistic representation (5.5) holds.

As mentioned earlier, the continuous semigroup $t \to u = u(t, u_0) = S(t)u_0$ given by Theorem 2.4 is the *nonlinear Fokker–Planck flow* associated with the nonlinear diffusion $\beta = \beta(u)$ and the drift term $u \to Db(u)u$. This means that the operator A_0 defined by (2.90) is the infinitesimal generator of the continuous semigroup of contractions $S(t)u_0$, that is, of the Fokker–Planck flow. Theorem 2.4 can be rephrased in terms of semigroup theory as follows: *Under Hypotheses* (i)–(iv), *the operator* A_0 *generates in a weak-mild sense a continuous semigroup of contractions in* L^1 *which leaves invariant the set* $\mathcal{P}$ *of all probability densities.*

As mentioned earlier, Theorem 2.4 is implied by the following key result.

Proposition 2.2 *Let* $1 \le d$. *Then, under Hypotheses* (i)–(iv), *the operator* A_0 *is m-accretive in* L^1. *Moreover, one has*

$$\overline{D(A_0)} = L^1, \tag{2.98}$$

$$(I + \lambda A_0)^{-1}(\mathcal{P}) \subset \mathcal{P}, \ \forall \lambda > 0. \tag{2.99}$$

(Here, $\overline{D(A_0)}$ *is the closure of* $D(A_0)$ *in* L^1.*)*

The following uniqueness lemma is one of the main steps of the proof.

Lemma 2.5 *For all* $f \in L^1$ *and* $\lambda > 0$, *there is at most one solution* $y \in D(A_0)$ *to the equation*

$$y + \lambda A_0(y) = f. \tag{2.100}$$

Proof We shall prove first that each solution $y \in D(A_0)$ to (2.100) satisfies

$$\beta(x, y) \in W^{1,q}_{\rm loc}(\mathbb{R}^d), \ \forall q \in \left[1, \frac{d}{d-1}\right), \ \text{if } d \ge 2, \tag{2.101}$$

$$\beta(x, y) \in W^{1,\infty}_{\rm loc}(\mathbb{R}^d), \ \text{if } d = 1, \tag{2.102}$$

and setting $\frac{d}{d-2} := \infty$ if $d = 2$

$$\|\nabla\beta(x, y)\|_{L^q(B_R)} + \|\beta(x, y)\|_{L^p(B_R)} \le C_R(|f|_1 + |y|_1), \ 1 < p < \frac{d}{d-2}, \quad \forall q \in \left[1, \frac{d}{d-1}\right), \tag{2.103}$$

for all $B_R = \{x \in \mathbb{R}^d; \ |x| < \mathbb{R}\}$. We note that (2.103) follows by (2.101) via the Sobolev–Gagliardo–Nirenberg theorem (see Theorem 6.12).

Consider first the case d=1. We have $\beta(x, y) \in L^1$ by Hypothesis (i) and

$$(\beta(x, y))'' - (Db^*(y))' = \frac{1}{\lambda}(y - f) \text{ in } \mathcal{D}'(\mathbb{R}).$$

Since $Db^*(y) \in L^1(\mathbb{R})$, it follows that $(\beta(x,y))' \in L^1_{loc}(\mathbb{R})$ and so $\beta(x,y) \in W^{1,1}_{loc}(\mathbb{R})$. Then, by (iv), $b^*(y) \in W^{1,1}_{loc}(\mathbb{R})$ and so, by (ii) we infer that $Db^*(y) \in W^{1,1}_{loc}(\mathbb{R})$ and $(\beta(x,y))'' \in L^1_{loc}(\mathbb{R})$. Hence, $\beta(x,y) \in W^{1,\infty}_{loc}(\mathbb{R})$, as claimed.

Consider now the case $2 \le d$. By (2.100), we have, for all $\varphi \in C_0^2(\mathbb{R}^d)$,

$$\begin{aligned}\frac{1}{\lambda}(y-f)\varphi &= \varphi\Delta\beta(x,y)-\varphi\,\mathrm{div}(Db^*(y))\\ &= \Delta(\varphi\beta(x,y))-\beta(x,y)\Delta\varphi-\mathrm{div}(D\varphi b^*(y))-2\nabla\beta(x,y)\cdot\nabla\varphi+(D\cdot\nabla\varphi)b^*(y)\\ &= \Delta(\varphi\beta(x,y))+\beta(x,y)\Delta\varphi+(D\cdot\nabla\varphi)b^*(y)-\mathrm{div}(2\beta(x,y)\nabla\varphi+D\varphi b^*(y)),\end{aligned}$$

and, therefore,

$$\Delta(\varphi\beta(x,y)) = f_1 + \mathrm{div}\, f_2 \ \text{ in } \mathscr{D}'(\mathbb{R}^d), \tag{2.104}$$

where

$$\begin{aligned} f_1 &= \frac{1}{\lambda}(y-f)\varphi - \beta(x,y)\Delta\varphi - (D\cdot\nabla\varphi)b^*(y)\\ f_2 &= 2\beta(x,y)\nabla\varphi + D\varphi b^*(y).\end{aligned} \tag{2.105}$$

We set $u=\varphi\beta(x,y)$, $u_\varepsilon = u * \Psi_\varepsilon$, $f_i^\varepsilon = f_i * \Psi_\varepsilon$, $i=1,2$, where Ψ_ε is a standard mollifier, that is,

$$\Psi_\varepsilon(x) = \frac{1}{\varepsilon^d}\Psi\left(\frac{x}{\varepsilon}\right),\ \Psi \in C_0^\infty(\mathbb{R}^d), \text{support}\,\Psi \subset \{x; |x|\le 1\}, \int_{\mathbb{R}^d}\Psi(x)dx = 1.$$

Let $\mathscr{O}$ and $\mathscr{O}'$ be open balls in $\mathbb{R}^d$ centered at zero such that $\overline{\mathscr{O}'} \subset \mathscr{O}$ and choose $\varphi \in C_0^2(\mathbb{R}^d)$ such that $\varphi=1$ on $\mathscr{O}'$ and $(\mathrm{supp}\,\varphi)_\varepsilon \subset \mathscr{O}$, $\varepsilon \in (0,1]$, where $(\mathrm{supp}\,\varphi)_\varepsilon$ denotes the closed ε-neighbourhood of $\mathrm{supp}\,\varphi$. Then, by (2.104) we have

$$\Delta u_\varepsilon = f_1^\varepsilon + \mathrm{div}\, f_2^\varepsilon \ \text{ in } \mathscr{O},\quad u_\varepsilon \in C_0^\infty(\mathscr{O}). \tag{2.106}$$

Hence, by the uniqueness of the solution u_ε to (2.106), we have $u_\varepsilon = u_\varepsilon^1 + u_\varepsilon^2$, where $u_\varepsilon^1, u_\varepsilon^2 \in C^\infty(\mathscr{O}) \cap C(\overline{\mathscr{O}})$ are the solutions to the boundary value problems

$$\Delta u_\varepsilon^1 = f_1^\varepsilon \text{ in } \mathscr{O},\ \ u_\varepsilon^1 = 0 \text{ on } \partial\mathscr{O}, \tag{2.107}$$

$$\Delta u_\varepsilon^2 = \mathrm{div}\, f_2^\varepsilon \text{ in } \mathscr{O},\ \ u_\varepsilon^2 = 0 \text{ on } \partial\mathscr{O}. \tag{2.108}$$

By the standard existence theory for elliptic equations, we know that (see, e.g., [34], Corollary 12)

$$\|u_\varepsilon^1\|_{W_0^{1,q}(\mathscr{O})} \le C\|f_1^\varepsilon\|_{L^1(\mathscr{O})} \le C(|y|_1 + |f|_1),\ \forall\,\varepsilon > 0, \tag{2.109}$$

where $1 \le q < \frac{d}{d-1}$ and where we used Hypotheses (i)–(iii) for the last inequality. By the Sobolev–Galiardo–Nirenberg theorem, it follows by (2.109) that we have

$$\|u_\varepsilon^1\|_p \le C(|f|_1 + |y|_1), \quad \forall\, p \in \left[1, \frac{d}{d-2}\right) \text{ if } d > 2, \tag{2.110}$$

$$\|u_\varepsilon^1\|_p \le C(|f|_1 + |y|_1), \quad \forall\, p \ge 1 \text{ if } d = 2. \tag{2.111}$$

(In the following, we shall denote by the same symbol C several positive constants independent of ε and $\|\cdot\|_p$ is the norm of $L^p(\mathcal{O})$.)

Consider now the solution u_ε^2 to Eq. (2.108).

If $\psi \in L^m(\mathcal{O})$, $m > d$, and $\theta \in W^{2,m}(\mathcal{O}) \cap W_0^{1,m}(\mathcal{O})$ is the solution to the Dirichlet problem

$$-\Delta\theta = \psi \text{ in } \mathcal{O};\quad \theta = 0 \text{ on } \partial\mathcal{O}, \tag{2.112}$$

we see by (2.112) and by the Morrey embedding theorem (see, e.g., [30], p. 282) that $\nabla\theta \in L^\infty(\mathcal{O})$ and, therefore, by Green's formula, since $u_\varepsilon^2 = \theta = 0$ on $\partial\mathcal{O}$,

$$\int_{\mathcal{O}} u_\varepsilon^2 \Delta\theta dx = -\int_{\mathcal{O}} f_2^\varepsilon \cdot \nabla\theta dx \le |f_2^\varepsilon|_1 \|\nabla\theta\|_\infty \le C(|f|_1 + |y|_1)\|\psi\|_m. \tag{2.113}$$

This yields

$$\left|\int_{\mathcal{O}} u_\varepsilon^2 \psi dx\right| \le C(|f|_1 + |y|_1)\|\psi\|_m,\ \forall \psi \in L^m(\mathcal{O}). \tag{2.114}$$

Then, if $\frac{1}{m'} = 1 - \frac{1}{m}$, by (2.114) it follows by duality that $u_\varepsilon^2 \in L^{m'}(\mathcal{O}) \subset L^q(\mathcal{O})$ for all $q \in \left[1, \frac{d}{d-1}\right)$ and

$$\|u_\varepsilon^2\|_q \le C(|f|_1 + |y|_1),\ \ \forall \varepsilon > 0,$$

and so, by (2.109), it follows also that $u_\varepsilon^i \in L^q(\mathcal{O})$, $i = 1, 2$, and

$$\|u_\varepsilon^i\|_q \le C(|f|_1 + |y|_1),\ \forall q \in \left[1, \frac{d}{d-1}\right),\ \ i = 1, 2.$$

Hence,

$$\|u_\varepsilon\|_q \le C(|f|_1 + |y|_1),\ \forall \varepsilon > 0,\ q \in \left[1, \frac{d}{d-1}\right). \tag{2.115}$$

Finally, taking into account that $u_\varepsilon = (\varphi\beta(x,y)) * \Psi_\varepsilon$, by letting $\varepsilon \to 0$ we see by (2.115) and (2.110)–(2.111) that

$$\|\varphi\beta(x,y)\|_q \le C(|f|_1 + |y|_1), \ \ \forall q \in \left[1, \frac{d}{d-1}\right).$$

Because φ and the corresponding ball $\mathscr{O}$ are arbitrary, we conclude that $y, \beta(x,y) \in L^q_{\rm loc}(\mathbb{R}^d)$ and that (for a possible larger C, still independent of ε)

$$\|\beta(x,y)\|_q \le C(|f|_1 + |y|_1), \ \forall q \in \left[1, \frac{d}{d-1}\right).$$

In particular, by Hypothesis (iv), this implies that

$$\|f_2\|_q \le C(|f|_1 + |y|_1)$$

and, therefore,

$$\|f_2^\varepsilon\|_q \le C(|f|_1 + |y|_1), \ \forall \varepsilon > 0, q \in \left[1, \frac{d}{d-1}\right). \tag{2.116}$$

Now, we shall improve the last estimate by invoking a bootstrap argument. Namely, we take in (2.114) $\psi \in L^\ell(\mathscr{O})$, where $\frac{d}{2} < \ell$. This yields as above that

$$\left|\int_{\mathscr{O}} u_\varepsilon^2 \psi \, dx\right| \le \int_{\mathscr{O}} |f_2^\varepsilon| \, |\nabla\theta| dx \le C\|f_2^\varepsilon\|_q \|\nabla\theta\|_{q'}$$

for all $q \in \left[1, \frac{d}{d-1}\right)$ and $q' = \frac{q}{q-1} > d$. Again by the Sobolev–Galiardo–Nirenberg theorem, we have, for all $\ell \in \left(\frac{d}{2}, d\right)$ that $q' := d\ell/(d-\ell) > d$ and

$$\|\nabla\theta\|_{q'} \le C\|\theta\|_{W^{2,\ell}(\mathscr{O})} \le C\|\psi\|_\ell. \tag{2.117}$$

This yields, for all $\ell \in \left(\frac{d}{2}, d\right)$,

$$\left|\int_{\mathscr{O}} u_\varepsilon^2 \psi \, dx\right| \le C\|f_2^\varepsilon\|_q \|\psi\|_\ell \le C(|f|_1 + |y|_1)\|\psi\|_\ell, \ \forall \psi \in L^\ell(\mathscr{O}), \tag{2.118}$$

and, therefore, putting $\frac{d}{d-2} := \infty$, if $d = 2$,

$$\|u_\varepsilon^2\|_r \le C(|f|_1 + |y|_1), \ \forall r \in \left[1, \frac{d}{d-2}\right), \tag{2.119}$$

Then, by (2.110) and (2.111), we get

$$\|u_\varepsilon\|_r \le C(|f|_1 + |y|_1),\ \forall r \in \left[1, \frac{d}{d-2}\right). \tag{2.120}$$

Letting $\varepsilon \to 0$, this yields the bound for $\|y\|_{L^p(B_R)}$ in (2.103), since $\varphi \in C_0^2(\mathbb{R}^d)$ was arbitrary. Furthermore, (2.120) implies that (2.116) is strengthened to

$$\|f_2^\varepsilon\|_\nu \le C(|f|_1 + |y|_1),\ \forall \nu \in \left[1, \frac{d}{d-2}\right),\ \varepsilon > 0. \tag{2.121}$$

For $\varepsilon \to 0$, this yields, since f_2 has compact support in $\mathscr{O}$,

$$|f_2|_\nu \le C(|f|_1 + |y|_1),\ \ \forall \nu \in \left[1, \frac{d}{d-2}\right). \tag{2.122}$$

Hence, for $d \in [2, 3]$, we get

$$\|f_2^\varepsilon\|_2 \le C(|f|_1 + |y|_1),\ \forall \varepsilon > 0, \tag{2.123}$$

and so

$$\|\mathrm{div}(f_2^\varepsilon)\|_{H^{-1}(\mathscr{O})} \le C(|f|_1 + |y|_1),\ \forall \varepsilon > 0.$$

Then, since $\frac{d}{d-1} \le 2$, by Eq. (2.108) it follows that $u_2^\varepsilon \in H_0^1(\mathscr{O})$ and

$$\|u_2^\varepsilon\|_{W^{1,q}(\mathscr{O})} \le \|u_2^\varepsilon\|_{H_0^1(\mathscr{O})} \le C(|f|_1 + |y|_1),\ \forall \varepsilon > 0. \tag{2.124}$$

Hence, by (2.109), we have

$$\|u_\varepsilon\|_{W_0^{1,q}(\mathscr{O})} \le C(|y|_1 + |f|_1),\ \forall \varepsilon > 0,\ \text{if } d = 2, 3. \tag{2.125}$$

Letting $\varepsilon \to 0$ we get the estimate

$$\|u\|_{W^{1,q}(\mathscr{O})} \le C(|f|_1 + |y|_1),\ \ d = 2, 3. \tag{2.126}$$

Recalling that $u = \varphi\beta(x, y)$, since φ and the corresponding ball $\mathscr{O}$ are arbitrary, this implies (2.101), for $d \in [2, 3]$, as claimed.

We shall consider now the case $d > 3$. To this end, we come back to Eq. (2.104) and note that $\varphi\beta(x, y) = u_1 + u_2$, where u_1, u_2 are solutions to the equations

$$\Delta u_1 = f_1 \text{ in } \mathscr{D}'(\mathbb{R}^d), \tag{2.127}$$

$$\Delta u_2 = \mathrm{div}(f_2) \text{ in } \mathscr{D}'(\mathbb{R}^d), \tag{2.128}$$

where f_1, f_2 are given by (2.105).

Since $f_1 \in L^1$, it follows that u_1 is given by the representation formula

$$u_1 = -E_d * f_1 \ \text{ in } \mathbb{R}^d,$$

where $E_d(x) \equiv \frac{1}{(d-2)\omega_d |x|^{d-2}}$ is the fundamental solution to Δ.

Hence (see (6.21)–(6.22) in Sect. 6.3), $u_1 \in M^{\frac{d}{d-2}}(\mathbb{R}^d) \subset L^p_{\text{loc}}(\mathbb{R}^d)$, $\forall p \in \left[1, \frac{d}{d-2}\right)$ and $|\nabla u_1| = |\nabla E_d * f_1| \in M^{\frac{d}{d-1}}(\mathbb{R}^d) \subset L^p_{\text{loc}}(\mathbb{R}^d)$, $\forall p \in \left[1, \frac{d}{d-1}\right)$ with

$$\|\nabla u_1\|_{L^p(B_R)} \le C(|y|_1 + |f|_1),\ \forall R > 0,\ p \in \left[1, \frac{d}{d-1}\right). \tag{2.129}$$

As regards the solution u_2 to Eq. (2.128), we note that from (2.122) it follows that $f_2 \in L^p(\mathbb{R}^d)$, $\forall p \in \left[1, \frac{d}{d-1}\right)$. Let $u_{2,\varepsilon} = u_2 * \Psi_\varepsilon$. Then

$$u_{2,\varepsilon} = -\nabla E_d * f_2^\varepsilon \ \text{in } \mathbb{R}^d.$$

Taking into account that $|\nabla^2 E_d(x)| \le C|x|^{-d}$, $\forall x \ne 0$, it follows by the Calderon–Zygmund theorem (see Theorem 6.13 and estimate (2.121)) that

$$|\nabla u_{2,\varepsilon}|_p \le C|f_2^\varepsilon|_p \le C(|f|_1 + |y|_1),\ \forall p \in \left[1, \frac{d}{d-1}\right)$$

and, after letting $\varepsilon \to 0$, together with (2.129) this yields

$$\|\nabla u\|_{L^p(B_R)} \le C_R(|f|_1 + |y|_1),\ \forall p \in \left[1, \frac{d}{d-1}\right),$$

and so (2.101) and (2.103) hold for all $d \ge 2$.

Now, let us prove the uniqueness of the solution to (2.100).

If $y_1, y_2 \in D(A_0)$ are two solutions, we have

$$y_1 - y_2 - \lambda\Delta(\beta(x, y_1) - \beta(x, y_2)) + \lambda\, \text{div}(D(b^*(y_1) - b^*(y_2))) = 0. \tag{2.130}$$

Then we introduce the cut-off function $\eta \in C^2([0, \infty))$ such that

$$\eta(r) \ge 0,\ \eta(r) = 1,\ \forall r \in [0, 1];\ \eta(r) = 0,\ \forall r \in [2, \infty).$$

We set

$$\varphi_n(x) = \eta\left(\frac{|x|^2}{n}\right),\ \forall x \in \mathbb{R}^d,\ n \in \mathbb{N},$$

and note that

$$|\nabla\varphi_n(x)| \le \frac{4}{\sqrt{n}}\,|\eta'|_\infty,\ \forall\, x \in \mathbb{R}^d, \tag{2.131}$$

$$|\Delta\varphi_n(x)| \le \frac{1}{n}\,(2d|\eta'|_\infty + 8|\eta''|_\infty),\ \forall\, x \in \mathbb{R}^d. \tag{2.132}$$

By (2.130), we have

$$\begin{aligned}
&\varphi_n(y_1 - y_2) - \lambda\Delta(\varphi_n(\beta(x, y_1) - \beta(x, y_2))) + \lambda\,\mathrm{div}(D\varphi_n(b^*(y_1) - b^*(y_2)))\\
&\quad = \lambda(\nabla\varphi_n \cdot D)(b^*(y_1) - b^*(y_2)) - \lambda(\beta(x, y_1) - \beta(x, y_2))\Delta\varphi_n\\
&\quad\quad - 2\lambda\nabla\varphi_n \cdot \nabla(\beta(x, y_1) - \beta(x, y_2)) \text{ in } \mathscr{D}'(\mathbb{R}^d).
\end{aligned} \tag{2.133}$$

Let $\mathscr{X}_\delta : \mathbb{R} \to \mathbb{R}$ be the function (2.41) and let

$$j_\delta(s) = \int_0^s \mathscr{X}_\delta(r)dr,\ \forall\, s \in \mathbb{R}.$$

We know by (2.101)–(2.103) that

$$\beta(x, y_i) \in L^p_{\mathrm{loc}},\ \forall p \in \left[1, \frac{d}{d-2}\right),\ \ \nabla\beta(x, y_i) \in L^q_{\mathrm{loc}},\ \forall q \in \left[1, \frac{d}{d-1}\right).$$

Moreover, by Hypothesis (iv) it follows that

$$|b^*(y_i)| \le \alpha_2|\beta(x, y_i)|,\ \ |\nabla b^*(y_i)| \le \alpha_2|\nabla\beta(x, y_i)|,\ \ \text{a.e. on } \mathbb{R}^d,\ i = 1, 2,$$

and, therefore,

$$b^*(y_i) \in L^p_{\mathrm{loc}},\ |\nabla b^*(y_i)| \in L^q_{\mathrm{loc}},\ i = 1, 2,\ \forall\, p \in \left[1, \frac{d}{d-2}\right), q \in \left[1, \frac{d}{d-1}\right).$$

This implies that

$$\mathrm{div}(Db^*(y_i)) = D \cdot \nabla b^*(y_i) + b^*(y_i)\mathrm{div}\, D \in L^1_{\mathrm{loc}},\ i = 1, 2,$$

because, by (ii), $\mathrm{div}(D) \in L^m_{\mathrm{loc}}$ for some $m > \frac{d}{2}$. Since $y_i \in D(A_0)$, $i = 1, 2$, we have therefore that $\Delta(\varphi_n(\beta(x, y_1) - \beta(x, y_2)))$ and $\varphi_n(\beta(x, y_1) - \beta(x, y_2))$ are in L^1.

This yields

$$
\begin{aligned}
&-\int_{\mathbb{R}^d} \Delta(\varphi_n(\beta(x,y_1)-\beta(x,y_2)))\mathscr{X}_\delta(\beta(x,y_1)-\beta(x,y_2))dx \\
&=\int_{\mathbb{R}^d} \nabla(\varphi_n(\beta(x,y_1)-\beta(x,y_2)))\nabla(\beta(x,y_1) \\
&-\beta(x,y_2))\mathscr{X}_\delta'(\beta(x,y_1)-\beta(x,y_2))dx \\
&\geq \frac{1}{\delta}\int_{[|\beta(x,y_1)-\beta(x,y_2)|\leq\delta]} (\nabla\varphi_n\cdot\nabla(\beta(x,y_1)-\beta(x,y_2)))(\beta(x,y_1)-\beta(x,y_2))dx,
\end{aligned}
$$

and so, by (2.131)–(2.133), we have

$$
\begin{aligned}
&\int_{\mathbb{R}^d} \varphi_n(y_1-y_2)\mathscr{X}_\delta(\beta(x,y_1)-\beta(x,y_2))dx \\
&\qquad+\frac{\lambda}{\delta}\int_{[|\beta(x,y_1)-\beta(x,y_2)|\leq\delta]} \varphi_n(b^*(y_1)-b^*(y_2))(D\cdot\nabla(\beta(x,y_1)-\beta(x,y_2)))dx \\
&\qquad\leq \frac{C\lambda}{\sqrt{n}}+I^\delta_{\lambda,n},
\end{aligned}
\tag{2.134}
$$

where

$$
I^\delta_{\lambda,n}\leq\lambda|D|_\infty\int_{[|\beta(x,y_1)-\beta(x,y_2)|\leq\delta]} |\varphi_n|\cdot|\nabla(\beta(x,y_1)-\beta(x,y_2))|dx\to 0 \text{ as } \delta\to 0,
$$

because

$$
|\nabla\beta(x,y_1)-\nabla\beta(x,y_2)|=0, \text{ a.e. on } \{x;\ |\beta(x,y_1)(x)-\beta(x,y_2)(x)|=0\}.
$$

To obtain (2.134), we have used the relation

$$
\begin{aligned}
-2\lambda&\int_{\mathbb{R}^d} (\nabla\varphi_n\cdot\nabla(\beta(x,y_1)-\beta(x,y_2)))\mathscr{X}_\delta(\beta(x,y_1)-\beta(x,y_2))dx \\
&=-2\lambda\int_{\mathbb{R}^d} \nabla\varphi_n\cdot\nabla j_\delta(\beta(x,y_1)-\beta(x,y_2))dx \\
&=2\lambda\int_{\mathbb{R}^d} \Delta\varphi_n j_\delta(\beta(x,y_1)-\beta(x,y_2))dx \\
&\leq\frac{2\lambda}{n}(2d|\eta'|_\infty+8|\eta''|_\infty)\int_{\mathbb{R}^d}(|\beta(x,y_1(x))|+|\beta(x,y_2(x))|)dx \\
&\leq\frac{4\lambda}{n}\alpha_1(d|\eta'|_\infty+4|\eta''|_\infty)(|y_1|_1+|y_2|_1)\leq\frac{C\lambda}{\sqrt{n}}|f|_1,
\end{aligned}
$$

where C is independent of n. On the other hand, recalling that

$$D \in L^\infty(\mathbb{R}^d),\ \nabla\beta(x, y_i) \in L^q_{\rm loc},\ 1 \le q < \frac{d}{d-1},$$

while by Hypothesis (iv), we have

$$|b^*(y_1) - b^*(y_2)| \le \alpha_2|\beta(x, y_1) - \beta(x, y_2)|, \text{ a.e. in } \mathbb{R}^d,$$

and so, by (2.134), it follows for $\delta \to 0$ that

$$\int_{\mathbb{R}^d} \varphi_n(y_1 - y_2)\text{sign}(\beta(x, y_1) - \beta(x, y_2))dx \le \frac{C\lambda}{\sqrt{n}}|f|_1.$$

This yields for $n \to \infty$ that $|y_1 - y_2|_1 = 0$, as claimed.

Lemma 2.6 *Assume that $d \ge 1$. Then, for each $f \in L^1$ and all $\lambda > 0$, Eq.* (2.100) *has a unique solution $y = J_\lambda(f)$. Moreover, one has*

$$|J_\lambda(f_1) - J_\lambda(f_2)|_1 \le |f_1 - f_2|_1,\ \forall f_1, f_2 \in L^1,\ \lambda > 0, \tag{2.135}$$

$$J_\lambda(f) \in \mathcal{P},\ \forall\, f \in \mathcal{P},\ \forall\lambda > 0, \tag{2.136}$$

$$J_\lambda(f) \in L^\infty,\ \forall f \in L^\infty, \tag{2.137}$$

and $\overline{D(A_0)} = L^1$.

Proof We assume first that $f \in L^1 \cap L^2$ and approximate Eq. (2.100) by

$$y + \lambda(\varepsilon I - \Delta)(\beta(x, y) + \varepsilon y) + \lambda\,\text{div}(D_\varepsilon b^*_\varepsilon(y)) = f, \tag{2.138}$$

where $b^*_\varepsilon \in C^1_b(\mathbb{R}) \cap C_b(\mathbb{R})$ is a smooth approximation of b^* such that $|b^*_\varepsilon(r)| \le C|r|$, $\lim\limits_{\varepsilon\to 0} b^*_\varepsilon(r) = b(r)r$ uniformly on compacts, and

$$D_\varepsilon = \eta_\varepsilon D,\ \eta_\varepsilon \in C^1_0(\mathbb{R}^d),\ 0 \le \eta_\varepsilon \le 1,\ |\nabla\eta_\varepsilon| \le 1,\ \eta_\varepsilon(x){=}1 \text{ if } |x| < \frac{1}{\varepsilon}. \tag{2.139}$$

Clearly, we have

$$\begin{aligned}&|D_\varepsilon| \in L^\infty \cap L^2,\ |D_\varepsilon| \le |D|,\ \lim_{\varepsilon\to\infty} D_\varepsilon(x) = D(x), \text{ a.e. } x \in \mathbb{R}^d.\\ &\text{div}\, D_\varepsilon \in L^1,\ (\text{div}\, D_\varepsilon)^- \le (\text{div}\, D)^- + \mathbf{1}_{\left[|x|>\frac{1}{\varepsilon}\right]}|D|.\end{aligned} \tag{2.140}$$

A typical example is

$$b_\varepsilon \equiv b * \varphi_\varepsilon, \quad b^*_\varepsilon(r) \equiv \frac{b_\varepsilon(r)r}{1+\varepsilon|r|}, \quad r \in \mathbb{R},$$
$$\varphi_\varepsilon(r) = \frac{1}{\varepsilon}\varphi\left(\frac{r}{\varepsilon}\right), \quad \varphi \in C_0^\infty(\mathbb{R}), \quad \int_{\mathbb{R}} \varphi(x)dx = 1. \tag{2.141}$$

where $*$ is the convolution product symbol.

We can rewrite (2.138) equivalently as the following equation on L^2:

$$\lambda(\beta(x,y)+\varepsilon y) + (\varepsilon I - \Delta)^{-1}y + \lambda(\varepsilon I - \Delta)^{-1}\mathrm{div}(D_\varepsilon b^*_\varepsilon(y)) = (\varepsilon I - \Delta)^{-1}f. \tag{2.142}$$

We set

$$F(y) = \lambda(\beta(x,y)+\varepsilon y) + (\varepsilon I - \Delta)^{-1}y + \lambda(\varepsilon I - \Delta)^{-1}\mathrm{div}(D_\varepsilon b^*_\varepsilon(y))$$

and note that

$$\begin{aligned}
(F(y_1)-F(y_2), y_1-y_2)_2 &= \varepsilon|(\varepsilon I-\Delta)^{-1}(y_1-y_2)|_2^2 \\
&\quad +\lambda(\beta(x,y_1)-\beta(x,y_2), y_1-y_2)_2 + \varepsilon\lambda|y_1-y_2|_2^2 \\
&\quad +|\nabla(\varepsilon I-\Delta)^{-1}(y_1-y_2)|_2^2 - \lambda(D_\varepsilon(b^*_\varepsilon(y_1)-b^*_\varepsilon(y_2)), \nabla(\varepsilon I-\Delta)^{-1}(y_1-y_2))_2 \\
&\geq \varepsilon\lambda|y_1-y_2|_2^2 + |(\varepsilon I-\Delta)^{-1}(y_1-y_2)|_2^2 + |\nabla(\varepsilon I-\Delta)^{-1}(y_1-y_2)|_2^2 \\
&\quad -\lambda|D_\varepsilon|_\infty|b^*_\varepsilon|_{\mathrm{Lip}}|y_1-y_2|_2|\nabla(\varepsilon I-\Delta)^{-1}(y_1-y_2)|_2 \geq 0, \ \text{ for } 0<\lambda<\lambda_\varepsilon.
\end{aligned}$$

It is also clear that $(F(y),y)_2 \geq \alpha_\varepsilon\lambda|y|_2^2$ and so, for $0<\lambda<\lambda_\varepsilon$, F is monotone, continuous and coercive on L^2. Hence, it is surjective and so Eq. (2.142) has a solution $y_\varepsilon \in L^2$ and the latter is then true for all $\lambda > 0$ (see Theorem 6.4). By (2.138), it follows also that $y_\varepsilon, \beta(x,y_\varepsilon) \in H^1$. If $f \in L^1 \cap L^\infty$, we have

$$|y_\varepsilon|_\infty \leq C|f|_\infty, \quad 0<\lambda<\lambda_0<1. \tag{2.143}$$

Indeed, by (2.138) we see that, for $M = |(\mathrm{div}\, D_\varepsilon)^-|_\infty^{\frac{1}{2}}|f|_\infty + \sup\{|\Delta(\beta_\varepsilon(x,r))|, (x,r) \in \mathbb{R}^d \times \mathbb{R}\}$ and $0<\lambda<\lambda_0$,

$$\begin{aligned}
&(y_\varepsilon - |f|_\infty - M) - \lambda\Delta(\beta_\varepsilon(x,y_\varepsilon) - \beta_\varepsilon(x,|f|_\infty+M)) \\
&\quad +\lambda\varepsilon(\beta_\varepsilon(x,u_\varepsilon) - \beta_\varepsilon(x,|f|_\infty+M)) + \lambda\,\mathrm{div}(D_\varepsilon(b^*_\varepsilon(u_\varepsilon) - b^*_\varepsilon(|f|_\infty+M))) \\
&\leq f - |f|_\infty - M - \lambda b^*_\varepsilon(M+|f|_\infty)\mathrm{div}\, D_\varepsilon - \lambda\Delta\beta_\varepsilon(x,|f|_\infty+M) \leq 0,
\end{aligned}$$

where $\beta_\varepsilon(x,r)=\beta(x,r)+\varepsilon r$. (Here we have also used the conditions

$$(\operatorname{div} D)^- \in L^\infty,\ b \ge 0, \tag{2.144}$$

from Hypotheses (ii) and (iii).)

Multiplying the above equation by $\mathscr{X}_\delta((y_\varepsilon-(|f|_\infty+M))^+)$ and integrating over $\mathbb{R}^d$, we get as above, for $\delta\to 0$, $|(y_\varepsilon-|f|_\infty-M)^+|_1\le 0$ and, therefore, by (2.140) and since $b\ge 0$, $y_\varepsilon\le |f|_\infty-M$, a.e. in $\mathbb{R}^d$. Similarly, one gets that $y_\varepsilon\ge M-|f|_\infty$, a.e. in $\mathbb{R}^d$, and so (2.143) follows.

Let us denote the solution to (2.138) by $y_\lambda^\varepsilon(f)$ and define $\beta_\varepsilon(x,r):=\beta(x,r)+\varepsilon r$, $r\in\mathbb{R}$. Then, we multiply the equation

$$\begin{aligned}(y_\lambda^\varepsilon(f_1)-y_\lambda^\varepsilon(f_2))+\lambda(\varepsilon I-\Delta)(\beta_\varepsilon(x,y_\lambda^\varepsilon(f_1))-\beta_\varepsilon(x,y_\lambda^\varepsilon(f_2)))\\ +\lambda\operatorname{div}(D_\varepsilon(b_\varepsilon^*(y_\lambda^\varepsilon(f_1))-b_\varepsilon^*(y_\lambda^\varepsilon(f_2)))=f_1-f_2\end{aligned}$$

by $\mathscr{X}_\delta(\beta_\varepsilon(x,y_\lambda^\varepsilon(f_1))-\beta_\varepsilon(x,y_\lambda^\varepsilon(f_2)))$ and integrate over $\mathbb{R}^d$.

We set for $\lambda\in(0,\lambda_1)$ and $\delta>0$

$$E_{\lambda,\delta}^\varepsilon=\{x\in\mathbb{R}^d;\ |\beta_\varepsilon(x,y_\lambda^\varepsilon(f_1)(x))-\beta_\varepsilon(x,y_\lambda^\varepsilon(f_2)(x))|\le\delta\}.$$

Since $|\beta_\varepsilon(x,r)-\beta_\varepsilon(x,\bar r)|\ge\varepsilon|r-\bar r|$, $r,\bar r\in\mathbb{R}$, and b_ε^* is Lipschitz, we have

$$\begin{aligned}&\int_{\mathbb{R}^d}(y_\lambda^\varepsilon(f_1)-y_\lambda^\varepsilon(f_2))\mathscr{X}_\delta(\beta_\varepsilon(x,y_\lambda^\varepsilon(f_1))-\beta_\varepsilon(x,y_\lambda^\varepsilon(f_2)))dx \qquad (2.145)\\
&\le |f_1-f_2|_1+\lambda\int_{\mathbb{R}^d}(b_\varepsilon^*(y_\lambda^\varepsilon(f_1))-b_\varepsilon^*(y_\lambda^\varepsilon(f_2)))\\
&\quad D_\varepsilon\cdot\nabla(\beta_\varepsilon(x,y_\lambda^\varepsilon(f_1)-\beta_\varepsilon(x,y_\lambda^\varepsilon(f_2)))\mathscr{X}_\delta'(\beta_\varepsilon(x,y_\lambda^\varepsilon(f_1))-\beta_\varepsilon(x,y_\lambda^\varepsilon(f_2)))dx\\
&\le |f_1-f_2|_1+\frac{C_\varepsilon\lambda}{\delta}\int_{E_{\lambda,\delta}^\varepsilon}|\beta_\varepsilon(x,y_\lambda^\varepsilon(f_1))-\beta_\varepsilon(x,y_\lambda^\varepsilon(f_2))|\cdot\\
&\quad\cdot|\nabla(\beta_\varepsilon(x,y_\lambda^\varepsilon(f_1)))-\beta_\varepsilon(x,y_\lambda^\varepsilon(f_2))||D_\varepsilon|dx\\
&\le |f_1-f_2|_1+C_\varepsilon\lambda|D_\varepsilon|_2\left(\int_{E_{\lambda,\delta}^\varepsilon}|\nabla(\beta_\varepsilon(x,y_\lambda^\varepsilon(f_1))-\beta_\varepsilon(x,y_\lambda^\varepsilon(f_2)))|^2dx\right)^{\frac12}.\end{aligned}$$

Then, letting $\delta\to 0$ and recalling that

$$\operatorname{sign}(\beta_\varepsilon(x,y_\lambda^\varepsilon(f_1))-\beta_\varepsilon(x,y_\lambda^\varepsilon(f_2)))=\operatorname{sign}(y_\lambda^\varepsilon(f_1)-y_\lambda^\varepsilon(f_2)),\ \text{a.e. in } \mathbb{R}^d,$$

we get by monotone convergence

$$|y_\lambda^\varepsilon(f_1)-y_\lambda^\varepsilon(f_2)|_1\le|f_1-f_2|_1. \tag{2.146}$$

In particular, for all $f \in L^1 \cap L^\infty$, we have $\sup\limits_{\lambda,\varepsilon>0} |y_\lambda^\varepsilon(f)|_1 \le |f|_1$. Recall also that by (2.143) we have, for some $C_D \in (0,\infty)$,

$$\sup_{\substack{\lambda\in(0,\lambda_0)\\ \varepsilon>0}} |y_\lambda^\varepsilon(t)|_\infty \le C_D |f|_\infty. \tag{2.147}$$

Hence, multiplying (2.138) by $\beta_\varepsilon(x, y_\varepsilon)$ and integrating over $\mathbb{R}^d$, we see that, for some $C \in (0,\infty)$ and all $\lambda \in (0,\lambda_0)$, $\varepsilon \in (0,1)$,

$$\lambda|\nabla\beta_\varepsilon(x, y_\lambda^\varepsilon(f))|_2^2 \le C\left(|f|_\infty + \sup_{|r|\le C_D|f|_\infty} |\beta(r)|\right)|f|_1. \tag{2.148}$$

Now, fix $\lambda \in (0,\lambda_0)$. Set $y_\varepsilon := y_\lambda^\varepsilon(f)$. Then, by (2.147) and (2.148), $\{\beta_\varepsilon(x, y_\varepsilon)\}$ is bounded in H^1 and $\{y_\varepsilon\}$ is bounded in L^2.

This implies that $\{\beta_\varepsilon(x, y_\varepsilon)\}$ is compact in L^2_{loc} and, therefore, along a subsequence $\{\varepsilon\} \to 0$, we have

$$\begin{aligned} y_\varepsilon &\to y \quad \text{weakly in } L^2,\\ \beta(x, y_\varepsilon) &\to v \quad \text{strongly in } L^2_{loc},\\ \nabla\beta_\varepsilon(x, y_\varepsilon) &\to \nabla v \ \text{ weakly in } L^2. \end{aligned} \tag{2.149}$$

Since the map $y \to \beta(x, y)$ is maximal monotone in $L^2(\mathcal{O})$ for every bounded, open $\mathcal{O} \subset \mathbb{R}^d$, it follows that $v(x) = \beta(x, y(x))$, a.e. $x \in \mathbb{R}^d$. Moreover, as $\beta(x,\cdot)$ is, for each $x \in \mathbb{R}^d$, by Hypothesis (i), continuous and strictly monotone, it follows that $y_\varepsilon \to y$ a.e. on $\mathbb{R}^d$, and, therefore, we have

$$b_\varepsilon(y_\varepsilon)y_\varepsilon \to b(y)y, \text{ a.e. in } \mathbb{R}^d, \quad b_\varepsilon(y_\varepsilon)y_\varepsilon \to b(y)y \text{ weakly in } L^2,$$

selecting another subsequence $\{\varepsilon\} \to 0$, if necessary. Then, letting $\varepsilon \to 0$ in (2.138), we see that

$$y - \lambda\Delta\beta(x, y) + \lambda\,\mathrm{div}(Db(y)y) = f \text{ in } H^{-1}. \tag{2.150}$$

(However, since as seen above the solution y to (2.144) is unique, it follows that (2.149) holds for all $\{\varepsilon\} \to 0$.) Moreover, we see that $\beta(x, y) \in H^1$ and (2.147) and (2.148) hold for y, $\beta(x, y)$ replacing $y_\lambda^\varepsilon(f)$ and $\beta(x, y_\lambda^\varepsilon(f))$, respectively. We denote this solution to Eq. (2.150) by $J_\lambda(f)$.

We shall now prove

$$|J_\lambda(f_1) - J_\lambda(f_2)|_1 \le |f_1 - f_2|_1, \ \forall f_1, f_2 \in L^1 \cap L^\infty, \ \lambda \in (0,\lambda_0), \tag{2.151}$$

which implies, in particular, that $J_\lambda(f) \in L^1$, $\forall f \in L^1 \cap L^\infty$.

Let $f \in L^1$ be arbitrary but fixed and let $\{f_n\} \subset L^1 \cap L^\infty$ be such that $f_n \to f$ in L^1 as $n \to \infty$. We set $y_n = y(\lambda, f_n)$, that is,

$$y_n - \lambda\Delta\beta(x, y_n) + \lambda \operatorname{div}(Db(y_n)y_n) = f_n \text{ in } H^{-1}. \tag{2.152}$$

Then, by (2.146), we have

$$|y_n - y_m|_1 \le |f_n - f_m|_1, \ \forall n, m \in \mathbb{N},$$

and so there is $y = \lim\limits_{n\to\infty} y_n$ in L^1. Since A_0 is closed on L^1, it follows by (2.152) that $y \in D(A_0)$ and $y + \lambda A_0 y = f$ and so $R(I + \lambda A_0) = L^1$, as claimed. The fact that y is the unique solution to $y+\lambda A_0 y = f$ follows by Lemma 2.5. Denoting this solution y by $J_\lambda(f)$, we obtain by (2.151) that (2.135) holds.

Now, taking into account that, as seen earlier, we have

$$J_{\lambda_2}(f) = J_{\lambda_1}\left(\frac{\lambda_1}{\lambda_2} f + \left(1 - \frac{\lambda_1}{\lambda_2}\right) J_{\lambda_2}(f)\right), \ \forall \lambda_1, \lambda_2 \in (0, \lambda_0), \tag{2.153}$$

it follows by (2.151) via the Banach fixed point theorem that λ extend to all $\lambda > 0$ such that (2.135) holds.

If $f \in \mathscr{P} \cap L^\infty$, it follows by (2.138) that $y_\varepsilon \in \mathscr{P}$. By (2.149), it follows that $y_\varepsilon \to J_\lambda(f)$ in $L^1_{\rm loc}$ as $\varepsilon \to 0$.

Let us prove that

$$y_\varepsilon \to J_\lambda(f) \text{ in } L^1 \text{ as } \varepsilon \to 0. \tag{2.154}$$

To this end, we shall argue as in the proof of Proposition 2.1. Namely, taking into account (2.135), it suffices to prove (2.154) for $f \in L^1 \cap L^\infty$.

Consider a function Φ given by (2.56). If we multiply (2.138) by $\Phi_\nu \mathscr{X}_\delta(\beta_\varepsilon(x, y_\varepsilon))$, where $\Phi_\nu = \Phi(x)\exp(-\nu\Phi(x))$, $\nu > 0$, and integrate over $\mathbb{R}^d$, we get as in Lemma 2.3 (see (2.59))

$$\int_{\mathbb{R}^d} |y_\varepsilon|\Phi_\nu dx \le \lambda \int_{\mathbb{R}^d} (|\beta_\varepsilon(x, y_\varepsilon)|\,|\Delta\Phi_\nu| + |D_\varepsilon \cdot \nabla\Phi_\nu|\,|b^*_\varepsilon(y_\varepsilon)|)dx + \int_{\mathbb{R}^d} |f|\Phi_\nu dx,$$

and, letting $\nu \to 0$, we get

$$\int_{\mathbb{R}} |y_\varepsilon|\Phi dx \le \int_{\mathbb{R}^d} |f|\Phi dx + \lambda \int_{\mathbb{R}^d} (|\beta_\varepsilon(x, y_\varepsilon)|\,|\Delta\Phi| + |D|_\infty |b|_\infty |y_\varepsilon|\,|\nabla\Phi|)dx.$$

Taking into account that, by (2.143) and (2.148) we have the estimate

$$|y_\varepsilon|_2 + |\beta_\varepsilon(x, y_\varepsilon)|_2 \le C(|f|_2 + |f|_1), \ \forall f \in L^1 \cap L^\infty,$$

we get

$$\int_{\mathbb{R}^d} |y_\varepsilon(x)|\Phi(x)dx \leq \int_{\mathbb{R}^d} |f(x)|\Phi(x)dx + C_\lambda(|f|_1 + |f_2|_2), \ \forall f \in L^1 \cap L^\infty.$$

Since $y_\varepsilon \to J_\lambda(f)$ in L^1_{loc}, we get (2.154) for $f \in L^1 \cap L^\infty$ and by density it extends to all of L^1. Next, by (2.154) and (2.149) it follows also that $J_\lambda(f) \in \mathcal{P}$ if $f \in \mathcal{P}$.

To prove that $\overline{D(A_0)} = L^1$, it suffices to note that, by Hypotheses (ii), (iii), $C_0^\infty(\mathbb{R}^d) \subset D(A_0)$ (because $\mathrm{div}\, D \in L^1_{\mathrm{loc}}$ and $\beta \in C^2(\mathbb{R}^d)$, $b \in C^1 \cap C_b$). This completes the proof of Lemma 2.6.

Proof of Theorem 2.4 (Continued) By Lemmas 2.5 and 2.6, it follows that A_0 is m-accretive in L^1 and $\overline{D(A_0)} = L^1$. Then, as mentioned earlier, the existence and uniqueness of a mild solution u to (2.80) follows by the Crandall and Liggett theorem (Theorem 6.7). Moreover, by (2.136) and (2.96) it follows that $u(t) \in \mathcal{P}$, $\forall t \geq 0$, if $u_0 \in \mathcal{P}$.

We shall show now that u is a distributional solution to (2.80). Since $u_h \to u$ in $L^1((0,T) \times \mathbb{R}^d)$ as $h \to 0$, we have along a subsequence $\{h\} \to 0$

$$\beta(u_h) \to \beta(u), \ \ b^*(u_h) \to b^*(u), \text{ a.e. in } (0,\infty) \times \mathbb{R}^d.$$

Then, taking into account that $|\beta(u_h)| \leq \alpha_1 |u_h|$, a.e. in $(0,T) \times \mathbb{R}^d$, it follows by a standard argument and by Hypothesis (iv) that

$$\beta(u_h) \to \beta(u), \ \ b^*(u) \to b^*(u) \text{ in } L^1((0,\infty) \times \mathbb{R}^d)$$

as $h \to 0$. By (2.90), we have

$$\begin{aligned} &\int_h^\infty \int_{\mathbb{R}^d} \Big(\frac{1}{h}(u_h(t,x) - u_h(t-h,x))\varphi(t,x) - \beta(x, u_h(t,x)) \cdot \Delta\varphi(t,x) \\ &-(D(x) \cdot \nabla\varphi(t,x)) b^*(u_h(t,x)) \Big) dx\, dt = 0 \, \forall \varphi \in C_0^\infty([0,\infty) \times \mathbb{R}^d). \end{aligned} \tag{2.155}$$

Taking into account that

$$\begin{aligned} \int_h^\infty \int_{\mathbb{R}^d} u_h(t-h,x)\varphi(t,x)dx\,dt &= \int_0^\infty \int_{\mathbb{R}^d} u_h(t,x)\varphi(t+h,x)dx\,dt \\ &+ \int_0^h \int_{\mathbb{R}^d} u_0(x)\varphi(t+h,x)dx\,dt, \end{aligned}$$

and, letting $h \to 0$ in (2.155), it follows that

$$\int_0^\infty \int_{\mathbb{R}^d} (u\varphi_t + \beta(x,u)\Delta\varphi + b(u)uD \cdot \nabla\varphi)dxdt + \int_{\mathbb{R}^d} \varphi(0,x)u_0(x)dx = 0$$

and, therefore, u is a distributional solution to (2.80), as claimed.

If $u_0 \in L^1 \cap L^\infty$, it follows by (2.147) and (2.96) that $u \in L^\infty((0,T) \times \mathbb{R}^d)$, $\forall T > 0$. This completes the proof. □

Analyzing the proof of Lemma 2.6, one sees that, under additional assumptions on β, condition (2.144) can be dispensed with. Indeed, we have the following existence result for Eq. (2.98).

Lemma 2.7 *Assume that Hypotheses* (i)–(iv) *hold excepting* (2.144) *and with* (2.85) *strengthen to*

$$\gamma_1|r| \le |\beta(x,r)| \le \alpha_1|r|, \ \forall r \in \mathbb{R}, \ x \in \mathbb{R}^d, \tag{2.156}$$

where $\gamma_1, \alpha_1 > 0$. *Then, for each* $f \in L^1$ *and* $\lambda > 0$, *Eq.* (2.100) *has a unique solution* $y = J_\lambda(f)$ *which satisfies* (2.135)–(2.137).

Proof If $y_\varepsilon = y_\lambda^\varepsilon(f)$ for $f \in L^1 \cap L^\infty$ is the solution to (2.138), it follows the estimate

$$|\beta(\cdot, y_\varepsilon)|_2^2 + |y_\varepsilon|_2^2 + \lambda|\nabla\beta(\cdot, y_\varepsilon)|_2^2 \le C|f|_2^2, \ \forall \varepsilon > 0,$$

and so, with reference to the above proof, (2.148) follows without assuming (2.144) and, in particular, that $b \ge 0$. Then, the existence of a solution $y = J_\lambda(f)$ to (2.98) follows as in the proof of Lemma 2.6. The uniqueness of this solution is given by Lemma 2.5.

In particular, it follows under the hypotheses of Lemma 2.7 that the operator A_0 is m-accretive and so Proposition 2.2 remains true in this case too and this yields

Theorem 2.5 *Assume that Hypotheses* (i)–(iv) *hold with* (2.83) *strengthen to* (2.156) *and with* (iii) *weaken to* $b \in C^1(\mathbb{R}) \cap C_b(\mathbb{R})$. *Then, for each* $u_0 \in L^1$, *there is a unique mild solution* $u = u(t, u_0) = S(t)u_0$ *to* (2.80) *satisfying* (2.94) *and* (2.95).

Remark 2.3 In the special case $b \equiv 0$, the operator (2.110), that is

$$A_0(u) = -\Delta\beta(u), \ \forall u \in D(A_0) = \{u \in L^1; \Delta\beta(u) \in L^1\}$$

is m-accretive in L^1 under the weaker hypothesis that $\beta : \mathbb{R} \to \mathbb{R}$ is continuous and monotonically nondecreasing or, more generally, if β is maximal monotone and $D(\beta) = \mathbb{R}$ (eventually multivalued). (See [5], p. 118.)

2.3 Uniqueness of Distributional Solutions to NFPE

As seen earlier in Theorem 2.4, under Hypotheses (i)–(iii) the *mild solution* ρ *is also a distributional solution to* (2.2). However, the uniqueness of distributional solutions to (2.2) is still an open problem under these hypotheses. For the porous media equation (that is, $D \equiv 0$) such a uniqueness result was established by H. Brezis and M.G. Crandall [31] and M. Pierre [84] in the class $L^\infty((0,\infty)\times\mathbb{R}^d)$ for distributional solutions.

In the limit case $\beta \equiv 0$, $Db(r) \equiv \{a_i(r)\}_{i=1}^d \equiv a(r)$, Eq. (2.1) (or (2.80)) reduces to

$$\begin{aligned} &u_t + \operatorname{div}(a(u)) = 0, \ \mathscr{D}'((0,\infty)\times\mathbb{R}^d), \\ &u(0) = u_0, \end{aligned} \tag{2.157}$$

which has a unique Kružkov's solution $u(t) = S_0(t)u_0$ defined as

$$|u-k|_t + \operatorname{div}((a(u)-a(k))\operatorname{sign}(u-k)) \le 0 \text{ in } \mathscr{D}'((0,\infty)\times\mathbb{R}^d),\ k \in \mathbb{Z}^+,$$

$u \in C([0,\infty); L^1)$, $u(0) = u_0$.

The uniqueness result for (2.157) is not in the class of distributional solutions but in that of entropic solutions and this fact emphasizes the role of the diffusion term $\Delta\beta(u)$ for the distributional uniqueness. The NFPE (2.2) involves two different models: diffusion and drift, and Hypotheses (jv) amounts to saying that, for uniqueness in a distributional sense, *the diffusion should dominate the drift.*

By virtue of the equivalence of Eqs. (2.80) and (2.83), these results have implications on the uniqueness of probabilistically weak solutions to the McKean-Vlasov equation (2.83).

Hypotheses

(j) $\beta \in C^1(\mathbb{R}^d\times\mathbb{R})$, $\beta_r \in L^\infty(\mathbb{R}^d\times(-N,N))$, $\forall N>0$, $\beta_r(x,r)\ge 0$, $\forall x\in\mathbb{R}^d$, $r\in\mathbb{R}$; $\beta(x,0)\equiv 0$.
(jj) $D \in L^\infty(\mathbb{R}^d;\mathbb{R}^d)$.
(jjj) $b \in C^1(\mathbb{R})$.
(jv) *For each compact* $K\subset\mathbb{R}$ *there exists* $\alpha_K \in (0,\infty)$ *such that* $|b'(r)r + b(r)| \le \alpha_K|\beta_r(x,r)|,\ \forall r\in K.$

We note that Hypotheses (j)–(jjj) are weaker than (i)–(iii), while (jv) is equivalent to (iv). If $\beta'(r) > 0$, $\forall r\in\mathbb{R}$, then (jv) is automatically fulfilled due to (jjj).

Theorem 2.6 below implies the uniqueness of distributional solutions y which are in $L^\infty((0,T)\times\mathbb{R}^d)\cap L^1((0,T)\times\mathbb{R}^d)$ and are weakly continuous in t. If $d=1$, then the uniqueness extends to all $L^2(0,T;L^1(\mathbb{R}^d))$.

Theorem 2.6 *Assume that Hypotheses* (j)–(jv) *hold. Let* $d \geq 1$, $T > 0$, *and let* $y_1, y_2 \in L^\infty((0,T)\times\mathbb{R}^d)$ *be two distributional solutions to* (2.2) *on* $(0,T)\times\mathbb{R}^d$ *(in the sense of* (2.154)*) such that* $y_1 - y_2 \in L^\infty(0,T;H^{-1})$ *and*

$$\lim_{t\to 0} \operatorname*{ess\,sup}_{s\in(0,t)} |(y_1(s)-y_2(s),\varphi)_2| = 0,\ \forall \varphi \in C_0^\infty(\mathbb{R}^d). \tag{2.158}$$

Then $y_1 \equiv y_2$. *If* $d = 1$, (j)–(jv) *hold for* $K = \mathbb{R}$ *and* β *is Lipschitzian, then, if* $y_1, y_2 \in L^2(0,T;L^1)$ *and* (2.158) *is satisfied, it follows that* $y_1 \equiv y_2$.

Proof (In the following, we shall simply write $\beta(r)$ instead of $\beta(x,r)$.) Replacing, if necessary, the functions β and b by

$$\beta_N(r) = \begin{cases} \beta(r) & \text{if } |r| \leq N, \\ \beta_r(N)(r-N)+\beta(N) & \text{if } r > N, \\ \beta_r(-N)(r+N)+\beta(-N) & \text{if } r < -N, \end{cases}$$

and

$$b_N(r) = \begin{cases} b(r) & \text{if } |r| \leq N, \\ b'(N)(r-N)+b(N) & \text{if } r > N, \\ b'(-N)(r+N)+b(-N) & \text{if } r < -N, \end{cases}$$

where $N \geq \max\{|y_1|_\infty, |y_2|_\infty\}$, we may assume in the following that

$$\beta \in C_b(\mathbb{R}^d\times\mathbb{R}),\ b' \in C_b(\mathbb{R}), \tag{2.159}$$

and, therefore, by (j) and (jv) we have

$$(\beta(x,r)-\beta(x,\bar r))(r-\bar r) \geq \alpha_1|\beta(x,r)-\beta(x,\bar r)|^2,\ \forall r,\bar r\in\mathbb{R},\ x\in\mathbb{R}^d, \tag{2.160}$$

$$|b(r)r-b(\bar r)\bar r| \leq \alpha_2|\beta(x,r)-\beta(x,\bar r)|,\ r,\bar r\in\mathbb{R},\ x\in\mathbb{R}^d, \tag{2.161}$$

where $\alpha_1^{-1} = \|\beta'\|_{L^\infty(-N,N)}$ and $\alpha_2 = \alpha_{[-N,N]}$. We set

$$\begin{aligned} &\Phi_\varepsilon(y) = (\varepsilon I - \Delta)^{-1}y,\ \forall\, y \in L^2, \\ &z = y_1 - y_2,\ w = \beta(x,y_1)-\beta(x,y_2),\ b^*(y_i) \equiv b(y_i)y_i,\ i=1,2. \end{aligned} \tag{2.162}$$

It is well known that $\Phi_\varepsilon : L^p \to L^p, \forall p \in [1,\infty]$ and

$$\varepsilon|\Phi_\varepsilon(y)|_p \leq |y|_p,\ \ \forall y \in L^p,\ \varepsilon > 0. \tag{2.163}$$

Moreover, $\Phi_\varepsilon(y) \in C_b(\mathbb{R}^d)$ if $y \in L^1 \cap L^\infty$.

We have

$$z_t - \Delta w + \operatorname{div} D(b^*(y_1) - b^*(y_2)) = 0 \text{ in } \mathscr{D}'((0,T)\times\mathbb{R}^d).$$

Since the functions z, w and y_1, y_2 are not smooth, we shall regularize the latter equation as follows. We set

$$z_\varepsilon = z * \theta_\varepsilon,\; w_\varepsilon = w * \theta_\varepsilon,\; \zeta_\varepsilon = (D(b^*(y_1) - b^*(y_2))) * \theta_\varepsilon,$$

where $\theta \in C_0^\infty(\mathbb{R}^d)$, $\theta_\varepsilon(x) \equiv \varepsilon^{-d}\theta\left(\frac{x}{\varepsilon}\right)$ is a standard mollifier.

We note that $z_\varepsilon, w_\varepsilon, \zeta_\varepsilon, \Delta w_\varepsilon, \operatorname{div}\zeta_\varepsilon \in L^2(0,T;L^2)$ and we have

$$(z_\varepsilon)_t - \Delta w_\varepsilon + \operatorname{div}\zeta_\varepsilon = 0 \text{ in } \mathscr{D}'(0,T;L^2). \tag{2.164}$$

This yields $\Phi_\varepsilon(z_\varepsilon), \Phi_\varepsilon(w_\varepsilon), \operatorname{div}\Phi_\varepsilon(\zeta_\varepsilon) \in L^2(0,T;L^2)$ and

$$\begin{aligned}(\Phi_\varepsilon(z_\varepsilon))_t = \Delta\Phi_\varepsilon(w_\varepsilon) - \operatorname{div}\Phi_\varepsilon(\zeta_\varepsilon) = \varepsilon\Phi_\varepsilon(w_\varepsilon) - w_\varepsilon - \operatorname{div}\Phi_\varepsilon(\zeta_\varepsilon)\\ \text{in } \mathscr{D}'(0,T;L^2).\end{aligned} \tag{2.165}$$

By (2.164) and (2.165) it follows that $(z_\varepsilon)_t = \frac{d}{dt}z_\varepsilon$, $(\Phi_\varepsilon(z))_t = \frac{d}{dt}\Phi_\varepsilon(z_\varepsilon) \in L^2(0,T;L^2)$, where $\frac{d}{dt}$ is taken in the sense of L^2-valued vectorial distributions on $(0,T)$ and this implies that $z_\varepsilon, \Phi_\varepsilon(z_\varepsilon) \in H^1(0,T;L^2)$ and

$$\Phi_\varepsilon(z_\varepsilon) \in L^2(0,T;H^2) \cap H^1(0,T;L^2), \tag{2.166}$$

and, since the spaces H^2, L^2 are in duality with the pivot space $H^1 = (H^2, L^2)_{\frac{1}{2}}$, modifying the function $t \to \Phi_\varepsilon(z_\varepsilon(t))$ on a subset of measure zero, we have

$$\Phi_\varepsilon(z_\varepsilon) \in C([0,T];H^1), \tag{2.167}$$

(see, e.g., [4], p. 25). We set

$$h_\varepsilon(t) = (\Phi_\varepsilon(z_\varepsilon(t)), z_\varepsilon(t))_2,\; t \in [0,T],$$

and get, therefore,

$$\begin{aligned} h'_\varepsilon(t) &= 2(z_\varepsilon(t), (\Phi_\varepsilon(z_\varepsilon(t)))_t)_2 \\ &= 2(\varepsilon\Phi_\varepsilon(w_\varepsilon(t)) - w_\varepsilon(t) - \operatorname{div}\Phi_\varepsilon(\zeta_\varepsilon(t)), z_\varepsilon(t))_2 \\ &= 2\varepsilon(\Phi_\varepsilon(z_\varepsilon(t)), w_\varepsilon(t))_2 + 2(\nabla\Phi_\varepsilon(z_\varepsilon(t)), \zeta_\varepsilon(t))_2 \\ &\quad -2(z_\varepsilon(t), w_\varepsilon(t))_2,\; \text{a.e. } t \in (0,T). \end{aligned} \tag{2.168}$$

By (2.168) it also follows that $t \to h_\varepsilon(t)$ is absolutely continuous on $[0, T]$. Since, by (2.160) and (2.161),

$$(z_\varepsilon(t), w_\varepsilon(t))_2 \geq \alpha_4 |w_\varepsilon(t)|_2^2 + \gamma_\varepsilon(t), \tag{2.169}$$

where

$$\gamma_\varepsilon(t) := (z_\varepsilon(t), w_\varepsilon(t))_2 - (z(t), w(t))_2,$$

we get, therefore, by (2.168) that

$$\begin{aligned} 0 \leq h_\varepsilon(t) \leq & \, h_\varepsilon(0+) + 2\varepsilon \int_0^t (\Phi_\varepsilon(z_\varepsilon(s)), w_\varepsilon(s))_2 ds - 2\alpha_4 \int_0^t |w_\varepsilon(s)|_2^2 ds \\ & + 2\alpha_3 |D|_\infty \int_0^t |\nabla \Phi_\varepsilon(z_\varepsilon(s))|_2 |w_\varepsilon(s)|_2 ds + 2 \int_0^t |\gamma_\varepsilon(s)| ds, \ \forall t \in [0, T]. \end{aligned} \tag{2.170}$$

Taking into account that $t \to \Phi_\varepsilon(z_\varepsilon(t))$ has an H^1-continuous version on $[0, T]$ (which we shall consider from now on), there exists $f \in H^1$ such that

$$\lim_{t \to 0} \Phi_\varepsilon(z_\varepsilon(t)) = f \ \text{ in } H^1.$$

Furthermore, for every $\varphi \in C_0^\infty(\mathbb{R}^d)$, $s \in (0, T)$,

$$0 \leq h_\varepsilon(s) \leq |\Phi_\varepsilon(z_\varepsilon(s)) - f|_{H^1} |z_\varepsilon(s)|_{H^{-1}} + |f - \varphi|_{H^1} |z_\varepsilon(s)|_{H^{-1}} + |(\varphi * \theta_\varepsilon, z(s))_2|.$$

Hence,

$$\begin{aligned} 0 \leq h_\varepsilon(0+) = & \lim_{t \downarrow 0} h_\varepsilon(t) = \lim_{t \to 0} \operatorname*{ess\,sup}_{s \in (0,t)} h_\varepsilon(s) \\ \leq & \left(\lim_{t \to 0} |\Phi_\varepsilon(z_\varepsilon(t)) - f|_{H^1} + |f - \varphi|_{H^1} \right) |z_\varepsilon|_{L^\infty(0,T;H^{-1})} \\ & + \lim_{t \to 0} \operatorname*{ess\,sup}_{s \in (0,t)} |(\varphi * \theta_\varepsilon, z(s))_2| = |f - \varphi|_{H^1} |z_\varepsilon|_{L^\infty(0,T;H^{-1})}. \end{aligned}$$

Since $C_0^\infty(\mathbb{R}^d)$ is dense in $H^1(\mathbb{R}^d)$, we get

$$h_\varepsilon(0+) = 0. \tag{2.171}$$

On the other hand, taking into account that, for a.e. $t \in (0, T)$,

$$\varepsilon \Phi_\varepsilon(z_\varepsilon(t)) - \Delta \Phi_\varepsilon(z_\varepsilon(t)) = z_\varepsilon(t), \tag{2.172}$$

and this yields

$$\varepsilon|\Phi_\varepsilon(z_\varepsilon(t))|_2^2+|\nabla\Phi_\varepsilon(z_\varepsilon(t))|_2^2=(z_\varepsilon(t),\Phi_\varepsilon(z_\varepsilon(t)))_2=h_\varepsilon(t),\ \text{a.e. } t\in(0,T). \tag{2.173}$$

Hence, h_ε has a continuous version on $[0, T]$, and (2.173) holds for all $t \in [0, T]$.

By (2.170) and (2.171), we get, for every $t \in [0, T]$,

$$\begin{aligned}0\le h_\varepsilon(t)\le{}& \varepsilon\int_0^t|\Phi_\varepsilon(z_\varepsilon(s))|_2^2ds-(\alpha_4-\varepsilon)\int_0^t|w_\varepsilon(s)|_2^2ds\\&+\frac{\alpha_3^2|D|_\infty^2}{\alpha_4}\int_0^t|\nabla\Phi_\varepsilon(z_\varepsilon(s))|_2^2ds+2\int_0^t|\gamma_\varepsilon(s)|ds,\end{aligned}$$

and so, by (2.173), this yields, for $\varepsilon \in (0, \alpha_4]$,

$$0\le h_\varepsilon(t)\le\max\left(1,\frac{\alpha_3^2|D|_\infty^2}{\alpha_4}\right)\int_0^t h_\varepsilon(s)ds+2\int_0^t|\gamma_\varepsilon(s)|ds,\ \forall t\in[0,T]. \tag{2.174}$$

On the other hand, we have that $\gamma_\varepsilon \to 0$ in $L^1(0, T)$ as $\varepsilon \to 0$. Then, (2.174) yields

$$0\le h_\varepsilon(t)\le\eta_\varepsilon(t)\exp\left(\max\left(1,\frac{\alpha_3^2|D|_\infty^2}{\alpha_4}\right),t\right),\ \forall t\in[0,T], \tag{2.175}$$

where $\lim\limits_{\varepsilon\to0}\eta_\varepsilon(T) = 0$ and hence $\lim\limits_{\varepsilon\to0} h_\varepsilon = 0$ in $C([0, T])$.

By (2.173) and (2.175), for every $t \in [0, T]$, it follows that, as $\varepsilon \to 0$, $\nabla\Phi_\varepsilon(z_\varepsilon(t)) \to 0$ in L^2 and $\varepsilon|\Phi_\varepsilon(z_\varepsilon(t))|_2^2 \to 0$, hence the left hand side of (2.172) converges to zero in $\mathscr{D}'(\mathbb{R}^d)$ for every $t \in [0, T]$, and so $0 = \lim\limits_{\varepsilon\to0} z_\varepsilon(t) = z(t)$ in $\mathscr{D}'(\mathbb{R}^d)$ for a.e. $t \in [0, T]$, which implies $y_1 \equiv y_2$.

Remark 2.4 If Hypotheses (j), (jv) are strenghten to (2.160) and (2.161), then, as easily follows by the proof, one can replace in Theorem 2.6 condition $y_1, y_2 \in L^\infty((0, T) \times \mathbb{R}^d)$ by $y_1, y_2 \in L^2((0, T) \times \mathbb{R}^d)$.

Remark 2.5 Let $u \in L^1((0, T) \times \mathbb{R}^d)$ such that $\beta(u) \in L^1((0, T) \times \mathbb{R}^d)$ and u is a solution to (2.80). Then, it is elementary to check that

$$\int_{\mathbb{R}^d} u(t,x)dx=\int_{\mathbb{R}^d} u_0(dx)\ \text{ for } dt\text{-a.e. } t\in(0,T).$$

Hence, if u_0 is nonnegative and $u \ge 0$, a.e. on $(0, T) \times \mathbb{R}^d$, it follows by Lemma 2.3 in [87] that there exists a $dt \otimes dx$-version $\widetilde{u}$ of u such that for $\widetilde{u}_t(dx) := \widetilde{u}(t, x)dx$, $t > 0$, and $\widetilde{u}_0(dx) := u_0(dx)$, the map $[0, T] \ni t \mapsto \widetilde{u}_t$ is narrowly continuous.

Then, Remark 2.5 implies the following consequence of Theorem 2.6.

Corollary 2.1 *Let $u_0 \in \mathcal{M}(\mathbb{R}^d)$, u_0 nonnegative, and $y_1, y_2 \in L^\infty((0,T)\times\mathbb{R}^d) \cap L^1((0,T)\times\mathbb{R}^d)$ be two nonnegative solutions to* (2.2). *Then, $y_1 \equiv y_2$.*

Proof Let $\widetilde{y}_1, \widetilde{y}_2$ be the $dt \otimes dx$-versions from Remark 2.5. Then, for every $\varphi \in C_0^\infty(\mathbb{R}^d)$,

$$\lim_{t\to 0} \operatorname*{ess\,sup}_{s\in(0,t)} |(y_1(s)-y_2(s),\varphi)_2| = \lim_{t\to 0} \sup_{s\in(0,t)} \left| \int_{\mathbb{R}} (\widetilde{y}_1(s,x)-\widetilde{y}_2(s,x))\varphi(x)dx \right| = 0.$$

So, (2.158) holds and Theorem 2.6 implies the assertion.

For the weak uniqueness of the McKean–Vlasov SDE (2.83) corresponding to the Fokker–Planck equation (2.2) (resp. (2.80)) to be treated in Chap. 5, we also need to prove the so-called "linearized uniqueness" for Eq. (2.83).

Theorem 2.7 (Linearized Uniqueness) *Assume that Hypotheses* (j)–(jv) *hold. Let $T > 0$ and $u \in L^\infty((0,T)\times\mathbb{R}^d)$. Let $y_1, y_2 \in L^2((0,T)\times\mathbb{R}^d)$ such that $y_1 - y_2 \in L^\infty(0,T;H^{-1})$ and y_1, y_2 are solutions to the following linearized version of* (2.83)

$$\int_0^T \int_{\mathbb{R}^d} \left(\varphi_t + \frac{\beta(u)}{u}\Delta\varphi + b(u)D\cdot\nabla\varphi\right) v\,dx\,dt + \int_{\mathbb{R}^d} \varphi(0,x)u_0(dx) = 0,$$
$$\forall \varphi \in C_0^\infty([0,T)\times\mathbb{R}^d), \tag{2.176}$$

for some $u_0 \subset \mathcal{M}(\mathbb{R}^d)$, where $\frac{\beta(0)}{0} := \beta'(0)$, such that (2.158) *holds. Then, $y_1 \equiv y_2$.*

Proof First, we note that by (j)–(jv) we have:

$$\frac{\beta(u)}{u},\, b(u) \in L^\infty((0,T)\times\mathbb{R}^d), \tag{2.177}$$

$$|Db(u)| \le \alpha_7 \frac{\beta(u)}{u}, \text{ a.e. on } (0,T)\times\mathbb{R}^d, \tag{2.178}$$

where $\alpha_7 := |D|_\infty \alpha_{[-|u|_\infty, |u|_\infty]}$. Now, we set

$$z := y_1 - y_2, \quad w := \frac{\beta(u)}{u}(y_1 - y_2). \tag{2.179}$$

Then, we have, since $\frac{\beta(u)}{u} \ge 0$,

$$wz = \frac{\beta(u)}{u}|y_1 - y_2|^2 \ge \left(\left|\frac{\beta(u)}{u}\right|_\infty + 1\right)^{-1} |w|^2, \text{ a.e. on } (0,T)\times\mathbb{R}^d, \tag{2.180}$$

and

$$|Db(u)z| \le \alpha_7 |w|. \tag{2.181}$$

We set

$$z_\varepsilon = z * \theta_\varepsilon,\ w_\varepsilon = w * \theta_\varepsilon,\ \zeta_\varepsilon = (Db(u)(y_1 - y_2)) * \theta_\varepsilon, \tag{2.182}$$

where θ_ε is as in the proof of Theorem 2.6. Now, by (2.177)–(2.181), we can repeat the proof of the latter line by line for z_ε, w_ε and ζ_ε in (2.182) to obtain $y_1 \equiv y_2$.

Corollary 2.2 *Let $u_0 \in \mathscr{M}(\mathbb{R}^d)$, $u_0 \ge 0$, and $y_1, y_2 \in (L^2 \cap L^1)((0,T) \times \mathbb{R}^d)$ be two nonnegative solutions to* (2.176). *Then, $y_1 \equiv y_2$.* □

Proof The assertion follows from Theorem 2.7 by analogous arguments as in the proof of Corollary 2.1. □

2.4 NFPEs with Superlinear Diffusion

Here, we shall consider the NFPE (2.2)

$$\begin{aligned} &u_t - \Delta(\beta(u)) + \operatorname{div}(Db(u)u) = 0,\ \forall (t,x) \in [0,\infty) \times \mathbb{R}^d, \\ &u(0,\cdot) = \mu,\ \text{in } \mathbb{R}^d, \end{aligned} \tag{2.183}$$

where μ is a bounded Radon measure on $\mathbb{R}^d$ and the functions $\beta : \mathbb{R} \to \mathbb{R}$, $D : \mathbb{R}^d \to \mathbb{R}^d$, $b : \mathbb{R} \to \mathbb{R}$, are assumed to satisfy the following hypotheses

(k) $\beta \in C^1(\mathbb{R})$, $\beta'(r) > 0$, $\forall r \ne 0$, $\beta(0) = 0$.
(kk) $D \in L^\infty(\mathbb{R}^d; \mathbb{R}^d)$, $\operatorname{div} D \in L^2 + L^\infty$, $(\operatorname{div} D)^- \in L^\infty$.
(kkk) $b \in C^1(\mathbb{R}) \cap C_b(\mathbb{R})$, $b \ge 0$.

Here,

$$L^2 + L^\infty = \{\eta = \eta_1 + \eta_2;\ \eta_1 \in L^2,\ \eta_2 \in L^\infty\}.$$

In particular, the above hypotheses cover the so-called *nonlinear Fokker–Planck equation with power-law diffusion*,

$$u_t - \Delta(|u|^{m-2}u) + \operatorname{div}(Db(u)u) = 0,$$

where $m > 1$, as well as other cases of nonlinear diffusions.

In Theorem 2.4, the existence (and uniqueness) of a mild solution u to (2.183) was proved for $u_0 \in L^1$, under stronger assumptions on β and b. In particular, for existence it was assumed that β is strictly monotone and sublinear, which is not the case in Hypothesis (i). Here, we shall prove first the existence under the weaker Hypotheses (k)–(kkk). Namely, we have

Theorem 2.8 *Let $d \geq 1$ and let Hypotheses* (k)–(kkk) *hold and additionally that*

$$\operatorname{div} D \in L^m_{\mathrm{loc}}(\mathbb{R}^d),\ m > \frac{d}{2} \ \text{for}\ d \geq 2,\ \text{and}\ m = 1\ \text{for}\ d = 1. \tag{2.184}$$

Then, there is a continuous semigroup $S(t)$ of contractions in L^1, defined on a closed subset $\mathscr{K} \subset L^1$ such that, for each $u_0 \in \mathscr{K}$, $u(t) = S(t)u_0$ is a mild solution to (2.183). *If, in addition, $\beta \in C^2(\mathbb{R})$, then $\mathscr{K} = L^1$ and, for every $u_0 \in L^1 \cap L^\infty$,*

$$|S(t)u_0|_\infty \leq \exp((1 + 2(|(\operatorname{div} D)^-|_\infty + |D|_\infty)|b|_\infty)t)|u_0|_\infty,\ \forall t \geq 0. \tag{2.185}$$

Moreover $S(t)(\mathscr{P}) \subset \mathscr{P}$, $\forall t \geq 0$, and, for $u_0 \in L^1$, $u(t) = S(t)u_0$ is a solution to (2.183) *in the sense of Schwartz distributions on* $(0, \infty) \times \mathbb{R}^d$, *that is,*

$$\begin{aligned} &\int_0^\infty \int_{\mathbb{R}^d} (u(\varphi_t + b(u)D \cdot \nabla\varphi) + \beta(u)\Delta\varphi)dxdt \\ &\qquad + \int_{\mathbb{R}^d} u_0(x)\varphi(0, x)dx = 0,\ \forall \varphi \in C_0^\infty([0, \infty) \times \mathbb{R}^d). \end{aligned} \tag{2.186}$$

Finally, if $u_0 \in L^1 \cap L^\infty$, and in addition to (k)–(kkk) *one assumes that Hypothesis* (jv) *of Theorem* 2.6 *holds, then the mild solution u is the unique distributional solution to* (2.183).

Remark 2.6 The additional assumption (2.184) can be dropped. We refer to [12] and the arguments in the proof of the uniqueness of solutions to (2.9) in $D(A_0) \cap H^1$ and the resulting derivation of the resolvent equation (2.52).

As a consequence of Theorems 2.8 and 5.1 below we obtain that the McKean–Vlasov SDE (2.83) has a probabilistically weak solution. If, in addition, Hypothesis (iv) of Theorem 2.4 holds, then by Theorem 5.2 and Remark 5.4 this solution is unique in law and the corresponding path laws $P_{s,\zeta}$, $(s, \zeta) \in \mathbb{R}_+ \times \mathscr{P}_0$ are a nonlinear Markov process in the sense of Definition 5.1.

We recall (see Definition 2.2) that a continuous function $u : [0, \infty] \to L^1$ is said to be a *mild solution* to Eq. (2.183) if, for each $T > 0$,

$$u(t) = \lim_{h \to 0} u_h(t) \text{ in } L^1 \text{ uniformly on each interval } [0, T],$$

$$\begin{aligned} &u_h(t) = u_h^i,\ \forall t \in (ih, (i+1)h],\ i = 0, 1, \ldots, N-1, \\ &u_h^0 = u_0,\ \beta(u_h^i) \in L^1_{\mathrm{loc}},\ u_h^i \in L^1,\ \forall i = 1, 2, \ldots, N,\ Nh = T, \\ &u_h^{i+1} - h\Delta(\beta(u_h^{i+1})) + h \operatorname{div}(Db(u_h^{i+1})u_h^{i+1}) = u_h^i,\ \text{in } \mathscr{D}'(\mathbb{R}^d), \end{aligned} \tag{2.187}$$

for all $i = 0, 1, \ldots, N-1$.

Proof of Theorem 2.8 As seen earlier, the idea of the proof is to associating with Eq. (2.183) an m-accretive operator A in L^1 and so to reduce it to the Cauchy problem

$$\frac{du}{dt} + Au = 0,\ \ t \geq 0, \qquad u(0) = u_0.$$

The operator A is constructed here as in Sect. 2.1 (see (2.18)). Namely, consider in L^1 the nonlinear operator

$$\begin{aligned} A_0 y &= -\Delta\beta(y) + \operatorname{div}(Db(y)y),\ \forall y \in D(A_0), \\ D(A_0) &= \{y \in L^1,\ \beta(y) \in L^1_{\text{loc}},\ -\Delta\beta(y) + \operatorname{div}(Db(y)y) \in L^1\}. \end{aligned} \tag{2.188}$$

We have (see Proposition 2.1) □

Proposition 2.3 *There is $\lambda_0 > 0$ such that*

$$R(I + \lambda A_0) = L^1,\ \forall \lambda \in (0, \lambda_0), \tag{2.189}$$

and there is a family of operators $J_\lambda : L^1 \to L^1$ such that, for all $\lambda, \lambda_1, \lambda_2 \in (0, \lambda_0)$,

$$\begin{aligned} &J_\lambda(f) \in (I + \lambda A_0)^{-1} f,\ \forall f \in L^1, \\ &|J_\lambda f - J_\lambda g|_1 \leq |f - g|_1,\ \forall f, g \in L^1, \end{aligned} \tag{2.190}$$

$$J_{\lambda_2} f = J_{\lambda_1}\left(\frac{\lambda_1}{\lambda_2} f + \left(1 - \frac{\lambda_1}{\lambda_2}\right) J_{\lambda_2}(f)\right), \tag{2.191}$$

$$\int_{\mathbb{R}^d} J_\lambda(f)(x)dx = \int_{\mathbb{R}^d} f(x)dx, \tag{2.192}$$

$$J_\lambda(f) \geq 0,\ \ \text{a.e. in } \mathbb{R}^d,\ \text{if } f \geq 0,\ \text{a.e. in } \mathbb{R}^d. \tag{2.193}$$

Moreover, for each compact set $K \subset \mathbb{R}^d$,

$$\|\beta(J_\lambda(f))\|_{L^q(K)} \leq C_K \left(\frac{1+\lambda}{\lambda}\right) |f|_1,\ q \in \left(1, \frac{d}{d-1}\right),\ \ \forall f \in L^1, \tag{2.194}$$

$$\begin{aligned} &|J_\lambda(f)|_\infty \leq (1 + 2\lambda(|(\operatorname{div} D)^-|_\infty + |D|_\infty)|b|_\infty)|f|_\infty,\ \forall f \in L^1 \cap L^\infty, \\ &0 < \lambda < \lambda_0 = (2(|(\operatorname{div} D)^-|_\infty + |D|_\infty)|b|_\infty)^{-1}, \end{aligned} \tag{2.195}$$

$$J_\lambda(\mathscr{P}) \subset \mathscr{P},\ \lambda \in (0, \lambda_0). \tag{2.196}$$

If $\beta \in C^2(\mathbb{R})$ holds, then

$$|J_\lambda(\varphi) - \varphi|_1 \le C\lambda\|\varphi\|_{H^2(\mathbb{R}^d)}, \ \forall \varphi \in C_0^\infty(\mathbb{R}^d). \tag{2.197}$$

Here $R(I + \lambda A_0)$ is the range of the operator $I + \lambda A_0$ and $(I + \lambda A_0)^{-1} : L^1 \to D(A_0)$ (which, in general, might be multivalues) is a right inverse of $(I + \lambda A_0)$. Before proving Proposition 2.3, let us briefly discuss some consequences.

Define the operator $A : D(A) \subset L^1 \to L^1$,

$$Au = A_0 u, \ \ \forall u \in D(A), \tag{2.198}$$

$$D(A) = \{J_\lambda(f), \ f \in L^1\}. \tag{2.199}$$

It is easily seen by (2.191) that $D(A) = \{u = J_\lambda(f), \ f \in L^1\}$ is independent of $\lambda \in (0, \lambda_0)$. By Proposition 2.3, we have

Lemma 2.8

(i) *J_λ coincides with the inverse $(I + \lambda A)^{-1}$ of $(I + \lambda A)$.*
(ii) *The operator A is m-accretive in L^1, that is, $R(I + \lambda A) = L^1$, $\forall \lambda > 0$, and*

$$|(I + \lambda A)^{-1}u - (I + \lambda A)^{-1}v|_1 \le |u - v|_1, \ \forall u, v \in L^1, \ \lambda > 0. \tag{2.200}$$

Moreover, (2.190)–(2.197) *hold with $(I+\lambda A)^{-1}$ instead of J_λ. If $\beta \in C^2(\mathbb{R})$ holds, then $D(A)$ is dense in L^1.*

Proof (i) follows immediately by (2.198) and (2.199). Except for the density of $D(A)$ in L^1, all assertions of (ii) are immediate by the definition of A and Proposition 2.3. If $\beta \in C^2(\mathbb{R})$, by (2.197) we have $C_0^\infty(\mathbb{R}^d) \subset \overline{D(A)}$ and, since $C_0^\infty(\mathbb{R}^d)$ is dense in L^1, so is $D(A)$. □

We recall that Lemma 2.8 implies via the Crandall and Liggett theorem that, for each $u_0 \in \overline{D(A)} = \mathscr{K}$ and $T > 0$, the Cauchy problem

$$\frac{du}{dt} + Au = 0, \ t \in (0, T), \qquad u(0) = u_0, \tag{2.201}$$

has a unique mild solution $u \in C([0, T]; L^1)$, that is,

$$u(t) = \lim_{h \to 0} u_h(t) \text{ in } L^1 \text{ uniformly on } [0, T], \tag{2.202}$$

where $u_h : [0, T] \to L^1$ is given by

$$\begin{aligned} &u_h(t) = u_h^{i+1}, \ t \in (ih, (i+1)h], \\ &u_h^{i+1} + hAu_h^{i+1} = u_h^i, \ i = 0, 1, \dots, N-1; \ Nh = T, \\ &u_h^0 = u_0. \end{aligned} \tag{2.203}$$

In fact, as mentioned earlier, the solution $u = u(t, u_0)$ defined by (2.202) and (2.203) is given by the exponential formula

$$S(t)u_0 = u(t, u_0) = \lim_{n\to\infty} \left(I + \frac{t}{n}A\right)^{-n} u_0 \text{ in } L^1,\ t \geq 0, \tag{2.204}$$

where the convergence is uniform in t on compact intervals $[0, T]$, and $S(t)$ is a semigroup of contractions on L^1, that is,

$$S(t+s) = S(t)S(s)u_0,\ \forall t, s \geq 0, \quad S(0) = I,$$
$$|S(t)u_0 - S(t)\bar{u}_0|_1 \leq |u - \bar{u}_0|_1,\ \forall u_0, \bar{u}_0 \in \overline{D(A)},\ t \geq 0.$$

Proof of Proposition 2.3 For each $f \in L^1$, consider the equation

$$u + \lambda A_0 u = f \tag{2.205}$$

or, equivalently,

$$u - \lambda\Delta\beta(u) + \lambda \operatorname{div}(Db(u)u) = f \text{ in } \mathscr{D}'(\mathbb{R}), \tag{2.206}$$

$$u \in L^1,\ \beta(u) \in L^1_{\text{loc}},\ -\Delta\beta(u) + \operatorname{div}(Db(u)u) \in L^1. \tag{2.207}$$

We shall assume first $f \in L^1 \cap L^2$ and approximate (2.206) by an equation of the form (2.138), that is,

$$u + \lambda(\varepsilon I - \Delta)\widetilde{\beta}_\varepsilon(u) + \lambda \operatorname{div}(D_\varepsilon b_\varepsilon(u)u) = f, \tag{2.208}$$

where $\widetilde{\beta}_\varepsilon(u) \equiv \beta_\varepsilon(u) + \varepsilon u$, and for $\varepsilon > 0$, $r \in \mathbb{R}$,

$$\beta_\varepsilon(r) \equiv \frac{1}{\varepsilon}(r - (I + \varepsilon\beta)^{-1}r) = \beta((I + \varepsilon\beta)^{-1}r),$$

$$D_\varepsilon = \begin{cases} D, & \text{if } |D| \in L^2, \\ \eta_\varepsilon D, & \text{else.} \end{cases}$$

$$b_\varepsilon(r) = \begin{cases} b, & \text{if } b \text{ is a constant}, \\ \dfrac{(b * \varphi_\varepsilon)(r)}{1 + \varepsilon|r|}, & \text{otherwise.} \end{cases}$$

Here $\varphi_\varepsilon(r) = \frac{1}{\varepsilon}\rho\left(\frac{r}{\varepsilon}\right)$, $\rho \in C_0^\infty(\mathbb{R})$, $\rho \geq 0$, is a standard modifier. Furthermore, $\eta_\varepsilon \in C_0^1(\mathbb{R}^d)$, $0 \leq \eta_\varepsilon \leq 1$, $|\nabla\eta_\varepsilon| \leq 1$ and $\eta_\varepsilon(x) = 1$ if $|x| \leq \frac{1}{\varepsilon}$. Clearly, we have

$$|D_\varepsilon| \in L^\infty \cap L^2,\ |D_\varepsilon| \leq |D|,\ \lim_{\varepsilon\to 0} D_\varepsilon(x) = D(x), \text{ a.e. } x \in \mathbb{R}^d, \tag{2.209}$$

$$(\operatorname{div} D_\varepsilon)^- \leq (\operatorname{div} D)^- + \mathbf{1}_{|x|\geq\frac{1}{\varepsilon}}\,|D|,\ \text{a.e. } x \in \mathbb{R}^d. \tag{2.210}$$

Define the operator

$$\begin{aligned} A_\varepsilon(u) &= -\Delta\widetilde{\beta}_\varepsilon(u) + \operatorname{div}(D_\varepsilon b_\varepsilon(u)u) \\ D(A_\varepsilon) &= \{u \in L^1;\ -\Delta\widetilde{\beta}_\varepsilon(u) + \operatorname{div}(D_\varepsilon b_\varepsilon(u)u) \in L^1\}. \end{aligned} \tag{2.211}$$

Since $\widetilde{\beta}_\varepsilon$ and $D_\varepsilon, b_\varepsilon$ satisfy Hypotheses (i)–(iv) and (2.156), it follows as in the proof of Lemma 2.6 that, for each $f \in L^1$ and all $\lambda > 0$, the equation

$$u_\varepsilon(f) + \lambda A_\varepsilon(u_\varepsilon(f)) = f, \tag{2.212}$$

has a unique solution $y_\varepsilon \in H^1$. We also have (see (2.146))

$$|u_\varepsilon(f_1) - u_\varepsilon(f_2)|_1 \leq |f_1 - f_2|_1,\ \forall f_1, f_2 \in L^1. \tag{2.213}$$

Moreover, if $u_\varepsilon = u_\varepsilon(\lambda, f)$ is our solution to (2.208), we have, for all $0 < \lambda_1, \lambda_2 < \infty$ and $f \in L^1 \cap L^2$, by definition

$$u_\varepsilon(\lambda_2, f) = u_\varepsilon\left(\lambda_1, \frac{\lambda_1}{\lambda_2} f + \left(1 - \frac{\lambda_1}{\lambda_2}\right) u_\varepsilon(\lambda_2, f)\right). \tag{2.214}$$

If $f \in L^1 \cap L^\infty$, we have

$$|u_\varepsilon(f)|_\infty \leq (1 + 2\lambda(|(\operatorname{div} D)^-|_\infty + |D|_\infty)|b|_\infty)|f|_\infty,\ \ 0 < \lambda < \lambda_0, \tag{2.215}$$

where λ_0 is given by (2.195) and $\varepsilon > 0$ is sufficiently small. The argument is that used in the proof of Lemma 2.6. Again, below we set $\frac{1}{0} := \infty$.

Indeed, by (2.208), we see that, for $M := 2\lambda|b|_\infty(|(\operatorname{div} D)^-|_\infty + |D|_\infty)|f|_\infty$, $u_\varepsilon = u_\varepsilon(f)$, $b_v^* p(r) = b_v p(r) r$ and $\lambda < \lambda_0 = (2|b|_\infty(|(\operatorname{div} D)^-|_\infty + |D|_\infty))^{-1}$, we have

$$\begin{aligned} &(u_\varepsilon - |f|_\infty - M) - \lambda\Delta(\widetilde{\beta}_\varepsilon(u_\varepsilon) - \widetilde{\beta}_\varepsilon(|f|_\infty + M)) + \lambda\varepsilon(\widetilde{\beta}_\varepsilon(u_\varepsilon) - \widetilde{\beta}(|f|_\infty + M)) \\ &\quad + \lambda \operatorname{div}(D_\varepsilon(b_\varepsilon^*(u_\varepsilon) - b_\varepsilon^*(|f|_\infty + M))) \\ &\quad \leq f - |f|_\infty - M - \lambda b_\varepsilon^*(M + |f|_\infty)\operatorname{div} D_\varepsilon \leq 0. \end{aligned}$$

Multiplying the above equation by $\mathscr{X}_\delta((u_\varepsilon - (|f|_\infty + M))^+)$ and integrating over $\mathbb{R}^d$, we get as above, for $\delta \to 0$, $|(u_\varepsilon - |f|_\infty - M)^+|_1 \leq 0$ and, therefore, by (2.210)

$$u_\varepsilon \leq (1 + 2\lambda|b|_\infty(|(\operatorname{div} D)^-|_\infty + |D|_\infty))|f|_\infty,\ \text{a.e. in } \mathbb{R}^d.$$

Similarly, one gets that

$$u_\varepsilon \geq -(1+2\lambda|b|_\infty(|(\operatorname{div} D)^-|_\infty + |D|_\infty))|f|_\infty, \quad \text{a.e. in } \mathbb{R}^d,$$

and so, by (2.210), (2.215) follows.

We note that, by (2.213) and (2.215) we have, for $0 < \lambda \leq \lambda_0$,

$$|u_\varepsilon(\lambda, f)|_1 + |u_\varepsilon(\lambda, f)|_\infty \leq |f|_1 + C_\lambda |f|_\infty, \ \forall f \in L^1 \cap L^\infty. \tag{2.216}$$

Also, if we multiply (2.208), where $u = u_\varepsilon(\lambda, f)$, by $\widetilde{\beta}(u_\varepsilon(\lambda, f))$ and integrate over $\mathbb{R}^d$, we get

$$\begin{aligned}
&(u_\varepsilon(\lambda, f), \widetilde{\beta}_\varepsilon(u_\varepsilon(\lambda, f)))_2 + \lambda|\nabla\widetilde{\beta}_\varepsilon(u_\varepsilon(\lambda, f))|_2^2 + \lambda\varepsilon|\widetilde{\beta}_\varepsilon(u_\varepsilon(\lambda, f))|_2^2 \\
&\leq \lambda|\nabla\widetilde{\beta}(u_\varepsilon(\lambda, f))|_2|D_\varepsilon b_\varepsilon^*(u_\varepsilon(\lambda, f))|_2 + |\widetilde{\beta}_\varepsilon(u_\varepsilon(\lambda, f))|_\infty|f|_1 \\
&\leq \frac{\lambda}{2}|\nabla\widetilde{\beta}(u_\varepsilon(\lambda, f))|_2^2 + \frac{\lambda}{2}|D_\varepsilon|_\infty|b_\varepsilon|_\infty|u_\varepsilon(\lambda, f)|_2 + |\beta_\varepsilon(u_\varepsilon(\lambda, f))|_\infty|f|_1, \\
&\qquad \forall f \in L^1 \cap L^\infty.
\end{aligned}$$

Taking into account that, by (2.216),

$$|\beta_\varepsilon(u_\varepsilon(\lambda, f))|_\infty \leq C_\lambda^1|f|_\infty, \ \forall \varepsilon > 0, \ \lambda \in (0, \lambda_0), \tag{2.217}$$

we get

$$\lambda(1+\varepsilon)|\nabla\widetilde{\beta}_\varepsilon(u_\varepsilon|_2^2 \leq C_\lambda^2(|f|_1 + |f|_2|), \ \forall \varepsilon > 0, \ \lambda \in (0, \lambda_0). \tag{2.218}$$

Now, on a subsequence $\{\varepsilon\} \to 0$, we have by (2.216)–(2.218), for $f \in L^1 \cap L^\infty$, that

$$\begin{array}{rll}
u_\varepsilon \to u = u(\lambda, f) & \text{weakly in } L^2 \text{ and weak-star in } L^\infty, \\
\widetilde{\beta}_\varepsilon(u_\varepsilon) \to \eta & \text{weak-star in } L^\infty \text{ and strongly in } L^2_{\text{loc}}, \\
\nabla\widetilde{\beta}_\varepsilon(u_\varepsilon) \to \nabla\eta & \text{weakly in } (L^2)^d, \\
b_\varepsilon(u_\varepsilon) \to \zeta & \text{weak-star in } L^\infty, \\
\varepsilon\widetilde{\beta}_\varepsilon(u_\varepsilon) \to 0 & \text{strongly in } L^2.
\end{array} \tag{2.219}$$

Since β is maximal monotone, the map $u \to \beta(u)$ is weakly-strongly closed in L^2_{loc} and so $\eta = \beta(u)$, a.e. in $\mathbb{R}^d$. We have, therefore,

$$u - \lambda\Delta\beta(u) + \lambda \operatorname{div}(D\zeta u) = f \text{ in } \mathscr{D}'(\mathbb{R}^d),$$

for all $\lambda \in (0, \lambda_0)$ and $f \in L^1 \cap L^\infty$. Moreover, by (2.216)–(2.218), we have

$$|u|_1 + |u|_\infty + |\beta(u)|_\infty + |\nabla\beta(u)|_2 \leq C_\lambda(|f|_1 + |f|_2), \ \forall\lambda \in (0, \lambda_0), \ f \in L^1 \cap L^\infty.$$

Let us prove now that $\zeta = b(u)$, a.e. in $\mathbb{R}^d$. Multiplying (2.212) by $u_\varepsilon = u_\varepsilon(f)$ and integrating over $\mathbb{R}^d$, since $u_\varepsilon, b_\varepsilon(u_\varepsilon)u_\varepsilon \in H^1 \cap L^1 \cap L^\infty$, $\widetilde{\beta}_\varepsilon(u_\varepsilon) \in H^1$, we obtain

$$
\begin{aligned}
&|u_\varepsilon|_2^2 + \lambda \int_{\mathbb{R}^d} \beta_\varepsilon'(u_\varepsilon)|\nabla u_\varepsilon|^2 dx \\
&\quad \le \lambda \int_{\mathbb{R}^d} (\nabla u_\varepsilon \cdot D_\varepsilon) b_\varepsilon(u_\varepsilon) u_\varepsilon dx + \frac{1}{2}\, |u_\varepsilon|_2^2 + \frac{1}{2}\, |f|_2^2.
\end{aligned}
\tag{2.220}
$$

Defining

$$
\psi(r) = \int_0^r b_\varepsilon(s) s\, ds, \ r \in \mathbb{R},
$$

we see that $\psi \ge 0$, hence the first integral on the right hand side of (2.220) is equal to

$$
-\int_{\mathbb{R}^d} \operatorname{div} D_\varepsilon \psi(u_\varepsilon) dx \le C < \infty, \tag{2.221}
$$

where (see (2.210))

$$
C := |b|_\infty |(\operatorname{div} D)^- + |D||_\infty \sup_{\varepsilon \in (0,1)} (|u_\varepsilon|_\infty |u_\varepsilon|_1)
$$

is by (2.213) and (2.216) finite, since $f \in L^1 \cap L^\infty$.

Define $g_\varepsilon(r) = (I + \varepsilon\beta)^{-1}(r)$, $r \in \mathbb{R}$, and

$$
a(r) = \int_0^r \frac{\beta'(s)}{1 + \beta'(s)}\, ds, \quad r \in \mathbb{R}.
$$

Since

$$
\beta_\varepsilon'(r) \ge \frac{\beta'(g_\varepsilon(r))}{1 + \beta'(g_\varepsilon(r))} \ge \frac{\beta'(g_\varepsilon(r))}{1 + \beta'(g_\varepsilon(r))}\, (g_\varepsilon'(r))^2, \ r \in \mathbb{R},
$$

and thus

$$
\beta_\varepsilon'(u_\varepsilon)|\nabla u_\varepsilon|^2 \ge |\nabla a(g_\varepsilon(u_\varepsilon))|^2,
$$

we obtain from (2.220) and (2.221)

$$
|u_\varepsilon|_2^2 + 2\lambda \int_{\mathbb{R}^d} |\nabla a(g_\varepsilon(u_\varepsilon))|^2 dx \le |f|_2^2. \tag{2.222}
$$

Since $|a(r)| \le |r|$ and $|g_\varepsilon(r)| \le |r|$, $r \in \mathbb{R}$, this implies that $\{a(g_\varepsilon(u_\varepsilon)),\ \varepsilon > 0\}$ is bounded in H^1, hence compact in L^2_{loc}, so along a subsequence $\varepsilon \to 0$

$$a(g_\varepsilon(u_\varepsilon)) \to v \text{ in } L^2_{loc} \text{ and a.e. on } \mathbb{R}^d.$$

Since a is strictly increasing and continuous, and thus so is its inverse function a^{-1}, it follows that $g_\varepsilon(u_\varepsilon) \to a^{-1}(v)$, a.e. , and so, as $\varepsilon \to 0$,

$$u_\varepsilon = g_\varepsilon(u_\varepsilon) \to a^{-1}(v), \text{ a.e. on } \mathbb{R}^d.$$

Therefore, by (2.215) and by (2.219)

$$u_\varepsilon \to u \text{ in } L^p_{loc}, \quad \forall p \in [1, \infty) \tag{2.223}$$

on a subsequence $\{\varepsilon\} \to 0$. Then, $b_\varepsilon(u_\varepsilon) \to b(u)$, a.e. in $\mathbb{R}^d$ and, therefore, $\zeta = b(u)$, a.e. in $\mathbb{R}^d$. Hence $u = u(\lambda, f)$ is a solution to (2.205).

To resume, we have shown that for $\lambda \in (0, \lambda_0)$ and $f \in L^1 \cap L^\infty$ there is a subsequence of $\{\varepsilon\} \to 0$ such that

$$u_\varepsilon(\lambda, f) \to u(\lambda, f) \text{ weakly in } L^2$$

where $u(\lambda, f)$ is a solution to (2.205) and, if $\beta'(r) > 0$, $\forall r \ne 0$, then by (2.223) it follows also that

$$u_\varepsilon(\lambda, f) \to u(\lambda, f) \text{ in } L^1_{loc}. \tag{2.224}$$

We define $J_\lambda : L^1 \cap L^\infty \to L^1$,

$$J_\lambda(f) = u(\lambda, f),\ f \in L^1 \cap L^\infty,\ \lambda \in (0, \lambda_0). \tag{2.225}$$

It should be recalled that (2.219) as well as (2.224) follows for a subsequence $\{\varepsilon'\} \subset \{\varepsilon\}$ which might depend of f and λ. However, if $\beta'(r) > 0$, $\forall r \in \mathbb{R}$, as easily seen for all $f \in L^1 \cap L^\infty$, Eq. (2.206) has at most one solution $u \in L^2 \cap L^\infty \subset L^1$ such that $\nabla\beta(u) \in L^2$ and this implies that $u_\varepsilon \to u(\lambda, f) = J_\lambda(f)$ weakly in L^2 and this implies that (2.190) and (2.191) hold for $f \in L^1 \cap L^\infty$. In the general case of Hypothesis (k), this does not happen but we have, however, the following lemma (see also Lemma 2.3).

Lemma 2.9 *Let J_λ be defined by* (2.224), *for a subsequence $\{\varepsilon = \varepsilon(f)\} \to 0$. Then, there is $\{\varepsilon'\} \subset \{\varepsilon\}$ independent of f and $\lambda \in (0, \lambda_0)$ such that*

$$u_{\varepsilon'} \to u = J_\lambda(f) \text{ in } L^1,\ \forall\, f \in L^1 \cap L^\infty. \tag{2.226}$$

Proof The proof is essentially the same as that of Lemma 2.3, so it will be outlined only. We fix $f \in C_0^\infty(\mathbb{R}^d)$. We know by (2.224) that we have a subsequence $\{\varepsilon'\} \subset$

$\{\varepsilon\}$, $u_{\varepsilon'} \to u = J_\lambda(f)$ strongly in L^1_{loc}. Let us prove first that $\{u_{\varepsilon'}\} \to u$ in L^1. For simplicity, denote again $\{\varepsilon\}$ this subsequence. It suffices to show that there is C independent of ε such that

$$\|u_\varepsilon\| = \int_{\mathbb{R}^d} |u_\varepsilon(x)|\Phi(x)dx \le C, \ \ \forall\, \varepsilon > 0, \tag{2.227}$$

where $\Phi(x) \equiv (1+|x|^2)^{\frac{1}{2}}$.

We shall proceed as in the proof of (2.57). Namely, we multiply (2.208), where $u = u_\varepsilon$, by $\Phi_\nu \mathscr{X}_\delta(\widetilde{\beta}_\varepsilon(u_\varepsilon))$, where $\Phi_\nu(x) = \Phi(x)\exp(-\nu\Phi(x))$, $\nu > 0$, and integrate over $\mathbb{R}^d$. Since $\mathscr{X}'_\delta \ge 0$, we get

$$\begin{aligned}
\int_{\mathbb{R}^d} u_\varepsilon \mathscr{X}_\delta(\widetilde{\beta}_\varepsilon(u_\varepsilon))\Phi_\nu\, dx &\le -\lambda \int_{\mathbb{R}^d} \nabla\widetilde{\beta}_\varepsilon(u_\varepsilon)\cdot\nabla(\mathscr{X}_\delta(\widetilde{\beta}_\varepsilon(u_\varepsilon))\Phi_\nu)dx \\
&\quad +\lambda \int_{\mathbb{R}^d} D_\varepsilon b^*_\varepsilon(u_\varepsilon)\cdot\nabla(\mathscr{X}_\delta(\widetilde{\beta}_\varepsilon(u_\varepsilon))\Phi_\nu)dx + \int_{\mathbb{R}^d} |f|\Phi_\nu dx \\
&\le -\lambda \int_{\mathbb{R}^d} (\nabla\widetilde{\beta}_\varepsilon(u_\varepsilon)\cdot\nabla\Phi_\nu)\mathscr{X}_\delta(\widetilde{\beta}_\varepsilon(u_\varepsilon))dx \\
&\quad +\lambda \int_{\mathbb{R}^d} D_\varepsilon b^*_\varepsilon(u_\varepsilon)\cdot\nabla\widetilde{\beta}_\varepsilon(u_\varepsilon)\mathscr{X}'_\delta(\widetilde{\beta}_\varepsilon(u_\varepsilon))\Phi_\nu dx \\
&\quad +\lambda \int_{\mathbb{R}^d} (D_\varepsilon\cdot\nabla\Phi_\nu)b^*_\varepsilon(u_\varepsilon)\mathscr{X}_\delta(\widetilde{\beta}_\varepsilon(u_\varepsilon))dx + \int_{\mathbb{R}^d} |f|\Phi_\nu dx.
\end{aligned} \tag{2.228}$$

Letting $\delta \to 0$, we get as above

$$\begin{aligned}
\int_{\mathbb{R}^d} |u_\varepsilon|\Phi_\nu dx &\le -\lambda \int_{\mathbb{R}^d} \nabla|\widetilde{\beta}_\varepsilon(u_\varepsilon)|\cdot\nabla\Phi_\nu dx \\
&\quad +\overline{\lim_{\delta\to 0}}\frac{\lambda}{\delta}\int_{[|\widetilde{\beta}_\varepsilon(u_\varepsilon)|\le\delta]} |D_\varepsilon|\,|b^*_\varepsilon(u_\varepsilon)|\,|\nabla\widetilde{\beta}_\varepsilon(u_\varepsilon)|\Phi_\nu dx \\
&\quad +\lambda \int_{\mathbb{R}^d} (\operatorname{sign} u_\varepsilon)b^*_\varepsilon(u_\varepsilon)(D_\varepsilon\cdot\nabla\Phi_\nu)dx + \int_{\mathbb{R}^d} |f|\Phi_\nu dx \\
&\le \lambda \int_{\mathbb{R}^d} (|\widetilde{\beta}_\varepsilon(u_\varepsilon)|\Delta\Phi_\nu + |b^*_\varepsilon(u_\varepsilon)|\,|D_\varepsilon\cdot\nabla\Phi_\nu|)dx + \int_{\mathbb{R}^d} |f|\Phi_\nu dx,
\end{aligned} \tag{2.229}$$

where in the last step we used that

$$|b^*_\varepsilon(u_\varepsilon)| \le \mathrm{Lip}(b^*_\varepsilon)|u_\varepsilon| \le \frac{1}{\varepsilon}\,\mathrm{Lip}(b^*_\varepsilon)|\widetilde{\beta}_\varepsilon(u_\varepsilon)|.$$

We have $\nabla\Phi, \Delta\Phi \in L^\infty$, and

$$\begin{aligned}
&\nabla\Phi_\nu(x) = (\nabla\Phi - \nu\Phi\nabla\Phi)\exp(-\nu\Phi),\\
&\Delta\Phi_\nu(x) = (\Delta\Phi - \nu|\nabla\Phi|^2 - \nu\Phi\Delta\Phi + \nu^2\Phi|\nabla\Phi|^2 - \nu|\nabla\Phi|^2)\exp(-\nu\Phi).
\end{aligned}$$

Then, letting $\nu \to 0$ in (2.229), since $M := \sup\limits_{\varepsilon>0} |u_\varepsilon|_\infty < \infty$, $|b_\varepsilon^*(r)| \le |b|_\infty |r|$, $|D_\varepsilon| \le |D|$ and $|\widetilde{\beta}_\varepsilon(r)| \le \Big(\sup\limits_{|r|\le M} \beta'(r) + \varepsilon\Big)|r|$, $\forall r \in [-M, M]$, we get

$$\|u_\varepsilon\| \le \|f\| + C\lambda(|\Delta\Phi|_\infty + |D|_\infty|\nabla\Phi|_\infty)|f|_1, \ \forall \varepsilon \in (0,1).$$

Hence (2.227) holds for all $f \in C_0^\infty$.

Let us show now that there is a subsequence $\{\varepsilon'\}$ independent of f such that (2.226) holds. To this end, we choose a sequence $\{f_n\} \subset C_0^\infty(\mathbb{R}^d)$ which is dense in L^1. Then, by the first part of the proof for each n there is $\{\varepsilon_j^n\}_{j=1}^\infty \to 0$ such that

$$u_{\varepsilon_j^n}(f_k) \overset{j\to\infty}{\longrightarrow} J_\lambda(f_k) \text{ in } L^1, \ \forall 1 \le k \le n.$$

Then, for the diagonal sequence $\{\varepsilon_j^j\}_{j=1}^\infty$, we have for $j \to \infty$

$$u_{\varepsilon_j^j}(f_n) \to J_\lambda(f_n), \ \forall n,$$

and taking into account that

$$\begin{aligned}
&|J_\lambda(f) - J_\lambda(\bar{f})|_1 \le |f - \bar{f}|_1, \ \forall f, \bar{f} \in L^1,\\
&|J_{\lambda_2}(f(-J_{\lambda_1}(f)|_1 \le 2\left(1 - \frac{\lambda_1}{\lambda_2}\right)|f|_1, \ \forall f \in L^1,
\end{aligned}$$

it follows by density of $\{f_n\}$ in L^1 that (2.226) holds for $\{\varepsilon'\} = \{\varepsilon_j^j\}_{j=1}^\infty$.

Proof of Proposition 2.3 (Continued) By Lemma 2.9 and (2.214), it follows on $\varepsilon = \{\varepsilon'\} \to 0$ that (2.191) holds for all $f \in L^1 \cap L^\infty$.

We are going to extend by density J_λ to all of L^1. Under the hypotheses of Theorem 2.4, this was an immediate consequence of (2.85), but in our case it is more delicate, and to this end we need some further estimates on $\widetilde{\beta}(u_\varepsilon)$. We have

$$\Delta\widetilde{\beta}_\varepsilon(u_\varepsilon) = \lambda^{-1}(u_\varepsilon - f) + \operatorname{div}(D_\varepsilon b_\varepsilon(u_\varepsilon)u_\varepsilon). \tag{2.230}$$

We shall assume first that $d \geq 3$. Let $E_d(x) = \omega_d|x|_d^{2-d}$, $x \in \mathbb{R}^d$, be the fundamental solution of the Laplace operator. We have

$$\widetilde{\beta}_\varepsilon(u_\varepsilon) = \frac{1}{\lambda}\, E_d * (-u_\varepsilon + f) + \nabla(E_d * (D_\varepsilon b_\varepsilon(u_\varepsilon)u_\varepsilon)), \text{ a.e. in } \mathbb{R}^d. \tag{2.231}$$

For $1 < p < \infty$, we denote by M^p the Marcinkiewicz space with the norm (see Sect. 6.3)

$$\|u\|_{M^p} = \inf\left\{\lambda > 0;\ \int_K |u(x)|dx \leq \lambda(\text{meas } K)^{\frac{1}{p'}} \right.$$
$$\left. \text{for all Borel sets } K \subset \mathbb{R}^d\right\} < \infty,\ \tfrac{1}{p} + \tfrac{1}{p'} = 1.$$

We have

$$E_d \in M^{\frac{d}{d-2}}(\mathbb{R}^d),\ |\nabla E_d| \in M^{\frac{d}{d-1}}(\mathbb{R}^d), \tag{2.232}$$

and, for $f \in L^1$, the solution $u \in L^1$ to the equation $-\Delta u = f$ in $\mathscr{D}'(\mathbb{R}^d)$ is given by the convolution product $u = E * f$ and satisfies (see (6.20) and (6.21), Sect. 6.3)

$$\|u\|_{M^{\frac{d}{d-2}}} \leq \|E_d\|_{M^{\frac{d}{d-2}}} |f|_1. \tag{2.233}$$

$$\|\nabla u\|_{M^{\frac{d}{d-1}}} \leq \|\nabla E_d\|_{M^{\frac{d}{d-1}}} |f|_1. \tag{2.234}$$

This yields

$$\begin{aligned}\|\widetilde{\beta}_\varepsilon(u_\varepsilon) - \nabla(E_d * (D_\varepsilon b_\varepsilon(u_\varepsilon)u_\varepsilon))\|_{M^{\frac{d}{d-2}}} &\leq \frac{1}{\lambda}\, \|E_d\|_{M^{\frac{d}{d-2}}} (|u_\varepsilon - f|_1 \\ &\leq \frac{2+3\lambda}{\lambda}\, \|E_d\|_{M^{\frac{d}{d-2}}} |f|_1,\ \forall f \in L^1 \cap L^\infty,\end{aligned} \tag{2.235}$$

$\forall \lambda > 0$, because $\varepsilon|\widetilde{\beta}_\varepsilon(u)| \leq (2+\varepsilon^2)|u|$. Taking into account that, for $d \geq 3$,

$$M^{\frac{d}{d-2}} \subset L^p_{\text{loc}},\ \forall p \in \left(1, \frac{d}{d-2}\right),\quad M^{\frac{d}{d-1}} \subset L^p_{\text{loc}},\ \forall p \in \left(1, \frac{d}{d-1}\right)$$

we see by (2.235) that, for $1 < p < \frac{d}{d-2}$, and, for all compacts $K \subset \mathbb{R}^d$, we have

$$\|\widetilde{\beta}_\varepsilon(u_\varepsilon) - \text{div}(E_d * (D_\varepsilon b_\varepsilon(u_\varepsilon)u_\varepsilon))\|_{L^p(K)} \leq \frac{1+\lambda}{\lambda}\, C_K|f|_1,\ \forall \lambda > 0,$$

and so, by (2.233), we have, for $1 < q < \frac{d}{d-1}$,

$$\begin{aligned}\|\widetilde{\beta}_\varepsilon(u_\varepsilon)\|_{L^q(K)} &\leq C_K\left(\|\mathrm{div}(E_d * (D_\varepsilon b_\varepsilon(u_\varepsilon)u_\varepsilon))\|_{L^q(K)} + \frac{1+\lambda}{\lambda}|f|_1\right)\\ &\leq C_K\left(\|\mathrm{div}(E_d * (D_\varepsilon b_\varepsilon(u_\varepsilon)u_\varepsilon))\|_{M^{\frac{d}{d-1}}} + \frac{1+\lambda}{\lambda}|f|_1\right)\\ &\leq C_K\left(|u_\varepsilon|_1 + \frac{1+\lambda}{\lambda}|f|_1\right) \leq C_K\left(\frac{1+\lambda}{\lambda}\right)|f|_1,\\ &\qquad \forall f \in L^1 \cap L^\infty,\ \lambda > 0,\end{aligned} \tag{2.236}$$

for any compact subset $K \subset \mathbb{R}^d$, where we used that $|b_\varepsilon|_\infty \leq |b|_\infty$, $|D_\varepsilon| \leq |D|_\infty$.

Assume now $d = 2$, $E_2(x) = \frac{1}{2\pi}\log|x|$, $x \neq 0$, and set $B_R = \{x;\ |x| \leq R\}$, $\forall R > 0$.

Taking into account that $|\nabla E_2(x)| \leq \frac{1}{|x|}$, $\forall x \neq 0$,, we have, for all $1 \leq q < 2$,

$$\|\nabla E_2 * D_\varepsilon b_\varepsilon(u_\varepsilon)u_\varepsilon\|_{L^q(B_R)} \leq C_p|D_\varepsilon b_\varepsilon(u_\varepsilon)u_\varepsilon|_1 \leq C_p^1|f|_1. \tag{2.237}$$

Similarly,

$$\|\nabla(E_2 * (f - u_\varepsilon))\|_{L^q(B_R)} \leq C_p|f - u_\varepsilon|_1 \leq C_p^1|f|_1. \tag{2.238}$$

This yields

$$\|E_2 * (f - u_\varepsilon)\|_{L^q(B_{\frac{R}{2}})} \leq C_p^1|f|_1. \tag{2.239}$$

Here is the argument. Let $\varphi \in C_0^\infty(B_R)$ such that $\varphi = 1$ on $B_{\frac{3R}{4}}$ and let $\theta \in C_0^\infty(B_R)$ be such that $\theta = 1$ on $B_\nu^c = \{x;\ |x| > \nu\}$. We have

$$\Delta(\theta E_2) = \delta + \eta \ \text{ in } \mathscr{D}'(\mathbb{R}^2), \tag{2.240}$$

where $\eta \in C_0^\infty(B_\nu)$, and δ is the Dirac measure. We set $v_\varepsilon = (E_2 * (f - u_\varepsilon))\varphi$ and get by (2.240)

$$\begin{aligned}v_\varepsilon &= -\eta * v_\varepsilon + \nabla(\theta E_2) * \nabla v_\varepsilon\\ &= -\eta * v_\varepsilon + \nabla(\theta E_2) * (\varphi\nabla(E_2 * (f - u_\varepsilon)) + \nabla\varphi(E_2 * (f - u_\varepsilon))\\ &= -\eta * v_\varepsilon + \nabla(\theta E_2) * (\varphi(\nabla E_2 * (f - u_\varepsilon))) \text{ on } B_{\frac{3R}{4}}\end{aligned} \tag{2.241}$$

if $\nu < \frac{R}{4}$. Taking into account that $E_2 \in L^1_{\mathrm{loc}}$, it follows that

$$|\eta * v_\varepsilon|_q \leq |(E_2 * (f - u_\varepsilon))\varphi * \eta|_q \leq |(E_2 * (f - u_\varepsilon))\varphi|_1|\eta|_q \leq C|f|_1.$$

Then, by (2.241) and (2.238), we get that

$$\begin{aligned}\|v_\varepsilon\|_{L^q(B_{\frac{R}{2}})} &\le C|f|_1 + \|\nabla(\theta E_2) * (\varphi(\nabla E_2 * (f - u_\varepsilon)))\|_{L^q(B_{\frac{3R}{4}})}\\ &\le C|f|_1 + |\nabla(\theta E_2)|_1 \|\varphi(\nabla(E_2 * (f - u_\varepsilon)))\|_{L^q(B_R)}\\ &\le C(|f|_1 + \|\nabla(E_2 * (f - u_\varepsilon))_{L^q(B_R)})\\ &\le C_1|f|_1,\end{aligned}$$

and so (2.239) follows.

Now, by (2.237)–(2.239) we get (2.236) for all $K \subset \mathbb{R}^2$ and $1 < d$.

Consider now the case $d = 1$ and note that by (2.230) we have

$$|(\widetilde{\beta}_\varepsilon(u_\varepsilon))'|_\infty \le \frac{1}{\lambda}(|f|_1 + |u_\varepsilon|_1) \le \frac{2}{\lambda}|f|_1,$$

which clearly implies that

$$\|\beta_\varepsilon(u_\varepsilon)\|_{L^\infty(K)} \le C|f|_1, \ \forall K \subset (-\infty, \infty).$$

Hence (2.236) holds for all dimensions $d \ge 1$.

We note that, by (2.236) and (2.224), it also follows that

$$\|\beta(u(\lambda, f))\|_{L^q(K)} \le C_K\left(1 + \frac{1}{\lambda}\right)|f|_1, \ \forall f \in L^1 \cap L^\infty, \ \lambda > 0. \tag{2.242}$$

Moreover, by (2.213) and (2.219) we have

$$|J_\lambda(f_1) - J_\lambda(f_2)|_1 \le |f_1 - f_2|_1, \ \forall f_1, f_2 \in L^1 \cap L^\infty. \tag{2.243}$$

Now, let $f \in L^1$ and $\{f_n\} \subset L^1 \cap L^\infty$ be such that $f_n \to f$ in L^1. If $u_n = u(\lambda, f_n)$, we see by (2.243) that $u_n \to u$ in L^1 while, by (2.242), $\{\beta(u_n)\}$ is bounded in $L^q(K)$ for each $K \subset \mathbb{R}^d$, $q \in \left(0, \frac{d}{d-1}\right)$. Hence, $\{\beta(u_n)\}$ is weakly compact in $L^q(K)$ and, therefore, $\beta(u_n) \to \eta$ weakly in $L^q(K)$ on a subsequence $\{u_n\} \to 0$ in L^1. Now, by Luzin's theorem, for each $\delta > 0$, there is a Lebesgue measurable subset $K' \subset K$ such that $m(K \setminus K') \le \delta$ and on a subsequence, $u_n \to u$ in $L^\infty(K')$. Since the operator $\beta : L^q(K') \to L^q(K')$, $\frac{1}{q} + \frac{1}{q'} = 1$ is maximal monotone, and therefore strongly-weakly closed, we infer that $\eta = \beta(u)$, a.e. in K' and, since K' can be chosen arbitrarily large, we have that $\eta = \beta(u)$, a.e. in $\mathbb{R}^d$.

Then, letting $n \to \infty$ in the equation $u_n + \lambda A_0(u_n) = f_n$, we see that

$$u = J_\lambda(f) = \lim_{n\to\infty} J_\lambda(f_n) \text{ in } L^1, \ f_n \to f \text{ in } L^1, \tag{2.244}$$

is a solution to $u + \lambda A_0(u) = f$, as claimed.

Then, we may extend (2.245) and (2.243) to all of $f \in L^1$ and hence (2.190) and (2.191) hold.

By (2.215) and (2.219), it follows that (2.195) holds, which by (2.242) implies (2.194). Moreover, if $f \geq 0$ on $\mathbb{R}^d$, then by (2.208) it is easily seen via the maximum principle that $u_\varepsilon \geq 0$ on $\mathbb{R}^d$, and so, $u \geq 0$, as claimed. Moreover, we have by (2.206)

$$\int_{\mathbb{R}^d} u\,dx = \int_{\mathbb{R}^d} f\,dx. \tag{2.245}$$

Indeed, by (2.206), we have

$$\int_{\mathbb{R}^d} u\varphi dx - \lambda \int_{\mathbb{R}^d} \beta(u)\Delta\varphi dx + \lambda \int_{\mathbb{R}^d} ub(u)D\cdot\nabla\varphi dx, = \int_{\mathbb{R}^d} f\varphi\,dx, \tag{2.246}$$

$\forall \varphi \in C_0^\infty$.

Now, we choose in (2.246) $\varphi = \varphi_\nu \in C_0^\infty$, where $\varphi_\nu \to 1$ on $\mathbb{R}^d$, $0 \leq \varphi_\nu \leq 1$, and

$$|\Delta\varphi_\nu|_\infty + |\nabla\varphi_\nu|_\infty \to 0 \text{ as } \nu \to 0.$$

(Such an example is $\varphi_\nu(x) = \exp\left(-\frac{\nu|x|^2}{1-\nu|x|^2}\right)$.) Then, for $\nu \to 0$, it follows (2.245), and so (2.192) and (2.196) hold.

It remains to prove (2.197). Let $g \in C_0^\infty(\mathbb{R}^d)$ be arbitrary but fixed. Coming back to Eq. (2.212), we can write

$$g - \lambda\Delta\widetilde{\beta}_\varepsilon(g) + \lambda\,\mathrm{div}(D_\varepsilon b_\varepsilon(g)g) = g + \lambda A_\varepsilon(g),$$

and so (2.213) yields that, for some positive constant C,

$$|u_\varepsilon(g) - g|_1 \leq \lambda|A_\varepsilon(g)|_1 \leq C\lambda\|g\|_{H^2(\mathbb{R}^d)},\ \forall\varepsilon \in (0,1),$$

because $\beta \in C^2$, $b \in C^1$, and hence $\widetilde{\beta}_\varepsilon$, $(\widetilde{\beta}_\varepsilon)'$, $(\widetilde{\beta}_\varepsilon)''$ and b'_ε are locally uniformly bounded in $\varepsilon \in (0,1)$ and since $|D|_\infty < \infty$, $\mathrm{div}\,D \in L^2_{\mathrm{loc}}$ and $|b_\varepsilon| \leq |b|_\infty$. This, together with (2.219), implies (2.197). This completes the proof of Proposition 2.3. □

Proof of Theorem 2.8 (Continued) As seen earlier, the solution u_h to the finite difference scheme (2.187) is uniformly convergent on every compact interval $[0, T]$ to $u \in C([0,\infty); L^1)$. By (2.195) and (2.204), by a standard argument we obtain that, for all $u_0 \in L^1 \cap L^\infty$,

$$|u(t)|_\infty = |S(t)u_0|_\infty \leq \exp(2|b|_\infty(|(\mathrm{div}\,D)^-|_\infty + |D|_\infty)t)|u_0|_\infty,\ \forall t \geq 0,$$

and so (2.195) follows. By (2.196) and (2.204) it follows that $S(t)(\overline{D(A)} \cap \mathcal{P}) \subset \mathcal{P}$, $\forall t \geq 0$.

Let us prove now that u is a distributional solution to (2.183). We note first that by (2.187) we have (setting $u_h(t) = u_0$ for $t \in (-\infty, 0)$)

$$\begin{aligned} &u_h(t) - h\Delta\beta(u_h(t)) + h\,\mathrm{div}(Db(u_h(t))u_h(t)) = u_h(t-h),\ t \geq 0,\\ &u_h(t) = u_0. \end{aligned} \tag{2.247}$$

By (2.194) and (2.247), it follows that

$$\|\beta(u_h(t))\|_{L^p(K)} \leq C_K|u_0|_1,\ \forall t \in [0, T],$$

for any compact subset $K \subset \mathbb{R}^d$. In particular, this yields

$$\beta(u_h) \to \eta(t) \text{ weakly in } L^p_{\text{loc}},\ \forall t \in [0, T], \tag{2.248}$$

Since $\lim\limits_{h\to 0} u_h(t) = S(t)u_0$ in L^1 uniformly on compact intervals of $t \in [0, \infty)$ and β is maximal monotone, it follows that $\eta(t, x) = \beta(u(t, x))$, a.e. $(t, x) \in (0, T) \times \mathbb{R}^d$.

Let $\varphi \in C_0^\infty([0, \infty) \times \mathbb{R}^d)$. Then, by (2.247) we have

$$\begin{aligned} \int_0^\infty \int_{\mathbb{R}^d} \frac{1}{h}\,(u_h(t,x) - u_h(t-h,x))\varphi(t,x) - \beta(u_h(t,x))\Delta\varphi(t,x)&\\ -b(u_h(t,x))u_h(t,x)D_\varepsilon(x)\cdot\nabla\varphi(t,x)dt\,dx = 0&, \end{aligned}$$

if we take $u_h(t, x) \equiv u_0(x)$ for $t \in (-h, 0]$. Then, replacing the first term by

$$\int_0^\infty \int_{\mathbb{R}^N} \frac{1}{h}\, u_h(t,x)(\varphi(t+h,x) - \varphi(t,x))dt\,dx + \frac{1}{h}\int_0^h \int_{\mathbb{R}^N} u_0(x)\varphi(t,x)dt\,dx$$

and, letting $h \to 0$, by (2.248) we get (2.187), as claimed. Finally, if beside (k)–(kkk), Hypothesis (iv) of Theorem 2.4 holds, then, as proven earlier, the mild solution u is unique. This completes the proof. □

We shall prove now a vanishing viscosity type result for NFPE (2.183). Namely, let $A_\varepsilon : L^1 \to L^1$ be the operator defined by

$$\begin{aligned} &A_\varepsilon(y) = -\Delta(\beta(y) + \varepsilon y) + \mathrm{div}(D_\varepsilon b_\varepsilon^*(y)),\ \forall y \in D(A_\varepsilon),\\ &D(A_\varepsilon) = \{y \in L^1;\ \beta(y) \in L^1_{\text{loc}},\ -\Delta(\beta(y) + \varepsilon y) + \mathrm{div}(D_\varepsilon b_\varepsilon^*(y)) \in L^1\}. \end{aligned} \tag{2.249}$$

As seen above, A_ε is m-accretive in L^1 and so it generates a C_0-continuous semigroup of contractions $S_\varepsilon(t) = \exp(-tA_\varepsilon)$. We have

Theorem 2.9 *Let assumptions* (k)–(kkk) *hold. If $S(t)$ is the semigroup provided by Theorem* 2.8 *we have, for all $u_0 \in \overline{D(A)}$ and on a subsequence $\{\varepsilon\} \to 0$, independent of u_0,*

$$S_\varepsilon(t)u_0 \to S(t)u_0 \ \text{in } L^1 \tag{2.250}$$

uniformly in t on compact intervals.

Proof It suffices to recall Lemma 2.9 and apply the Trotter–Kato theorem for nonlinear semigroups (Theorem 6.9).

Theorem 2.9, which obviously remains true under assumptions (i)–(iii) as well, entitles to call a mild solution to NFPE (2.183) a *viscosity-weak solution.*

We shall prove now a version of Theorem 2.9 for possible degenerate diffusion term β. Namely, consider Eq. (2.183) under the following hypotheses

(k)′ $\beta \in C^1(\mathbb{R}),\ \beta'(r) > 0,\ \forall r \neq 0,\ \beta(0) = 0,$ *and*

$$\mu_1 \min\{|r|^\nu, |r|\} \le |\beta(r)| \le \mu_2 |r|,\ \forall r \in \mathbb{R},$$

where $\mu_1, \mu_2 > 0$ and $\nu \in (0, 1]$.

(kk)′ $D \in L^\infty$.

(kkk)′ $b \in C_b(\mathbb{R})$.

We have

Theorem 2.10 *Let $d \ge 1$. Then, under assumptions* (k)′–(kkk)′ *there is a continuous semigroup of contractions $S(t) : L^1 \to L^1$ defined on a closed subset $C \subset L^1$ such that, for each $u_0 \in C$, $u(t) = S(t)u_0$ is a mild solution to* (2.183). *Moreover, $S(t)\mathcal{P} \subset \mathcal{P}$, $\forall t \ge 0$, u is a distributional solution to* (2.183) *and there is a subsequence $\{\varepsilon'\} \subset \{\varepsilon\}$ such that*

$$u(t) = \lim_{\varepsilon' \to 0} u_{\varepsilon'}(t) \ \text{in } L^1, \tag{2.251}$$

uniformly in t on compact intervals, where $u_\varepsilon = \exp(-tA_\varepsilon)u_0$, that is,

$$\frac{du_\varepsilon}{dt} + A_\varepsilon(u_\varepsilon) = 0,\ t \ge 0;\ \ u_{\varepsilon'}(0) = u_0. \tag{2.252}$$

Here, $A_\varepsilon : L^1 \to L^1$ is defined by (2.249).

Proof We shall prove first

Lemma 2.10 *The operator A_ε is m-accretive in L^1 and*

$$\lim_{\varepsilon' \to 0} (I + \lambda A_{\varepsilon'})^{-1} f = J_\lambda(f) \ \text{in } L^1,\ \forall f \in L^1,\ \lambda > 0, \tag{2.253}$$

on a subsequence $\{\varepsilon'\} \subset \{\varepsilon\}$ independent of f and $\lambda \in (0, \lambda_0)$, where $J_\lambda : L^1 \to L^1$ satisfies (2.189)–(2.191).

Proof Consider the equation

$$u + \lambda A_\varepsilon(u) = f,\ \lambda > 0, \tag{2.254}$$

where $f \in L^1 \cap L^\infty$. Then, by Proposition 2.3, it follows that there is a solution $u_\varepsilon, \beta(u_\varepsilon) \in H^1(\mathbb{R}^d)$. Moreover, we have

$$(\beta(u_\varepsilon), u_\varepsilon)_2 + \lambda|\nabla\beta(u_\varepsilon)|_2^2 \le (f, \beta(u_\varepsilon))_2 + \lambda(D_\varepsilon b_\varepsilon(u_\varepsilon)u_\varepsilon, \nabla\beta(u_\varepsilon))_2,$$

and this yields

$$\begin{aligned}\alpha_1|u_\varepsilon|_2^2 + \tfrac{\lambda}{2}|\nabla\beta(u_\varepsilon)|_2^2 &\le (f, \beta(u_\varepsilon))_2 + \frac{\lambda}{2}|u_\varepsilon|_\infty^2|b_\varepsilon(u_\varepsilon)|_2^2|u_\varepsilon|_2^2\\ &\le |(f, \beta(u_\varepsilon))_2| + C_1\lambda|u_\varepsilon|_\infty^2|u_\varepsilon|_2^2.\end{aligned} \tag{2.255}$$

If $d \ge 3$, then by the Sobolev–Gagliardo–Nirenberg theorem, we have

$$|\beta(u_\varepsilon)|_{p^*} \le C_2|\nabla\beta(u_\varepsilon)|_2^2,\ \ p^* = \frac{2d}{d-2},$$

and so, (2.255) yields

$$\alpha_1|u_\varepsilon|_2^2 + \frac{\lambda}{2}|\nabla\beta(u_\varepsilon)|_2^2 \le \frac{C_1}{\lambda}|f|_2^2 + C_1\lambda|u_\varepsilon|_\infty^2|u_\varepsilon|_2^2.$$

Then, for $0 < \lambda \le \lambda_0$, we have

$$|u_\varepsilon|_2^2 + \lambda|\nabla\beta(u_\varepsilon)|_2^2 + \lambda|\beta(u_\varepsilon)|_{p^*}^2 \le C|f|_2^2.$$

For $d = 1, 2$, it follows that

$$|u_\varepsilon|_2^2 + \lambda|\nabla\beta(u_\varepsilon)|_2^2 \le C|f|_2^2,\ \forall\lambda \in (0, \lambda). \tag{2.256}$$

Let us denote the solution u_ε to (2.254) by $u_\lambda^\varepsilon(f)$ and set $\beta_\varepsilon(r) := \beta(r) + \varepsilon r$, $r \in \mathbb{R}$. Now, we multiply the equation

$$\begin{aligned}(u_\lambda^\varepsilon(f_1) - u_\lambda^\varepsilon(f_2)) - \lambda\Delta(\beta_\varepsilon(u_\lambda^\varepsilon(f_1)) - \beta_\varepsilon(u_\lambda^\varepsilon(f_2)))\\ +\lambda\,\mathrm{div}(u_\varepsilon(b_\varepsilon^*(f_1)) - b_\varepsilon^*(u_\lambda^\varepsilon(f_2))) = f_1 - f_2\end{aligned}$$

by $\mathcal{X}_\delta(\beta_\varepsilon(u_\lambda^\varepsilon(f_1)) - \beta(u_\lambda^\varepsilon(f_2)))$ and integrate over $\mathbb{R}^d$ and we get as above

$$|u_\lambda^\varepsilon(f_1) - u_\lambda^\varepsilon(f_2)|_1 \le |f_1 - f_2|_1.$$

Now, fix $\lambda \in (0, \lambda_0)$ and set $u_\varepsilon = u_\lambda^\varepsilon(f)$. Then, by (2.255) and (2.256) it follows that $\{u_\varepsilon\}$ and $\{\nabla \beta_\varepsilon(u_\varepsilon)\}$ are bounded in L^2. Moreover, for $d \geq 3$, $\{\beta(u_\varepsilon)\}$ is bounded in L^{p^*}, while for $d = 1, 2$ it is bounded in L^2. Hence, $\{\beta(u_\varepsilon)\}$ is bounded in W^{1,p^*} for $d \geq 3$ and in H^1 for $d = 1, 2$. This implies that $\{\beta(u_\varepsilon)\}$ is compact in L^2_{loc} and $\{u_\varepsilon\}$ is weakly compact in L^2.

Therefore, along a subsequence $\{\varepsilon\} \to 0$, we have

$$\begin{aligned} u_\varepsilon &\to u \quad \text{weakly in } L^2,\\ \beta(u_\varepsilon) &\to v \quad \text{strongly in } L^2_{\mathrm{loc}},\\ \nabla\beta_\varepsilon(u_\varepsilon) &\to \nabla v \ \ \text{wakly in } L^2. \end{aligned} \tag{2.257}$$

Since the map $u \to \beta(u)$ is maximal monotone in $L^2(\mathcal{O})$ for every bounded, open set $\mathcal{O} \subset \mathbb{R}^d$, it follows that $v(x) = \beta(u(x))$, a.e. $x \in \mathbb{R}^d$. Moreover, as β is continuous and strictly monotone, it follows that $u_\varepsilon \to u$, a.e. on $\mathbb{R}^d$ and, therefore, we have for $\varepsilon \to 0$,

$$b_\varepsilon(u_\varepsilon)u_\varepsilon \to b(u)u, \text{ a.e. in } \mathbb{R}^d \text{ and weakly in } L^2,$$

selecting another subsequence $\{\varepsilon\} \to 0$, if necessary. Then, letting $\varepsilon \to 0$, we see that

$$u - \lambda\Delta\beta(u) + \lambda\,\mathrm{div}(Db(u)u) = f \ \text{ in } H^{-1}.$$

Arguing as in Lemma 2.3, it follows that the subsequence $\{\varepsilon\} \to 0$ can be chosen independent of λ and f.

Also, it follows that $\nabla\beta(u) \in L^2$ and we have, for $0 < \lambda \leq \lambda_0$ and $f \in L^1 \cap L^2$,

$$|u|_2^2 + \lambda|\nabla\beta(u)|_2^2 + \lambda|\beta(u)|_{p^*}^2 \leq C_4|f|_{\frac{d}{2}}^2$$

if $d \geq 3$ and

$$|u|_2^2 + \lambda|\nabla\beta(u)|_2^2 + \lambda|\beta(u)|_2^2 \leq C_5|f|_2^2,$$

if $d = 1, 2$.

Let $f \in L^1$ be arbitrary but fixed and let $\{f_n\} \subset l^1 \cap L^\infty$ be such that $f_n \to f$ in L^1 as $n \to \infty$. We set $u_n = u(\lambda, f_n)$ and recall that

$$|u_n - u_m|_1 \leq |f_n - f_m|_1, \ \forall n, m \in \mathbb{N},$$

and so there is $u = \lim\limits_{n\to\infty} u_n$ in L^1. This implies that $u + \lambda A_0(u) = f$, and so $R(I + \lambda A_0) = L^1$, as claimed. By (2.257), it also follows that

$$u_\varepsilon = (I + \lambda A_\varepsilon)^{-1} f \to J_\lambda(f) \ \text{ in } L^1_{\mathrm{loc}} \ \text{ as } \varepsilon \to 0, \ \text{for all } \lambda > 0 \text{ and } f \in L^1.$$

Finally, by Lemma 2.9 it follows (2.253).

Now, we define the operator $A : D(A) \subset L^1 \to L^1$ as in (2.198) and (2.199). Then, by Lemma 2.10 we have

Lemma 2.11 *The operator A is m-accretive in* L^1 *and* $(I + \lambda A)^{-1} f = J_\lambda(f)$, $\forall f \in L^1$, $\lambda > 0$. *Moreover, there is a subsequence* $\{\varepsilon'\} \subset \{\varepsilon\}$ *independent of f and* $\lambda \in (0, \lambda_0)$ *such that*

$$(I + \lambda A)^{-1} f = \lim_{\varepsilon' \to 0} (I + \lambda A_{\varepsilon'})^{-1} f \text{ in } L^1, \ \forall f \in L^1, \ \lambda > 0. \tag{2.258}$$

Then, Theorem 2.10 follows as in the previous cases by the Crandall and Liggett theorem (Theorem 6.6).

About the Density in L^1 of the domain $D(A)$ of A

In the situations treated above for the density in L^1 of the domain $D(A)$ of the infinitesimal generator A of the Fokker–Planck semigroup $S(t)$, one should assume that $\beta \in C^2(\mathbb{R})$, which in many cases is an excessive condition. For instance, as seen in Proposition 2.1, part (2.19) (see also Lemma 4.1), a sufficient condition for $\overline{D(A)} = L^1$ is

$$\beta \in C^1(\mathbb{R}), \ \beta(0) = 0, \ 0 \le \beta'(r) \le \gamma_1 < \infty, \ b \in C_b(\mathbb{R}) \cap C^1(\mathbb{R}). \tag{2.259}$$

We have also

Under Hypotheses (k)–(kkk) *of Theorem* 2.8, *it follows that* $\overline{D(A)}=L^1$ *if* $D \equiv$ constant.

We outline the argument. For a fixed $f \in L^1$, consider the equation $u_\lambda + \lambda A u_\lambda = f$ and check first via the Kolmogorov compactness theorem that $\{u_\lambda\}$ is compact in $L^1_{\rm loc}$ and afterwards, by an argument similar to the one used in Lemma 2.9, that $\{u_\lambda\}$ is compact in L^1. Then, for $\{\lambda_n\} \to 0$, $u_{\lambda_n} \to u$ in L^1 and, since $\lambda_n A u_{\lambda_n} \to 0$ in $\mathcal{D}'(\mathbb{R}^d)$, it follows that $u_{\lambda_n} \to f$ in L^1 and so $f \in \overline{D(A)}$, as claimed. Such a result is, in particular, applicable for nonlinear porous media equations.

2.5 Smoothing Effect of NFP Flow on Initial Data

We shall study here Eq. (2.183) under the following hypotheses.

(ℓ) $\beta \in C^2(\mathbb{R})$, $\beta'(r) \ge a|r|^{\alpha-1}$, $\forall r \in \mathbb{R}$; $\beta(0) = 0$, where $\alpha \ge 1$, $d \ge 3$, $a > 0$.

($\ell\ell$) $D \in L^\infty(\mathbb{R}^d; \mathbb{R}^d)$, div $D \in L^2 + L^\infty$, div $D \ge 0$, a.e.

($\ell\ell\ell$) $b \in C_b(\mathbb{R}) \cap C^1(\mathbb{R})$, $b \ge 0$.

The next theorem is a *smoothing effect* property on initial data of a nonlinear Fokker–Planck semigroup $S(t)$.

Theorem 2.11 *Let $d \geq 3$. Then, under Hypotheses (ℓ), ($\ell\ell$), ($\ell\ell\ell$), the mild solution u to (2.183) given by Theorem 2.8 for $\mu = u_0 dx$, $u_0 \in L^1$, satisfies*

$$u(t) \in L^\infty, \ \forall t > 0, \tag{2.260}$$

$$|u(t)|_\infty \leq C t^{-\frac{d}{2+(\alpha-1)d}} |u_0|_1^{\frac{2}{2+d(\alpha-1)}}, \ \forall t \in (0,\infty), \ u_0 \in L^1, \tag{2.261}$$

where C is independent of u_0.

Proof We shall first prove the following lemma.

Lemma 2.12 *Let $u_\lambda = (I + \lambda A)^{-1} f$, where A is the operator defined by (2.198) and (2.199). Then, for each $p > 1$ and $\lambda > 0$, we have*

$$|u_\lambda|_p^p + \lambda a C \frac{p(p-1)}{(p+\alpha-1)^2} |u_\lambda|^{p+\alpha-1}_{\frac{(p+\alpha-1)d}{d-2}} \leq |f|_p^p, \forall f \in L^p \cap L^1. \tag{2.262}$$

where C is independent of p and λ.

Proof Let $p \in (1, \infty)$. By density, we may assume that $f \in L^1 \cap L^\infty$. Let us first explain the proof by heuristic computations in the case $b \equiv 1$, that is, for the equation

$$u_\lambda - \lambda \Delta \beta(u_\lambda) + \lambda \operatorname{div}(D u_\lambda) = f. \tag{2.263}$$

We multiply (2.208) by $|u_\lambda|^{p-2} u_\lambda$ and integrate over $\mathbb{R}^d$ and get

$$\begin{aligned} &|u_\lambda|_p^p + (p-1)\lambda \int_{\mathbb{R}^d} \beta'(u_\lambda) |\nabla u_\lambda|^2 |u_\lambda|^{p-2} dx \\ &\quad = \int_{\mathbb{R}^d} f |u_\lambda|^{p-2} u_\lambda dx - \frac{(p-1)}{p} \lambda \int_{\mathbb{R}^d} |u_\lambda|^p \operatorname{div} D\, dx \\ &\quad \leq |f|_p |u_\lambda|_p^{p-1} \leq \frac{1}{p} |f|_p^p + \left(1 - \frac{1}{p}\right) |u_\lambda|_p^p. \end{aligned} \tag{2.264}$$

Taking into account (ℓ) yields

$$|u_\lambda|_p^p + a p (p-1) \lambda \int_{\mathbb{R}^d} |u_\lambda|^{p+\alpha-3} |\nabla u_\lambda|^2 dx \leq |f|_p^p.$$

On the other hand, by the Sobolev–Gagliardo–Nirenberg theorem,

$$\begin{aligned} \int_{\mathbb{R}^d} |u_\lambda|^{p+\alpha-3} |\nabla u_\lambda|^2 dx &= \left(\frac{2}{p+\alpha-1}\right)^2 \int_{\mathbb{R}^d} \left|\nabla \left(|u_\lambda|^{\frac{p+\alpha-1}{2}}\right)\right|^2 dx \\ &\geq C \left(\frac{2}{p+\alpha-1}\right)^2 \left(\int_{\mathbb{R}^d} |u_\lambda|^{\frac{(\alpha-1+p)d}{d-2}} dx\right)^{\frac{d-2}{d}}. \end{aligned}$$

This yields

$$|u_\lambda|_p^p + \lambda a C \frac{p(p-1)}{(p+\alpha-1)^2} |u_\lambda|^{\alpha-1+p}_{\frac{(\alpha-1+p)d}{d-2}} \le |f|_p^p, \ \forall \lambda > 0,$$

as claimed.

To make the proof rigorous, we recall that the solution $u := u_\lambda = J_\lambda(f)$ to (2.206) constructed in Proposition 2.3 is an L^1-limit of solutions u_ε, $\varepsilon > 0$, to the approximating Eq. (2.212) (with A_ε as in (2.211)). So, we shall start with (2.212) (instead of (2.263)) and with its solution u_ε. Then we know that $u_\varepsilon, b(u_\varepsilon)u_\varepsilon \in H^1 \cap L^1 \cap L^\infty$, $\widetilde{\beta}_\varepsilon(u_\varepsilon) \in H^2$. We have, for all $r \in \mathbb{R}$,

$$\widetilde{\beta}'_\varepsilon(r) = \frac{\beta'(g_\varepsilon(r))}{(1+\varepsilon\beta')(g_\varepsilon(r))} + \varepsilon \ge h_\varepsilon(g_\varepsilon(r)),$$

where

$$h_\varepsilon(r) = \frac{a|r|^{\alpha-1}}{1+\varepsilon a|r|^{\alpha-1}}, \ r \in \mathbb{R}, \ g_\varepsilon = (I+\varepsilon\beta)^{-1}. \tag{2.265}$$

Define $\varphi_\delta : \mathbb{R} \to \mathbb{R}$ by

$$\varphi_\delta(r) = (|r|+\delta)^{p-2} r, \ \ r \in \mathbb{R}.$$

Then, $\varphi_\delta \in C_b^1$, $\lim\limits_{\delta\to 0} \varphi'_\delta(r) = (p-1)|r|^{p-2}$ and $\varphi'_\delta(r) \ge \min(1, p-1)(|r|+\delta)^{p-2}$. Now, we multiply (2.212) by $\varphi_\delta(u_\varepsilon)$ and obtain

$$\begin{aligned} &\int_{\mathbb{R}^d} u_\varepsilon \varphi_\delta(u_\varepsilon)dx + \lambda \int_{\mathbb{R}^d} \widetilde{\beta}'_\varepsilon(u_\varepsilon)|\nabla u_\varepsilon|^2 \varphi'_\delta(u_\varepsilon)dx \\ &\quad = \lambda \int_{\mathbb{R}^d} (D_\varepsilon \cdot \nabla u_\varepsilon)\varphi'_\delta(u_\varepsilon) b_\varepsilon(u_\varepsilon) u_\varepsilon dx + \int_{\mathbb{R}^d} f \varphi_\delta(u_\varepsilon)dx. \end{aligned} \tag{2.266}$$

Defining

$$\psi(r) = \int_0^r \varphi'_\delta(s) b_\varepsilon(s) s \, ds,$$

we see that $\psi \ge 0$, hence the first integral in the right hand side of (2.266) is equal to

$$-\int_{\mathbb{R}^d} (\operatorname{div} D_\varepsilon)\psi(u_\varepsilon)dx \le C_\varepsilon, \tag{2.267}$$

where (see (2.210))

$$C_\varepsilon=\frac{1}{2}\,|b|_\infty|D|_\infty\sup_{\varepsilon\in(0,1)}|u_\varepsilon|_\infty\int_{\mathbb{R}^d}\mathbf{1}_{|x|\ge\frac{1}{\varepsilon}}|u_\varepsilon|dx.$$

Furthermore, we deduce that the second integral on the left hand side of (2.266) dominates

$$\begin{aligned}\int_{\mathbb{R}^d}h_\varepsilon(g_\varepsilon(u_\varepsilon))|\nabla u_\varepsilon|^2\varphi_\delta'(u_\varepsilon)dx&=\int_{\mathbb{R}^d}|\nabla\psi_{\varepsilon,\delta}(u_\varepsilon)|^2dx\\&\ge C\left(\int_{\mathbb{R}^d}|\psi_{\varepsilon,\delta}(u_\varepsilon)|^{\frac{2d}{d-2}}dx\right)^{\frac{d-2}{d}},\end{aligned}\tag{2.268}$$

where

$$\psi_{\varepsilon,\delta}(r)=\int_0^r\sqrt{h_\varepsilon(g_\varepsilon(s))\varphi_\delta'(s)}\,ds,\ \ r\in\mathbb{R}.\tag{2.269}$$

(Here, we used the Sobolev embedding.) Combining (2.266)–(2.268) and letting $\delta\to0$, we obtain by Fatou's lemma and (2.215)

$$|u_\varepsilon|_p^p+\frac{\lambda C}{p}\left(\int_{\mathbb{R}^d}|\psi_\varepsilon(u_\varepsilon)|^{\frac{2d}{d-2}}dx\right)^{\frac{d-2}{2}}\le\frac{1}{p}\,|f|_p^p+\left(1-\frac{1}{p}\right)|u_\varepsilon|_p^p+C_\varepsilon,\tag{2.270}$$

where

$$\psi_\varepsilon(r)=\sqrt{p-1}\int_0^r\sqrt{h_\varepsilon(g_\varepsilon(s))|s|^{p-2}}\,ds,\ \ r\in\mathbb{R}.\tag{2.271}$$

Obviously, $\psi_\varepsilon(u_\varepsilon)$, $\varepsilon>0$, are equicontinuous, hence by (2.215) and (2.223)

$$(\psi_\varepsilon(u_\varepsilon))^2\to a(p-1)\left(\frac{2}{p+\alpha-1}\right)^2|u|^{p+\alpha-1},\ \text{ a.e. on } \mathbb{R}^d,$$

and, since $u_\varepsilon\to u$ in L^1 by Lemma 2.9, $\lim\limits_{\varepsilon\to0}C_\varepsilon=0$. Therefore, by Fatou's lemma, (2.270) implies (2.262).

Proof of Theorem 2.11 We choose $p=p_n$, where $\{p_n\}$ are defined by

$$p_{n+1}=\frac{d}{d-2}\,(p_n+\alpha-1),\ p_0>1.$$

Then, by (2.262), we get

$$|u_\lambda|_{p_n}^{p_n}+Ca\frac{p_n(p_n-1)}{(p_n+\alpha-1)^2}\,\lambda|u_\lambda|_{p_{n+1}}^{p_n+\alpha-1}\le|f|_{p_n}^{p_n},\ n=0,1,\dots$$

Now, we apply Theorem 5.2 in [83], where $\varphi_n(u) = |u|_{p_n}^{p_n}$, $\beta_n = \frac{d-2}{d}$, $C_n = Ca \inf_{p\in[p_0,\infty)} \frac{p(p-1)}{(p+\alpha-1)^2}$, and conclude that (see Proposition 6.5 in [83]) that

$$|u(t)|_\infty \le C_{p_0} t^{-\frac{d}{2p_0+(\alpha-1)d}} |u_0|_{p_0}^{\frac{2p_0}{2p_0+d(\alpha-1)}}, \ \forall t > 0,\ u_0 \in L^{p_0}. \tag{2.272}$$

Define

$$C_{\alpha,d} := \frac{d+2}{2d} + \sqrt{(\alpha-1)\left(\alpha+\frac{2}{d}\right) + \left(\frac{d+2}{2d}\right)^2}. \tag{2.273}$$

Note that, since $\alpha > 1 - \frac{2}{d}$, the value under the root is strictly bigger than $\left(\frac{d-2}{2d}\right)^2$, hence $C_{\alpha,d} > 1$. □

Lemma 2.13 *Let $p_0 \in (1, C_{\alpha,d})$. Then, for some constant $C_{p_0} > 0$,*

$$|u(t)|_{p_0} \le C_{p_0}\, t^{-\frac{p_0-\gamma}{p_0(\gamma+\alpha-1)}} |u_0|_1^{\frac{\gamma(p_0+\alpha-1}{p_0(\gamma+\alpha-1)}}, \ \forall t > 0,\ u_0 \in L^1 \cap L^{p_0}, \tag{2.274}$$

$$\gamma = \frac{2p_0 + (\alpha-1)d}{(p_0+\alpha-2)d+2} \in (0, 1). \tag{2.275}$$

Proof We may assume that $u_0 \in L^\infty$. We shall use the approximating scheme (2.202) and (2.203). By estimate (2.262), we have, for $\lambda = h$,

$$|u_h^{i+1}|_{p_0}^{p_0} + Ch|u_h^{i+1}|_{\frac{(p_0+\alpha-1)d}{d-2}}^{p_0+\alpha-1} \le |u_h^i|_{p_0}^{p_0},\ i = 0, 1, \dots$$

By the summation by parts formula, this yields, for all $t > 0$,

$$t|u_h(t)|_{p_0}^{p_0} + C\int_0^t s|u_h(s)|_{\frac{(p_0+\alpha-1)d}{d-2}}^{p_0+\alpha-1} ds \le \int_0^t |u_h(s)|_{p_0}^{p_0} ds + h|u_0|_{p_0}^{p_0}, \tag{2.276}$$

where u_h is given by (2.203) and where, here and below, by C we denote various constants independent of t and u_0, but depending on p_0.

On the other hand, by Hölder's inequality we have

$$|u_h(s)|_{p_0}^{p_0} \le |u_h(s)|_1^\gamma |u_h(s)|_{\frac{(p_0+\alpha-1)d}{d-2}}^{p_0-\gamma},\ s > 0.$$

Then, substituting into (2.276), we get

$$t|u_h(t)|_{p_0}^{p_0}+C\int_0^t s|u_h(s)|^{p_0+\alpha-1}_{\frac{(p_0+\alpha-1)d}{d-2}}\,ds-h|u_0|_{p_0}^{p_0}$$

$$\le |u_0|_1^\gamma\int_0^t |u_h(s)|^{p_0-\gamma}_{\frac{(p_0+\alpha-1)d}{d-2}}\,ds$$

$$\le |u_0|_1^\gamma\int_0^t s^{\frac{p_0-\gamma}{p_0+\alpha-1}}|u_h(s)|_{\frac{(p_0+\alpha-1)d}{d-2}})^{p_0-\gamma}\,s^{\frac{\gamma-p_0}{p_0+\alpha-1}}ds$$

$$\le |u_0|_1^\gamma\left(\int_0^t s|u_h(s)|^{p_0+\alpha-1}_{\frac{(p_0+\alpha-1)d}{d-2}}\,ds\right)^{\frac{p_0-\gamma}{p_0+\alpha-1}}\left(\int_0^t s^{\frac{\gamma-p_0}{\gamma+\alpha-1}}\,ds\right)^{\frac{\gamma+\alpha-1}{p_0+\alpha-1}}.$$

Since $p_0<C_{\alpha,d}$, it follows that $(\gamma-p_0)(\gamma+\alpha-1)^{-1}>-1$, and so, the above is dominated by

$$C|u_0|_1^\gamma\,t^{\frac{2\gamma+\alpha-p_0-1}{p_0+\alpha-1}}\left(\int_0^t s|u_h(s)|^{p_0+\alpha-1}_{\frac{(p_0+\alpha-1)d}{d-2}}\,ds\right)^{\frac{p_0-\gamma}{p_0+\alpha-1}}.$$

This yields, for $t>0$,

$$t|u_h(t)|_{p_0}^{p_0}+C\int_0^t s|u_h(s)|^{p_0+\alpha-1}_{\frac{(p_0+\alpha-1)d}{d-2}}ds\le C|u_0|_1^{\frac{\gamma(p_0+\alpha-1)}{\gamma+\alpha-1}}\,t^{\frac{2\gamma+\alpha-p_0-1}{\gamma+\alpha-1}}+h|u_0|_{p_0}^{p_0}.$$

Hence, dropping the integral on the left hand side and letting $h\to 0$, we obtain that

$$|u(t)|_{p_0}\le C_{p_0}|u_0|_1^{\frac{\gamma(p_0+\alpha-1)}{p_0(\gamma+\alpha-1)}}\,t^{\frac{\gamma-p_0}{p_0(\gamma+\alpha-1)}},\ t>0,$$

and (2.274) is proved.

Proof of Theorem 2.11 (Continued) By approximation, we may assume that $u_0\in L^1\cap L^\infty$. Then, combining estimates (2.272) and (2.274), we get, for $t>0$,

$$|u(t)|_\infty\le C_{p_0}\left(\frac{t}{2}\right)^{-\frac{d}{2p_0+(\alpha-1)d}}\left|u\left(\frac{t}{2}\right)\right|_{p_0}^{\frac{2p_0}{2p_0+d(\alpha-1)}}$$

$$\le C_{p_0}\,t^{-\left(\frac{d}{2p_0+(\alpha-1)d}+\frac{(p_0-\gamma)2p_0}{p_0(\gamma+\alpha-1)(2p_0+d(\alpha-1))}\right)}\,|u_0|_1^{\frac{\gamma(p_0+\alpha-1)2p_0}{p_0(\gamma+\alpha-1)(2p_0+d(\alpha-1))}}$$

$$=C_{p_0}\,t^{-\frac{d}{2+(\alpha-1)d}}\,|u_0|_1^{\frac{2}{2+(\alpha-1)d}},$$

which completes the proof. □

2.6 NFPE with a Measure as Initial Datum

We shall consider Eq. (2.183) with the initial datum $u(0) = \mu \in \mathcal{M} = \mathcal{M}(\mathbb{R}^d)$, the space of signed bounded Radon measures on $\mathbb{R}^d$.

Definition 2.3 The function $u : [0, \infty) \to \mathcal{M}$ is a distributional solution to (2.183) if

$$u, \beta(u) \in L^1_{loc}([0, \infty) \times \mathbb{R}^d), \tag{2.277}$$

$$\begin{aligned}\int_0^\infty \int_{\mathbb{R}^d} & u(t,x)(\varphi_t(t,x) + b(u(t,x))D(x) \cdot \nabla\varphi(t,x)) \\ & +\beta(u(t,x)\Delta\varphi(t,x))dt\,dx + \mu(\varphi(0,\cdot)) = 0, \\ & \forall \varphi \in C_0^\infty([0,\infty) \times \mathbb{R}^d).\end{aligned} \tag{2.278}$$

Theorem 2.12 *Assume that Hypotheses* (ℓ), ($\ell\ell$), ($\ell\ell\ell$) *hold and, in addition, that* $D \in L^2(\mathbb{R}^d; \mathbb{R}^d)$ *and*

$$|\beta(r)| \le C|r|^\alpha, \ \forall r \in \mathbb{R}. \tag{2.279}$$

Let $\mu \in \mathcal{M}$. *Then,* (2.183) *has a distributional solution u which satisfies, for dt-a.e.* $t \in (0, \infty)$,

$$u(t,x) \ge 0, \text{ a.e. on } \mathbb{R}^d, \text{ provided } \mu \ge 0, \tag{2.280}$$

$$\int_{\mathbb{R}^d} u(t,x)dx = \int_{\mathbb{R}^d} d\mu, \tag{2.281}$$

$$|u(t)|_\infty \le C\,t^{-\frac{d}{2+(\alpha-1)d}}\,\|\mu\|_{\mathcal{M}}^{\frac{2}{2+d(\alpha-1)}}, \tag{2.282}$$

$$|u(t)|_1 \le \|\mu\|_{\mathcal{M}}. \tag{2.283}$$

Furthermore, for all $p \in \left[1, \alpha + \frac{2}{d}\right)$,

$$u \in L^p((0,T) \times \mathbb{R}^d), \ \forall T > 0, \tag{2.284}$$

$$\beta(u) \in L^1((0,T) \times \mathbb{R}^d), \ \forall T > 0. \tag{2.285}$$

The map $t \mapsto u(t,x)dx \in \mathcal{M}$ *has a* $\sigma(\mathcal{M}, C_b)$*-continuous version on* $(0, \infty)$, *denoted by* $S(t)\mu$, $t > 0$, *for which* (2.280)–(2.283) *hold for all* $t > 0$. *Furthermore,*

$$\lim_{t\to 0} \int_{\mathbb{R}^d} (S(t)\mu)(x)\psi(x)dx = \mu(\psi), \ \forall \psi \in C_b. \tag{2.286}$$

Defining $S(0)\mu = \mu$, *then* $S(t), t \geq 0$, *restricted to* L^1 *coincides with the semigroup from Theorem* 2.8 *and we have* $\|S(t)\mu_1 - S(t)\mu_2\|_{\mathcal{M}} \leq \|\mu_1 - \mu_2\|_{\mathcal{M}}$, $\forall t \geq 0$, $\mu_1, \mu_2 \in \mathcal{M}$.

As a consequence of Theorems 2.12 and 5.1 in Chap. 5, we obtain that the McKean–Vlasov SDE (2.83) has a probabilistically weak solution. If, in addition, Hypothesis (iv) in Theorem 2.4 holds, then Theorem 5.3 and Remark 5.4 imply that this solution is unique in law and that the corresponding path laws $P_{s,\zeta}$, $(s, \zeta) \in \mathbb{R}_+ \times \mathcal{P}$, form a nonlinear Markov process in the sense of Definition 5.1.

Proof of Theorem 2.12 Consider a smooth approximation μ_ε of $u_0 = \mu$ of the form

$$\mu_\varepsilon(x) = (\mu * \varphi_\varepsilon),\ \varepsilon > 0,$$

where $\varphi_\varepsilon(x) \equiv \frac{1}{\varepsilon^d}\varphi\left(\frac{x}{\varepsilon}\right)$, is a standard mollifier. Then, by Theorem 2.11, the equation

$$\begin{aligned} &(u_\varepsilon)_t - \Delta\beta(u_\varepsilon) + \operatorname{div}(Db(u_\varepsilon)u_\varepsilon) = 0 \text{ in } (0,\infty) \times \mathbb{R}^d, \\ &u_\varepsilon(0) = \mu_\varepsilon, \end{aligned} \tag{2.287}$$

has, for each $\varepsilon > 0$, a mild solution $u_\varepsilon = S(t)\mu_\varepsilon \in C([0,\infty); L^1) \cap L^\infty((\delta,\infty) \times \mathbb{R}^d)$, $\forall \delta > 0$. More precisely, we have

$$|u_\varepsilon(t)|_\infty \leq C\, t^{-\frac{d}{2+(\alpha-1)d}}\, |\mu_\varepsilon|_1^{\frac{2}{2+d(\alpha-1)}} \leq C\, t^{-\frac{d}{2+(\alpha-1)d}}\, \|\mu\|_{\mathcal{M}}^{\frac{2}{2+d(\alpha-1)}},\ t > 0. \tag{2.288}$$

Everywhere in the following, C is a positive constant independent of t and μ possibly changing from line to line. Also, for simplicity, we set $\|\mu\| = \|\mu\|_{\mathcal{M}}$, and $\|\mu\|^{\frac{2}{2+(\alpha-1)d}}\, t^{-\frac{d}{2+(\alpha-1)d}} = \nu(t,\mu)$, $\forall t > 0$, $\mu \in \mathcal{M}$. We also have

$$|u_\varepsilon(t)|_1 \leq |\mu_\varepsilon|_1 \leq \|\mu\|,\ \forall t \geq 0,\ \forall \varepsilon > 0. \tag{2.289}$$

If we formally multiply (2.287) by $\beta(u_\varepsilon)$ and integrate over $(\delta, t) \times \mathbb{R}^d$, for $\psi(r) \equiv \int_0^r \beta'(s)b(s)s\,ds$, since $\psi, \operatorname{div} D \geq 0$, we get

$$\begin{aligned} &\int_{\mathbb{R}^d} g(u_\varepsilon(t,x))dx + \int_\delta^t \int_{\mathbb{R}^d} |\nabla\beta(u_\varepsilon(s))|_2^2 dx\,ds \\ &\quad = \int_\delta^t \int_{\mathbb{R}^d} \nabla(\psi(u_\varepsilon)) \cdot D\,dx\,ds + \int_{\mathbb{R}^d} g(u_\varepsilon(\delta,x))dx \\ &\quad \leq \int_{\mathbb{R}^d} g(u_\varepsilon(\delta,x))dx \leq C\|\mu\|(\nu(\delta,\mu))^\alpha,\ \forall t > \delta, \end{aligned} \tag{2.290}$$

where $g(r) \equiv \int_0^r \beta(s)ds \geq 0$ and the last inequality follows by (2.279).

Estimate (2.290) can be derived rigorously by using the finite difference scheme (2.202) and (2.203) corresponding to the resolvent of the regularized version (2.211) and (2.212) of Eq. (2.287). Indeed, by Theorem 2.9, for each $\varepsilon > 0$,

$$u_\varepsilon(t) = \lim_{\nu\to 0}\lim_{n\to\infty}\left(I + \frac{t}{n}A_\nu\right)^{-n}\mu_\varepsilon,$$

where A_ν, $\nu > 0$, is the operator defined by (2.211) and both limits are in L^1, locally uniformly in $t \in [0, \infty)$. Hence,

$$u_\varepsilon(t) = \lim_{\nu\to 0}\lim_{h\to 0} u_{\nu,h}(t),\ t \in [0, T], \tag{2.291}$$

where

$$\begin{aligned} &u_{\nu,h}(t) = u^{i+1}_{\nu,h},\ t \in (ih, (i+1)h], \\ &u^{i+1}_{\nu,h} + hA_\nu u^{i+1}_{\nu,h} = u^i_{\nu,h},\ i = 0, 1, \dots, N-1;\ \ Nh = T,\ u^0_{\nu,h} = \mu_\varepsilon. \end{aligned} \tag{2.292}$$

We know by the proof of Proposition 2.3 that, if $v \subset L^1 \cap L^\infty$, then for the solution u_h to the equation $u_h + hA_\nu u_h = v$, we have $u_h, b_\nu(u_h)u_h \in H^1 \cap L^1 \cap L^\infty$, $\widetilde{\beta}_\nu(u_h) \in H^2$ and $|u_h|_\infty \le |v|_\infty$. Hence, if we multiply (2.292) by $\widetilde{\beta}_\nu(u^{i+1}_{\nu,h})$ and integrate over $\mathbb{R}^d$, we get as above

$$\begin{aligned} &\int_{\mathbb{R}^d} g_\nu(u^{i+1}_{\nu,h}(x))dx + h\int_{\mathbb{R}^d} |\nabla\widetilde{\beta}_\nu(u^{i+1}_{\nu,h})|^2 dx \\ &\quad\le \int_{\mathbb{R}^d} g_\nu(u^i_{\nu,h})dx, i = 0, 1, \dots, N_1, Nh = T, \end{aligned}$$

where $g_\nu(r) = \int_0^r \widetilde{\beta}_\nu(s)ds$ and we used that $\operatorname{div} D_\nu \ge 0$, since $D_\nu = D$ because $D \in L^2(\mathbb{R}^d; \mathbb{R}^d)$. Summing over from $j = [N\delta/T]+1$ to $k-1 = [Nt/T]$, we get

$$\int_{\mathbb{R}^d} g_\nu(u^k_{\nu,h})dx + \frac{h}{2}\sum_{i=j}^{k-1}\int_{\mathbb{R}^d} |\nabla\widetilde{\beta}_\nu(u^{i+1}_{\nu,h})|^2 dx \le \int_{\mathbb{R}^d} g_\nu(u^j_{\nu,h})dx,\ \forall k.$$

Then, letting $h \to 0$ and afterwards $\nu \to 0$, by (2.291) and, since $|u^i_{\nu,h}|_\infty \le |\mu_\varepsilon|_\infty$, the closedness of the gradient on $L^2(0, T; L^2)$ and the weak lower semicontinuity implies (2.290), as claimed. Multiplying (2.287) by $|u_\varepsilon|^{q-2}u_\varepsilon$, $q \ge 2$, and integrating over $(\delta, t) \times \mathbb{R}^d$, we get by Hypothesis (k) that

$$\begin{aligned} &a(q-1)\left(\frac{2}{q+\alpha-1}\right)^2\int_\delta^t\int_{\mathbb{R}^d} |\nabla|u_\varepsilon|^{\frac{q+\alpha-1}{2}}|^2 ds\,dx + \frac{1}{q}\int_{\mathbb{R}^d} |u_\varepsilon(t, x)|^q dx \\ &\quad\le \frac{1}{q}\int_{\mathbb{R}^d} |u_\varepsilon(\delta, x)|^q dx \le C\|\mu\|(\nu(\delta, \mu))^{q-1}. \end{aligned} \tag{2.293}$$

As in the previous case, the above calculus can be made rigorous if we replace (2.287) by its discrete version (2.292), which we multiply by $|u^{i+1}_{\nu,h}|^{q-2}u^{i+1}_{\nu,h}$ and integrate over $\mathbb{R}^d$. Indeed, noting that, since $u^{i+1}_{\nu,h}, b_\nu(u^{i+1}_{\nu,h})u^{i+1}_{\nu,h} \in H^1 \cap L^1 \cap L^\infty$ and $\widetilde{\beta}'_\nu(u^{i+1}_{\nu,h}) \in H^2$, we get similarly as in the proof of Lemma 2.12 for $\psi_\nu(r) = \sqrt{p-1}\int_0^r \sqrt{h_\nu(g_\nu(s))|s|^{q-2}}\,ds$, $r \in \mathbb{R}$, where h_ε, g_ε are as in (2.265),

$$\frac{1}{q}\int_{\mathbb{R}^d}|u^{i+1}_{\nu,h}|^q dx + h\int_{\mathbb{R}^d}|\nabla\psi_\nu(u^{i+1}_{\nu,h})|^2 dx$$
$$\leq \frac{1}{q}\int_{\mathbb{R}^d}|u^i_{\nu,h}|^q dx,\ i = 0,\dots,N-1,\ Nh = T.$$

Summing over i from $j = [N\delta/T]+1$ to $k-1 = [Nt/T]$, we get

$$\frac{1}{q}\int_{\mathbb{R}^d}|u^k_{\nu,h}|^q dx + h\sum_{i=j}^{k-1}\int_{\mathbb{R}^d}|\nabla\psi_\nu(u^{i+1}_{\nu,h})|^2 dx \leq \frac{1}{q}\int_{\mathbb{R}^d}|u^j_{\nu,h}|^q dx,$$
$$i = 0,\dots,N-1,\ Nh = T.$$

Letting $h \to 0$, and afterwards $\nu \to 0$, (2.293) follows from (2.291) and the closedness of the gradient on $L^2(0,T;L^2)$.

Now, taking into account that by (2.293), with $q = 2p-1-\alpha$, we get

$$\int_\delta^t |\nabla(|u_\varepsilon|^{p-1})|_2^2 ds \leq \|\mu\|(\nu(\delta,\mu))^{2(p-1)-\alpha}),\ \ \forall t \geq \delta, \forall p \geq \frac{\alpha+3}{2}. \tag{2.294}$$

Moreover, by (2.290), $\{\nabla\beta(u_\varepsilon)\}_{\varepsilon>0}$ is bounded in $L^2(\delta,T,L^2)$, and so

$$\|\Delta\beta(u_\varepsilon) - \operatorname{div}(Db(u_\varepsilon)u_\varepsilon)\|_{L^2(\delta,T;H^{-1})} \leq C,\ \forall\varepsilon > 0. \tag{2.295}$$

Note also that, by (2.294), it follows that

$$\int_\delta^t |\nabla(|u_\varepsilon|^{p-1})|_2^2 ds \leq C,\ \forall\varepsilon > 0. \tag{2.296}$$

Hence, $\{|u_\varepsilon|^{p-1}\}_{\varepsilon>0}$ is bounded in $L^2(\delta,T;H^1)$ and so, by (2.295), we infer that, for $m \geq 4$,

$$\|\,|u_\varepsilon|^{p-1}(\Delta\beta(u_\varepsilon)\|_{L^1(\delta,T;H^{-m})} + \|\operatorname{div}(Db(u_\varepsilon)u_\varepsilon))\|_{L^2(\delta,T;H^{-1})} \leq C.$$

This implies that the set

$$\left\{\frac{\partial}{\partial t}(|u_\varepsilon|^{p-1}u_\varepsilon)\right\}_{\varepsilon>0} = \left\{p|u_\varepsilon|^{p-1}(\Delta\beta(u_\varepsilon) - \operatorname{div}(Db(u_\varepsilon)u_\varepsilon))\right\}_{\varepsilon>0}$$

is bounded in $L^1(\delta, T; H^{-m})$, and note that by (2.294) applied to $p+1$ replacing p, we have that also $\{|u_\varepsilon|^{p-1}u_\varepsilon\}_{\varepsilon>0}$ is bounded in $L^2(\delta, T; H^1)$.

Then, by the Aubin-Lions-Simon compactness theorem (see [4, 5]), the set $\{|u_\varepsilon|^{p-1}u_\varepsilon\}_{\varepsilon>0}$ is relatively compact in $L^2(\delta, T; L^2_{loc})$ for all $0 < \delta < T < \infty$. Hence, along a subsequence $\{\varepsilon\} \to 0$, we have for $\gamma(r) := |r|^{p-1}r$, $r \in \mathbb{R}$,

$$\gamma(u_\varepsilon) \to v, \quad \text{a.e. on } (0, \infty) \times \mathbb{R}^d. \tag{2.297}$$

Then, since γ has a continuous inverse and since β is continuous, we have

$$u_\varepsilon \to u = \gamma^{-1}(v) \text{ and } \beta(u_\varepsilon) \to \beta(u), \text{ a.e. on } (0, \infty) \times \mathbb{R}^d. \tag{2.298}$$

By (2.293) and (2.294), we have, for all $p \geq 1$, $\forall t > 0$, $\varepsilon > 0$,

$$|u_\varepsilon(t)|_p \leq |u_\varepsilon(t)|_1^{\frac{1}{p}} |u_\varepsilon(t)|_\infty^{\frac{p-1}{p}} \leq C\|\mu\|^{\frac{2p+(\alpha-1)d}{(2+(\alpha-1)d)p}} t^{-\frac{d(p-1)}{(2+(\alpha-1)d)p}}. \tag{2.299}$$

We have $\frac{d(p-1)}{2+(\alpha-1)d} < 1$ for every $p \in \left[1, \alpha + \frac{2}{d}\right)$. Then, for such a p, (2.299) implies that, for every $T > 0$,

$$\int_0^T |u_\varepsilon(t)|_p^p dt \leq C, \tag{2.300}$$

and, therefore, if in addition $p > 1$, along a subsequence $\varepsilon \to 0$,

$$u_\varepsilon \to u \text{ weakly in } L^p((0, T) \times \mathbb{R}^d). \tag{2.301}$$

Moreover, by (2.279), (2.298), and (2.300), it follows that $\{\beta(u_\varepsilon)\}$ is bounded in $L^q((0, T) \times \mathbb{R}^d)$ for all $q \in \left(1, 1 + \frac{2}{\alpha d}\right)$, and so (along a subsequence)

$$\beta(u_\varepsilon) \to \beta(u) \text{ weakly in } L^q((0, T) \times \mathbb{R}^d). \tag{2.302}$$

Since, by (2.186), we have

$$\int_0^\infty \int_{\mathbb{R}^d} (u_\varepsilon(\varphi_t + D \cdot \nabla\varphi) + \beta(u_\varepsilon)\Delta\varphi) dt\, dx + \int_{\mathbb{R}^d} \mu_\varepsilon(x)\varphi(0, x)dx = 0$$

for any $\varphi \in C_0^\infty([0, \infty) \times \mathbb{R}^d)$, letting $\varepsilon \to 0$, we see by (2.301) and (2.302) that u satisfies (2.278). As regards (2.280)–(2.283), by (2.298) they immediately follow from the corresponding properties of u_ε and (2.293). Furthermore, (2.284) follows from (2.300) and Fatou's Lemma. Taking $p = \alpha$ in (2.284) and (2.285) follows by (2.279).

By (2.281) and (2.285), we may apply Lemma 8.1.2 in [1], to conclude that $t \mapsto u(t, x)dx \in \mathcal{M}$ has a $\sigma(\mathcal{M}, C_b)$-continuous version on $(0, \infty)$, denoted by

$\mu_t, t > 0$. To show (2.286), we apply (2.278) with $\varphi(t,x) = \psi(t)\zeta(x)$, $\psi \in C_0^\infty([0,\infty))$ and $\zeta \in C_0^\infty(\mathbb{R}^d)$. Then, for

$$L\zeta(t,x) = \beta(u(t,x))\Delta\zeta(x) + D(x)\cdot\nabla\zeta(x),$$

we have from (2.278)

$$\int_0^\infty \psi(t)\int_{\mathbb{R}^d} L\zeta\, d\mu_t\, dt + \psi(0)\int_{\mathbb{R}^d}\zeta\, d\mu = -\int_0^\infty \frac{d}{dt}\psi(t)\int_{\mathbb{R}^d}\zeta\, d\mu_t\, dt, \tag{2.303}$$

hence, choosing $\psi \in C_0^\infty((0,\infty))$, we obtain for dt-a.e. $t \in (0,\infty)$,

$$\int_{\mathbb{R}^d}\zeta\, d\mu_t = C + \int_0^t\int_{\mathbb{R}^d} L\zeta\, d\mu_s ds. \tag{2.304}$$

By (2.285), the right hand side is continuous in $t \in [0,\infty)$ and equal to C at $t = 0$, while, as seen above, also the left hand side is continuous in $t \in (0,\infty)$. Hence, we obtain that (2.304) holds for all $t \in (0,\infty)$ and

$$\lim_{t\to 0}\int_{\mathbb{R}^d}\zeta\, d\mu_t = C.$$

Plugging (2.304) into the right hand side of (2.303), with $\psi \in C_0^\infty([0,\infty))$ such that $\psi(0) = 1$ and integrating by parts, we find

$$\int_0^\infty \psi\int_{\mathbb{R}^d} L\zeta\, d\mu_t\, dt + \int_{\mathbb{R}^d}\zeta\, d\mu = C + \int_0^\infty \psi\int_{\mathbb{R}^d} L\zeta\, d\mu_t\, dt$$

and (2.286) follows, because (2.281) holds for all $t > 0$, as we shall see below. It is obvious that, for the $\sigma(\mathcal{M}, C_b)$-continuous version $t \mapsto \mu_t$ of $t \mapsto u(t,x)dx$ on $(0,\infty)$, properties (2.280), (2.281), and (2.283) hold for all $t > 0$. For this version, it is also easily seen that $t \mapsto |u(t)|_\infty$ is lower semicontinuous, hence also (2.282) follows for all $t > 0$.

It remains to prove the last assertion in Theorem 2.12. To express the dependence of our $\sigma(\mathcal{M}, C_b)$-continuous version $[0,\infty) \ni t \mapsto \mu_t \in \mathcal{M}$ with $\mu_0 = \mu$ of our solution to (2.278), we set, for $\mu \in \mathcal{M}$, $P(t)\mu = \mu_t$, $t \geq 0$, and recall that μ_t has a density in L^1 for $t > 0$, which we identify with μ_t, i.e., $\mu_t \in L^1$, $\forall t > 0$. Let $T > 0$. By construction, we know that (along a subsequence depending on μ) $\varepsilon \to 0$

$$S(\cdot)(\mu * u_\varepsilon) \to P(\cdot)\mu, \text{ a.e. on } (0,T)\times\mathbb{R}^d \text{ and weakly in } L^p((0,T)\times\mathbb{R}^d) \tag{2.305}$$

as functions of (t, x) for $p \in \left(1, \alpha + \frac{2}{d}\right)$ (see (2.298) and (2.301), respectively). (Here $S(t)$, $t \geq 0$, is the semigroup from Theorem 2.12.)

Claim If $\mu \in L^1$, *then* $S(t)\mu = P(t)\mu$ for all $t \geq 0$.

To prove the claim we recall that, since $\mu \in L^1$, we have $\mu * u_\varepsilon \to \mu$ in L^1. Hence, by (2.305),

$$S(t)\mu = P(t)\mu \text{ in } \mathcal{M} \text{ for } dt - \text{a.e. } t \in [0, T].$$

Since both sides are $\sigma(L^1, C_b)$-continuous in $t \in [0, T]$, this equality holds $\forall t \in [0, T]$, $T > 0$, and the Claim is proved.

Therefore, we may rename $P(t) : \mathcal{M} \to \mathcal{M}$, $t \geq 0$, and set $S(t) = P(t)$, $t \geq 0$, since it is an extension of $S(t) : L^1 \to L^1$ for every $t \geq 0$.

Finally, for $\mu, \widetilde{\mu} \in \mathcal{M}$ with corresponding solutions $u_\varepsilon, \widetilde{u}_\varepsilon$ to (2.287), we have for all $t \geq 0$,

$$|u_\varepsilon(t) - \widetilde{u}_\varepsilon(t)|_1 \leq |(\mu - \widetilde{\mu}) * u_\varepsilon|_1 \leq \|\mu - \widetilde{\mu}|_{\mathcal{M}}.$$

Hence, for all $\varphi \in C_b([0, \infty))$, $\varphi \geq 0$, by (2.298) and Fatous's lemma, letting $\varepsilon \to 0$ we get

$$\int_0^\infty \varphi(t)|S(t)\mu - S(t)\widetilde{\mu}|_{L^1} dt \leq \int_0^\infty \varphi(t)\|\mu - \widetilde{\mu}\|_{\mathcal{M}} dt,$$

so,

$$|S(t)\mu - S(t)\widetilde{\mu}|_1 \leq \|\mu - \widetilde{\mu}\|_{\mathcal{M}} \text{ for } dt - \text{a.e. } t \in (0, \infty).$$

But the left hand side is lower semicontinuous in $t \in [0, \infty)$, since $t \mapsto S(t)\mu$ is $\sigma(L^1, C_b)$-continuous, hence

$$\|S(t)\mu - S(t)\widetilde{\mu}\|_{\mathcal{M}} \leq \|\mu - \widetilde{\mu}\|_{\mathcal{M}}, \ \forall t \in [0, \infty),$$

as claimed. □

2.7 NFPEs with Discontinuous Coefficients

Consider the equation

$$\begin{aligned} & u_t - \Delta\beta(u) + \operatorname{div}(b(u)u) = 0, \ \text{ in } (0, \infty) \times \mathbb{R}^d, \\ & u(0, x) = u_0(x), \ x \in \mathbb{R}^d, \end{aligned} \tag{2.306}$$

under the following assumptions on the functions $\beta : \mathbb{R} \to 2^{\mathbb{R}}$ and $b : \mathbb{R} \to \mathbb{R}^d$:

(m) β is maximal monotone in $\mathbb{R} \times \mathbb{R}$, $0 \in \beta(0)$, and

$$\sup\{|\eta|;\ \eta \in \beta(r)\} \le \alpha_1 |r|,\ \forall r \in \mathbb{R}.$$

(mm) $b \in L^\infty(\mathbb{R}; \mathbb{R}^d)$.

Recall (see Sect. 6.1) that the multivalued function $\beta : \mathbb{R} \to 2^{\mathbb{R}}$ is said to be maximal monotone if, for all $\lambda > 0$, the range of mapping $r \to r + \lambda\beta(r)$ is all of $\mathbb{R}$ and $(I + \lambda\beta)^{-1}$ is Lipschitz with Lipschitz constant 1.

In particular, Hypothesis (m) holds if β is monotonically-nondecreasing, continuous and $|\beta(r)| \le \alpha_1 |r|$, $\forall r \in \mathbb{R}$. Such a situation arises, for instance, in the case of nonlinear Fokker–Planck equations of the form

$$u_t - \Delta\beta_0(u) + \operatorname{div}(b(u)u) = 0 \text{ in } (0, \infty) \times \mathbb{R}^d, \tag{2.307}$$

where $\beta_0 : \mathbb{R} \to \mathbb{R}$ is a discontinuous and monotonically nondecreasing function. This equation can be reduced to one of the form (2.306) with maximal monotone (multi-valued) β by filling the jumps of β_0 in discontinuity points r_i, that is,

$$\beta(r) = \begin{cases} \beta_0(r), & \forall r \ne r_i,\ i = 1, 2, \\ [\beta_0(r_i + 0), \beta_0(r_{i+1} - 0)] & \text{if } r = r_i. \end{cases}$$

This is the case of the most mathematical models of phase-transition of diffusion processes and, in particular, of the famous Bak–Tang–Wiesenfeld model of self-organized criticallity (see, e.g., [5], p. 178). In this case, β is of the form

$$\beta(r) \equiv \alpha r H(r - u_c),\ \forall r \in \mathbb{R},$$

where $\alpha > 0$, $u_c > 0$ and H is the Heviside function. Other problems arising in phase transition, in particular, the Stefan two phase problem with convection can be written in a similar way (see [5], p. 159).

As for the function b, to get existence in (2.307), one should extend it to a multivalued upper–semicontinuous function on $\mathbb{R}$. More precisely, we replace b by the multivalued Filipov mapping $F_b : \mathbb{R} \to 2^{\mathbb{R}^d}$

$$F_b(r) = \bigcap_{\delta > 0} \bigcap_{\mu(S)=0} \overline{\operatorname{conv}}\, b([r - \delta, r + \delta] \setminus S),\ r \in \mathbb{R}, \tag{2.308}$$

where $S \subset \mathbb{R}^d$, μ is the Lebesgue measure, and $\overline{\operatorname{conv}}$ denotes the closed convex hull.

Definition 2.4 A function $u : [0, \infty) \to L^1(\mathbb{R}^d) = L^1$ is said to be a *mild solution* to NFPE (2.306) if $u \in C([0, \infty); L^1)$ and, for each $T > 0$, we have

$$u(t) = \lim_{h\to 0} u_h(t) \text{ in } L^1,\ \forall\, t \in [0, T) \text{ uniformly on } [0, T],$$

where $u_h : (0, T) \to L^1$ is the step function,

$$\begin{aligned}
&u_h(t) = u_h^j,\ \forall\, t \in [jh, (j+1)h),\ j = 0, 1, \ldots, N = \left[\tfrac{T}{h}\right],\\
&u_h^{j+1} - h\Delta\eta_h^{j+1} + h\,\mathrm{div}(\xi_h^{j+1}u_h^{j+1}) = u_h^j \text{ in } \mathscr{D}'(\mathbb{R}^d),\\
&u_h^{j+1}, \eta_h^{j+1} \in L^1,\ \xi_h^{j+1} \in (L^1)^d,\ \eta_h^{j+1}(x) \in \beta(u_h^{j+1}(x)),\ \text{a.e. } x \in \mathbb{R}^d,\\
&u_h^0(t, x) = u_0(x),\ \text{a.e. in } \mathbb{R}^d,\ \xi_h^{j+1}(x) \in F_b(u_h^{j+1}(x)),\ \text{a.e. } x \in \mathbb{R}^d.
\end{aligned}$$

Such a function can be seen as a Filipov's solution to NFPE (2.306) and is obtained by replacing in the standard definition of mild solution the drift function b by F_b.

We denote by $A_0 : D(A_0) \subset L^1 \to L^1$ the multivalued operator

$$\begin{aligned}
A_0(y) &= -\Delta\eta + \mathrm{div}(\xi y) \text{ in } \mathscr{D}'(\mathbb{R}^d),\ \forall\, y \in D(A_0),\\
D(A_0) &= \{y \in L^1;\ -\Delta\eta + \mathrm{div}(\xi y) \in L^1,\ \xi, \eta \in L^1_{\mathrm{loc}},\\
&\qquad \xi(x) \in F_b(y(x)),\ \eta(x) \in \beta(y(x)),\ \text{a.e. } x \in \mathbb{R}^d\}.
\end{aligned}$$

As in the previous cases, we shall replace A_0 by an m-accretive version of $A : L^1 \to L^1$ such that

$$(I + \lambda A)^{-1}y \in (I + \lambda A_0)^{-1}y,\ \forall\, y \in L^1,\ \lambda > 0. \tag{2.309}$$

We have the following lemma which will be proved later on.

Lemma 2.14 *Under Hypotheses* (m)–(mm), *we have*

$$R(I + \lambda A_0) = L^1,\ \forall \lambda > 0,$$

and there is a family $\{J_\lambda\}_{\lambda>0}$ *of operators from* L^1 *to itself such that* (2.190) *and* (2.191) *hold.*

Then, starting from A_0, we define the operator $A : D(A) \subset L^1 \to L^1$

$$\begin{aligned}
A(y) &= A_0(y),\ \ \forall y = J_\nu(f),\ \forall\, f \in L^1,\\
D(A) &= J_\nu(L^1),
\end{aligned}$$

where $\nu > 0$ is arbitrary but fixed. It follows by Lemma 2.14 that A is m-accretive in L^1 and J_λ is the resolvent of A, that is,

$$(I + \lambda A)^{-1}(y) = J_\lambda(y), \ \forall y \in L^1, \lambda > 0.$$

The domain $D(A)$ of A is, of course, nonempty and if, besides (m)–(mm), one assumes that

$$b \in C_b^1(\mathbb{R}; \mathbb{R}^d), \ \beta^{-1} \in \mathrm{Lip}(\mathbb{R}^d), \tag{2.310}$$

then $\overline{D(A)} = L^1$. Indeed, in this case we have

$$\begin{aligned}&\{y \in L^1, -\Delta\eta + \mathrm{div}(b(y)y) \in L^1; \eta \in L^1, \eta(x) \in \beta(y(x)), \ \text{a.e. } x \in \mathbb{R}^d\}\\ &\quad = \{z \in L^1; -\Delta z + \mathrm{div}(b(\beta^{-1}(z))\beta^{-1}(z)) \in L^1\}\end{aligned}$$

and then, since by (2.310) $\mathrm{div}(b(\beta^{-1}(z))\beta^{-1}(z)) \in L^1$ for $z \in C_0^\infty(\mathbb{R}^d)$, the latter set is dense in L^1 and, therefore, so is $D(A)$.

Then, the Cauchy problem

$$\begin{aligned}&\frac{du}{dt} + A(u) = 0, \ \ \forall t > 0,\\ &u(0) = u_0,\end{aligned}$$

has a unique mild solution for each $u_0 \in \overline{D(A)}$. Taking into account (2.309), it follows that $u(t)$ is just a mild solution to (2.306) in the sense of Definition 2.4. We have, therefore,

Theorem 2.13 *Under Hypotheses* (m)–(mm), *for each* $u_0 \in \overline{D(A)}$ *there is a mild solution* u *to NFPE* (2.306) *and, if* $u_0 \in \mathcal{P}$, *then* $u(t) \in \mathcal{P}$, $\forall t \geq 0$. *Moreover,*

$$u(t) = S(t)u_0, \ \ \forall t \geq 0,$$

where $S(t) : \overline{D(A)} \to \overline{D(A)}$ *is a continuous semigroup of nonlinear contractions in* L^1. *If* (2.310) *also holds, then* $\overline{D(A)} = L^1$.

As mentioned in the previous cases, though the mild solution u to (2.306) is not unique, the operator A generates a unique semigroup (flow) of selections in the class of mild solutions.

In particular, it follows by Theorem 2.13 the existence of a probabilistically weak solutions to the McKean–Vlasov equation

$$\begin{aligned}&dX(t) = F_b\left(\frac{d\mathcal{L}_{X(t)}}{dx}(X(t))\right)dt + \sqrt{2\sigma\left(\frac{d\mathcal{L}_{X(t)}}{dx}(X(t))\right)}\,dw(t),\\ &X(0) = X_0,\end{aligned}$$

where $\sigma(x) \equiv \frac{\beta(r)}{r}$, and β, b satisfy Hypotheses (m), (mm) and F_b is defined by (2.308). Since F_b is multivalued in the definition of the probabilistically weak solution to this equation, one takes instead of $b(y)$ measurable sections of $F_b(y)$.

Proof of Lemma 2.14 For $\lambda > 0$ and $f \in L^1$, consider the equation

$$y + \lambda A_0(y) = f. \tag{2.311}$$

We approximate (2.311) by

$$y - \lambda\Delta(\beta_\varepsilon(y) + \varepsilon y) + \lambda\,\mathrm{div}(b_\varepsilon^*(y)) = f \text{ in } \mathscr{D}'(\mathbb{R}^d), \tag{2.312}$$

where

$$\begin{aligned} b_\varepsilon^*(r) &\equiv \frac{b_\varepsilon(r)r}{1+\varepsilon|r|}, \quad b_\varepsilon = b * \varphi_\varepsilon, \\ \beta_\varepsilon(r) &\equiv \tfrac{1}{\varepsilon}(r - (1+\varepsilon\beta)^{-1}r) \in \beta(1+\varepsilon\beta)^{-1}r, \ \forall r \in \mathbb{R}. \end{aligned} \tag{2.313}$$

As seen earlier, it follows that (2.312) has a unique solution $y_\varepsilon = y_\varepsilon(f)$ which satisfies

$$|y_\varepsilon(f_1) - y_\varepsilon(f_2)|_1 \le |f_1 - f_2|_1, \ \forall f_1, f_2 \in L^1.$$

Moreover, if $f \in L^1 \cap L^\infty$, then $y_\varepsilon \in H^1$, and

$$|y_\varepsilon|_2^2 + \lambda|\nabla\beta_\varepsilon(y_\varepsilon)|_2^2 + \varepsilon\lambda|\nabla y_\varepsilon|_2^2 \le C|f|_2^2, \ \forall \varepsilon > 0, \ \lambda \in (0, \lambda_1), \tag{2.314}$$

for λ_1 small enough but independent of ε. Hence, on a subsequence $\{\varepsilon\} \to 0$, we have

$$\beta_\varepsilon(y_\varepsilon) \to \eta \ \text{ weakly in } H^1 \text{ and strongly in } L^2_{\mathrm{loc}} \tag{2.315}$$

$$y_\varepsilon \to y \ \text{ weakly in } L^2. \tag{2.316}$$

Since the map $y \to \beta(y)$ is maximal monotone in $L^2(\mathscr{O})$ for every bounded, open set $\mathscr{O} \subset \mathbb{R}^d$, it follows that $\eta(x) = \beta(y(x))$, a.e. $x \in \mathbb{R}^d$.

Arguing as in the proof of Proposition 2.3, we also get that (see (2.242))

$$\|\beta_\varepsilon(y_\varepsilon)\|_{L^q(K)} \le C_K(|f|_1 + 1),$$

for any compact set $K \subset \mathbb{R}^d$. Hence

$$\|\eta\|_{L^q(K)} \le C_K(|f|_1 + 1). \tag{2.317}$$

We also have that the set $\{y_\varepsilon\}$ is compact in $L^1_{\rm loc}$. Here is the argument. We set $y^h_\varepsilon(x) \equiv y_\varepsilon(x+h) - y_\varepsilon(x)$, $f^h(x) \equiv f(x+h) - f(x)$, $x, h \in \mathbb{R}^d$, and we get

$$y^h_\varepsilon - \lambda\Delta((\beta^\varepsilon(y_\varepsilon))^h + \lambda\,{\rm div}((b^*_\varepsilon(y_\varepsilon))^h) = f^h \text{ in } \mathbb{R}^d,$$

where $\beta^\varepsilon(r) \equiv \beta_\varepsilon(r) + \varepsilon r$.

Now, we multiply the latter by $\mathscr{X}_\delta((\beta^\varepsilon(y_\varepsilon))^h)$ and integrate on $\mathbb{R}^d$. We have

$$\int_{\mathbb{R}^d} \Delta(\beta^\varepsilon(y_\varepsilon))^h \mathscr{X}_\delta(\beta^\varepsilon(y_\varepsilon))^h)dx \le 0,\ \forall \delta, \varepsilon > 0,$$

and

$$\begin{aligned}&\lim_{\delta\to 0}\int_{\mathbb{R}^d} {\rm div}(b^*_\varepsilon(y_\varepsilon))^h \mathscr{X}_\delta(\beta_\varepsilon(y_\varepsilon))^h)dx\\ &\quad = -\lim_{\delta\to 0}\frac{1}{\delta}\int_{[|(\beta^\varepsilon(y_\varepsilon))^h|\le\delta]} (b^*_\varepsilon(y_\varepsilon))^h\cdot\nabla\mathscr{X}_\delta((\beta^\varepsilon(y_\varepsilon))^h)dx\\ &\quad\quad -\lim_{\delta\to 0}\int_{[|(\beta^\varepsilon(y_\varepsilon))^h|\le\delta]} (b^*_\varepsilon(y_\varepsilon))^h\cdot\nabla(\beta^\varepsilon(y_\varepsilon))^h dx = 0.\end{aligned}$$

This yields

$$\begin{aligned}&\int_{[|(\beta^\varepsilon(y_\varepsilon))^h|\le\delta]} |(b^*_\varepsilon(y_\varepsilon))^h|\,|\nabla(\beta^\varepsilon(y_\varepsilon))^h|dx \le C^1_\varepsilon\int_{[|(\beta^\varepsilon(y_\varepsilon))^h|\le\delta]} |y_\varepsilon|\,|\nabla(\beta^\varepsilon(y_\varepsilon))^h|dx\\ &\quad\le C^1_\varepsilon|y_\varepsilon|_2\left(\int_{[|(\beta^\varepsilon(y_\varepsilon))^h|\le\delta]}|\nabla(\beta^\varepsilon(y_\varepsilon))^h|^2dx\right)^{\frac12} \to 0 \text{ as } \delta\to 0,\end{aligned}$$

for all $\varepsilon > 0$ and, therefore,

$$\int_{\mathbb{R}^d} y^h_\varepsilon(x)\mathscr{X}_\delta((\beta^\varepsilon(y_\varepsilon))^h)(x))dx \le \int_{\mathbb{R}^d} |f^h(x)|dx + \psi(\delta),\ \ \forall\varepsilon,\delta > 0,$$

where $\psi(\delta) \to 0$ as $\delta \to 0$. This yields

$$|y^h_\varepsilon|_1 \le |f^h|_1,\ \forall\,\varepsilon > 0,\ h \in \mathbb{R}^d.$$

Then, by the Riesz–Kolmogorov compactness theorem in L^1, it follows that the sequence $\{y_\varepsilon\}$ is compact in $L^1_{\rm loc}$. Hence, on a subsequence, again denoted $\{\varepsilon\} \to 0$, we have

$$y_\varepsilon \to y \text{ strongly in } L^1_{\rm loc} \text{ and a.e. on } \mathbb{R}^d, \tag{2.318}$$

and

$$b_\varepsilon(y_\varepsilon) \to \xi \text{ weak-star in } L^\infty(\mathbb{R}^d;\mathbb{R}^d). \tag{2.319}$$

Now, arguing as in the proof of Lemma 2.9, it follows by (2.318) that on a subsequence $\{\varepsilon\} \to 0$ independent of $f \in L^1$, we have

$$y_\varepsilon \to y_\lambda(f) \text{ in } L^1.$$

By (2.315) we have

$$\begin{aligned} &y - \lambda\Delta\eta + \lambda\,\mathrm{div}(\xi y) = f \text{ in } \mathscr{D}'(\mathbb{R}^d),\\ &\xi(x) \in F_b(y(x)),\ \eta(x) \in \beta(y(x)),\ \text{a.e. } x \in \mathbb{R}^d. \end{aligned} \tag{2.320}$$

Moreover, by (2.314), we see that $\eta \in H^1(\mathbb{R}^d)$, $y \in L^2$, and

$$|y|_2^2 + \lambda|\nabla\eta|_2^2 \le C|f|_2^2,\ \forall\lambda \in (0,\lambda_1). \tag{2.321}$$

If $B_N = \{x \in \mathbb{R}^d;\ |x| \le N\}$, it follows by (2.318) and the Egorov theorem that, for each $\nu > 0$, there is a Lebesgue measurable set $\mathscr{O}_\nu \subset B_N$ such that, for $\varepsilon \to 0$, $y_\varepsilon \to y$ uniformly on $B_N \setminus \mathscr{O}_\nu$ and $\mu(\mathscr{O}_\nu) \le \nu$. Hence, for each $\delta > 0$ there is $\varepsilon(\nu) > 0$ such that

$$y_\varepsilon(x) \in [y(x)-\delta, y(x)+\delta],\ x \in B_N \setminus \mathscr{O}_\nu,\ 0 < \varepsilon < \varepsilon(\nu).$$

Moreover, by Luzin's theorem, for each $\gamma > 0$ there is a measurable set $S_\gamma \subset \mathbb{R}^d$ such that $\mu(S_\gamma) \le \gamma$ and b is continuous on $\left(\bigcup_{x\in\mathscr{O}_\nu}[y(x)-\delta, y(x)+\delta]\right) \setminus S_\gamma$. Hence, for $0 < \varepsilon < \varepsilon(\nu)$, we have by (2.313) that

$$b_\varepsilon(y_\varepsilon(x)) \in \mathrm{conv}[b(y(x)-\delta), b(y(x)+\delta)] \setminus S_\gamma,\ \forall x \in B_N \setminus \mathscr{O}_\nu.$$

Then, by (2.319) we infer that

$$\xi(x) \in \bigcap_{\delta>0}\bigcap_{\gamma>0} \overline{\mathrm{conv}}([b(y(x)-\delta), b(y(x)+\delta)] \setminus S_\gamma)$$

and, therefore,

$$\xi(x) \in \bigcap_{\delta>0}\bigcap_{\gamma>0} \mathrm{conv}([b(y(x)-\delta), b(y(x)+\delta)] \setminus S).$$

Since ν is arbitrarily small while N is arbitrarily large, we get

$$\xi(x) \in F_b(y(x)),\ \text{a.e. } x \in \mathbb{R}^d. \tag{2.322}$$

We denote by $J_\lambda(f) = y_\lambda(f)$ and note that

$$|J_\lambda(f_1) - J_\lambda(f_2)|_1 \le |f_1 - f_2|_1, \ \forall f_1, f_2 \in L^1 \cap L^2, \ \lambda \in (0, \lambda_1). \tag{2.323}$$

To prove (2.323), we denote by $y_\lambda^\varepsilon(f)$ the solution to (2.312) and set $\beta^\varepsilon(r) := \beta_\varepsilon(r) + \varepsilon r$, $r \in \mathbb{R}$. Then, arguing as above, we multiply the equation

$$\begin{aligned}(y_\lambda^\varepsilon(f_1) - y_\lambda^\varepsilon(f_2)) - \lambda\Delta(\beta_\varepsilon(y_\lambda^\varepsilon(f_1)) - \beta_\varepsilon(y_\lambda^\varepsilon(f_2)))\\ + \lambda \operatorname{div}(b_\varepsilon^*(y_\lambda^\varepsilon(f_1)) - b_\varepsilon^*(y_\lambda^\varepsilon(f_2))) = f_1 - f_2\end{aligned}$$

by $\mathscr{X}_\delta(\beta^\varepsilon(y_\lambda^\varepsilon(f_1) - \beta^\varepsilon(y_\lambda^\varepsilon(f_2)))$ and integrate over $\mathbb{R}^d$. We get, as above, that

$$\begin{aligned}&\int_{\mathbb{R}^d} (y_\lambda^\varepsilon(f_1) - y_\lambda^\varepsilon(f_2))\mathscr{X}_\delta(\beta^\varepsilon(y_\lambda^\varepsilon(f_1)) - \beta^\varepsilon(y_\lambda^\varepsilon(f_2)))dx\\ &\quad \le |f_1 - f_2|_1 + C\lambda\left(\int_{E_{\lambda,\delta}^\varepsilon} |\nabla(\beta^\varepsilon(y_\lambda^\varepsilon(f_1)) - \beta^\varepsilon(y_\lambda^\varepsilon(f_2)))|^2 dx\right)^{\frac{1}{2}}\\ &\quad \le |f_1 - f_2|_1 + C\lambda\nu(\delta), \ \text{ where } \nu(\delta) \to 0 \text{ as } \delta \to 0,\\ &|y_\lambda^\varepsilon(f_1) - y_\lambda^\varepsilon(f_2)|_1 \le |f_1 - f_2|_1,\end{aligned}$$

and so, for $\varepsilon \to 0$, it follows that (2.323) holds.

Let $f \in L^1$ be arbitrary but fixed and let $\{f_n\} \subset L^1 \cap L^2$ be such that $f_n \to f$ in L^1 as $n \to \infty$. We set $y_n = y(\lambda, f_n)$, that is,

$$y_n - \lambda\Delta\eta_n + \lambda \operatorname{div}(\xi_n y_n) = f_n \text{ in } H^{-1}(\mathbb{R}^d),$$

where $\eta_n \in \beta(y_n)$, $\xi_n \in F_b(y_n)$, a.e. in $\mathbb{R}^d$ and

$$\|\eta_n\|_{L^q(K)} + \|\xi_n\|_\infty \le C(|f|_1 + 1),$$

for any compact set $K \subset \mathbb{R}^d$ and $q > 1$, and by (2.323),

$$|y_n - y_m|_1 \le |f_n - f_m|_1, \ \forall n, m \in \mathbb{N}.$$

Hence, there is $y = \lim\limits_{n\to\infty} y_n$ in L^1.

Moreover, we have $\eta_n \to \eta$ weakly in $L^q_{\rm loc}$, $\xi_n \to \xi$ weak-star in L^∞ and this yields by the strong-weak closedness of the mapping $y \to \beta(y)$ that $\eta \in \beta(y)$, a.e. in $\mathbb{R}^d$.

On the other hand, arguing as above, it follows that

$$\xi(x) \in \bigcap_{\delta>0}\bigcap_{\gamma>0} \overline{\operatorname{conv}}([b(y(x) - \delta), b(y(x) + \delta)] \setminus S_\gamma), \ \forall x \in B_N \setminus \mathscr{O}_\nu,$$

where $\mu(\mathcal{O}_\nu) \le \nu$, $\mu(S_\gamma) \le \gamma$. Hence, $\xi(x) \in F_b(y(x))$, a.e. $x \in \mathbb{R}^d$. Therefore, $y \in D(A_0)$, $y + \lambda A_0 y = f$, and so $R(I + \lambda A_0) = L^1$, as claimed.

As regard (2.190) and (2.191), they follow for $\varepsilon \to 0$ as in the proof of Proposition 2.3 by the corresponding properties of $y_\varepsilon^\lambda(f)$.

2.8 NFPE with Fractional Laplacian

The existence and uniqueness theory developed in Sects. 2.2 and 2.3 extends *mutatis-mutandis* to NFPEs (2.80) with fractional Laplacian. Namely,

$$
\begin{aligned}
&u_t + (-\Delta)^s \beta(u) + \operatorname{div}(Db(u)u) = 0, \quad \text{in } (0,\infty) \times \mathbb{R}^d,\\
&u(0,x) = u_0(x), \ x \in \mathbb{R}^d,
\end{aligned}
\tag{2.324}
$$

where $\beta : \mathbb{R} \to \mathbb{R}$, $D : \mathbb{R}^d \to \mathbb{R}^d$, $d \ge 2$, and $b : \mathbb{R} \to \mathbb{R}$ are given functions, while $(-\Delta)^s$, $0 < s < 1$, is the fractional Laplace operator defined as follows. Let $S' := S'(\mathbb{R}^d)$ be the dual of the Schwartz test function space $S := S(\mathbb{R}^d)$. Define

$$
\begin{aligned}
&D_s := \{u \in S';\ \mathcal{F}(u) \in L^1_{\mathrm{loc}},\ |\xi|^{2s}\mathcal{F}(u) \in S'\}\ (\supset L^1)\\
&\mathcal{F}((-\Delta)^s u)(\xi) = |\xi|^{2s}\mathcal{F}(u)(\xi),\ \xi \in \mathbb{R}^d,\ u \in D_s,
\end{aligned}
$$

where $\mathcal{F}$ stands for the Fourier transform in $\mathbb{R}^d$, that is,

$$
\mathcal{F}(u)(\xi) = (2\pi)^{-d/2} \int_{\mathbb{R}^d} e^{ix\cdot\xi} u(x)dx,\ \xi \in \mathbb{R}^d,\ u \in L^1.
$$

NFPE (2.324) is used for modelling the dynamics of anomalous diffusion of particles in disordered media. The solution u may be viewed as the transition density corresponding to a distribution dependent stochastic differential equation with Lévy forcing term. (See Eq. (2.325) below.)

Hypotheses

(n) $\beta \in C^1(\mathbb{R}) \cap \mathrm{Lip}(\mathbb{R})$, $\beta'(r) > 0$, $\forall r \in \mathbb{R}$, $\beta(0) = 0$.
(nn) $D \in L^\infty(\mathbb{R}^d;\mathbb{R}^d) \cap C^1(\mathbb{R}^d;\mathbb{R}^d)$, $\operatorname{div} D \in L^2_{\mathrm{loc}}$, $(\operatorname{div} D)^- \in L^\infty$, $b \ge 0$.
(nnn) $b \in C_b(\mathbb{R}) \cap C^1(\mathbb{R})$.

We shall represent (2.324) as an abstract differential equation in L^1 of the form

$$
\begin{aligned}
&\frac{du}{dt} + A(u) = 0, \quad t \ge 0,\\
&u(0) = u_0,
\end{aligned}
\tag{2.325}
$$

where A is an m-accretive realization in L^1 of the operator

$$\begin{aligned} A_0(u) &= (-\Delta)^s\beta(u) + \operatorname{div}(Db(u)u),\ u \in D(A_0),\\ D(A_0) &= \left\{u \in L^1;\ (-\Delta)^s\beta(u) + \operatorname{div}(Db(u)u) \in L^1\right\}. \end{aligned} \tag{2.326}$$

As in the previous cases, a function $u \in C([0,\infty); L^1)$ is said to be a *mild solution* to (2.324) if, for each $0 < T < \infty$,

$$u(t) = \lim_{h\to 0} u_h(t) \text{ in } L^1,\ t \in [0,T),$$

where

$$u_h(t) = u_h^j,\ \forall t \in [jh,(j+1)h),\ j = 0,1,\dots,N = \left[\tfrac{T}{h}\right],$$

$$u_h^{j+1} + hA_0(u_h^{j+1}) = u_h^j,\ j = 0,1,\dots,N_h,$$

$$u_h^j \in D(A_0),\ \forall j = 0,\dots,N_h;\ u_h^0 = u_0.$$

We note that, if u is a mild solution to (2.324), then it is also a *Schwartz distributional solution*, that is,

$$\begin{aligned} &\int_0^\infty\int_{\mathbb{R}^d}(u(t,x)\varphi_t(t,x) - (-\Delta)^s\varphi(t,x)\beta(u(t,x))\\ &\qquad + b(u(t,x))u(t,x)D(x)\cdot\nabla\varphi(t,x))dxdt\\ &\qquad + \int_{\mathbb{R}^d}\varphi(0,x)u_0(dx) = 0,\ \forall\varphi \in C_0^\infty([0,\infty)\times\mathbb{R}^d), \end{aligned} \tag{2.327}$$

where u_0 is a measure of finite variation on $\mathbb{R}^d$. This equation is related to the following McKean–Vlasov SDE on $\mathbb{R}^d$

$$\begin{aligned} dX_t &= D(X_t)b(u(t,X_t))dt + \left(\frac{\beta(u(t,X_{t-}))}{u(t,X_{t-})}\right)^{\frac{1}{2s}} dL_t,\\ \mathcal{L}_{X_t}(dx) &:= \mathbb{P}\circ X_t^{-1}(dx) = u(t,x)dx,\ t \in [0,T], \end{aligned} \tag{2.328}$$

where L is a d-dimensional isotropic $2s$-stable process with Lévy measure $dz/|z|^{d+2s}$ (see Chap. 5).

As in the previous case ($s = 1$) to construct the operator $A : D(A) \subset L^1 \to L^1$, we need the following lemma.

Lemma 2.15 *Assume that* $\frac{1}{2} < s < 1$. *Under Hypotheses* (n)–(nnn) *there is a family of operators* $\{J_\lambda : L^1 \to L^1; \lambda > 0)\}$, *which for all* $\lambda \in (0, \lambda_0)$, $\lambda_0 > 0$, *satisfies*

$$(I + \lambda A_0)(J_\lambda(f)) = f,\ \forall f \in L^1,$$

$$|J_\lambda(f_1) - J_\lambda(f_2)|_1 \le |f_1 - f_2|_1,\ \forall f_1, f_2 \in L^1,$$

$$J_{\lambda_2}(f) = J_{\lambda_1}\left(\frac{\lambda_1}{\lambda_2} f + \left(1 - \frac{\lambda_1}{\lambda_2}\right) J_{\lambda_2}(f)\right),\ \forall f \in L^1,\ \lambda_1, \lambda_2 \in (0, \lambda_0),$$

$$\int_{\mathbb{R}^d} J_\lambda(f)dx = \int_{\mathbb{R}^d} f\,dx,\ \forall f \in L^1,$$

$$J_\lambda(f) \ge 0,\ \ a.e.\ on\ \mathbb{R}^d,\ \ if\ f \ge 0,\ \ a.e.\ on\ \mathbb{R}^d,$$

$$|J_\lambda(f)|_\infty \le (1 + ||D| + (\mathrm{div}\, D)^-|_\infty^{\frac{1}{2}})|f|_\infty,\ \forall f \in L^1 \cap L^\infty,$$

$$\beta(J_\lambda(f)) \in H^s \cap L^1 \cap L^\infty,\ \forall f \in L^1 \cap L^\infty.$$

The proof of Lemma 2.15 is similar with that of Proposition 2.3, so it will be outlined only, emphasizing however some technical specific arguments. For details, we refer to the work [17].

First, we shall prove the existence of a solution $y = y_\lambda \in D(A_0)$ to the equation

$$y + \lambda A_0(y) = f \text{ in } S', \tag{2.329}$$

for $f \in L^1$. For $\varepsilon \in (0, 1]$ we consider the equation

$$y + \lambda(\varepsilon I - \Delta)^s(\beta_\varepsilon(y)) + \lambda\, \mathrm{div}(D_\varepsilon b_\varepsilon(y)y) = f \text{ in } S', \tag{2.330}$$

where, for $r \in \mathbb{R}$, $\beta_\varepsilon(r) := \beta(r) + \varepsilon r$ and

$$D_\varepsilon := \eta_\varepsilon D,\ \eta_\varepsilon \in C_0^1(\mathbb{R}^d),\ 0 \le \eta_\varepsilon \le 1,\ |\nabla \eta_\varepsilon| \le 1,\ \eta_\varepsilon(x) = 1 \text{ if } |x| < \tfrac{1}{\varepsilon},$$

$$b_\varepsilon(r) \equiv \frac{(b * \varphi_\varepsilon)(r)}{1 + \varepsilon|r|},\ \forall r \in \mathbb{R},$$

and $\varphi_\varepsilon(r) = \frac{1}{\varepsilon}\,\varphi\left(\frac{r}{\varepsilon}\right)$ is a standard mollifier.

Assume first that $f \in L^2$ and consider the approximating equation

$$F_{\varepsilon,\lambda}(y) = f \text{ in } S', \tag{2.331}$$

where $F_{\varepsilon,\lambda} : L^2 \to S'$ is defined by

$$F_{\varepsilon,\lambda}(y) := y + \lambda(\varepsilon I - \Delta)^s \beta_\varepsilon(y) + \lambda \operatorname{div}(D_\varepsilon b_\varepsilon^*(y)), \; y \in L^2.$$

We recall that the Bessel space of order $s \in \mathbb{R}$ is defined as

$$H^s := \{u \in S'; \; (1 + |\xi|^2)^{\frac{s}{2}} \mathcal{F}(u) \in L^2\}$$

and the Riesz space as

$$\dot{H}^s := \{u \in S'; \; \mathcal{F}(u) \in L^1_{\mathrm{loc}} \text{ and } |\xi|^s \mathcal{F}(u) \in L^2\}$$

with the respective norms

$$|u|^2_{H^s} := \int_{\mathbb{R}^d} (1 + |\xi|^2)^s |\mathcal{F}(u)|^2(\xi) d\xi = \int_{\mathbb{R}^d} |(I - \Delta)^{\frac{s}{2}} u|^2 d\xi,$$

$$|u|^2_{\dot{H}^s} := \int_{\mathbb{R}^d} |\xi|^{2s} |\mathcal{F}(u)|^2(\xi) d\xi = \int_{\mathbb{R}^d} |(-\Delta)^{\frac{s}{2}} u|^2 d\xi.$$

We rewrite (2.331) as

$$(\varepsilon I - \Delta)^{-s} y + \lambda \beta_\varepsilon(y) + \lambda(\varepsilon I - \Delta)^{-s} \operatorname{div}(D_\varepsilon b_\varepsilon^*(y)) = (\varepsilon I - \Delta)^{-s} f. \quad (2.332)$$

Clearly, since $D_\varepsilon b_\varepsilon^*(y) \in L^2$, hence $\operatorname{div}(D_\varepsilon b_\varepsilon^*(y)) \in H^{-1}$, we have

$$(\varepsilon I - \Delta)^{-s} F_{\varepsilon,\lambda}(y) \in L^2, \; \forall y \in L^2,$$

because $s > \frac{1}{2}$. Now, it is easy to see that (2.332) has a unique solution, $y_\varepsilon \in L^2$, because $(\varepsilon I - \Delta)^{-s} F_{\varepsilon,\lambda} : L^2 \to L^2$ is strictly monotone.

It follows so that the solution y_ε to (2.332) belongs to H^{2s-1}, hence $b_\varepsilon^*(y_\varepsilon) \in H^{2s-1}$. Since $s > \frac{1}{2}$ and $D \in C^1(\mathbb{R}^d; \mathbb{R}^d)$, by a bootstrapping argument it follows that $y_\varepsilon \in H^1$.

If $f \geq 0$, a.e. on $\mathbb{R}^d$, then we have $y_\varepsilon \geq 0$, a.e. on $\mathbb{R}^d$.

Here is the argument. For $\delta > 0$, consider the function

$$\eta_\delta(r) = \begin{cases} -1 & \text{for } r \leq -\delta, \\ \dfrac{r}{\delta} & r \in (-\delta, 0), \\ 0 & \text{for } r \geq 0. \end{cases}$$

If we multiply Eq. (2.330), where $y = y_\varepsilon$, by $\eta_\delta(\beta_\varepsilon(y_\varepsilon))$ ($\in H^1$) and integrate over $\mathbb{R}^d$, we get

$$\int_{\mathbb{R}^d} y_\varepsilon \eta_\delta(\beta_\varepsilon(y_\varepsilon))dx + \lambda \int_{\mathbb{R}^d} (\varepsilon I - \Delta)^s (\beta_\varepsilon(y_\varepsilon))\eta_\delta(\beta_\varepsilon(y_\varepsilon))dx$$
$$= \int_{\mathbb{R}^d} f \eta_\delta(\beta_\varepsilon(y_\varepsilon))dx + \lambda \int_{\mathbb{R}^d} D_\varepsilon b_\varepsilon^*(y_\varepsilon)\eta_\delta'(\beta_\varepsilon(y_\varepsilon)) \cdot \nabla\beta_\varepsilon(y_\varepsilon)dx,$$

and so, by Lemma 5.2 in [45] we have

$$\int_{\mathbb{R}^d} (\varepsilon I - \Delta)^s u\Psi(u)dx \geq \int_{\mathbb{R}^d} |(\varepsilon I - \Delta)^{\frac{s}{2}} \widetilde{\Psi}(u)|^2 dx \geq 0, \ u \in H^1(\mathbb{R}^d),$$

for any pair of functions $\Psi, \widetilde{\Psi} \in \mathrm{Lip}(\mathbb{R})$ such that $\Psi'(r) \equiv (\widetilde{\Psi}'(r))^2,\ r \in \mathbb{R}$. This yields

$$\int_{\mathbb{R}^d} (\varepsilon I - \Delta)^s \beta_\varepsilon(y_\varepsilon)\eta_\delta(\beta_\varepsilon(y_\varepsilon))dx \geq 0.$$

Taking into account that $|\beta_\varepsilon(y_\varepsilon)| \geq \varepsilon|y_\varepsilon|$, we have

$$\left|\int_{\mathbb{R}^d} D_\varepsilon b_\varepsilon(y_\varepsilon) y_\varepsilon \eta_\delta'(\beta_\varepsilon(y_\varepsilon))\nabla\beta_\varepsilon(y_\varepsilon)dx\right|$$
$$\leq \frac{1}{\varepsilon}|b|_\infty \|D_\varepsilon\|_{L^2} \left(\int_{\widetilde{E}_\varepsilon^\delta} |\nabla\beta_\varepsilon(y_\varepsilon)|^2 dx\right)^{\frac{1}{2}} \to 0 \text{ as } \delta \to 0,$$

where $\widetilde{E}_\varepsilon^\delta = \{-\delta < \beta_\varepsilon(y_\varepsilon) \leq 0\}$.

If $y_\varepsilon = y_\varepsilon(\lambda, f)$ is the solution to (2.330), we have for $f_1, f_2 \in L^1 \cap L^2$

$$y_\varepsilon(\lambda, f_1) - y_\varepsilon(\lambda, f_2) + \lambda(\varepsilon I - \Delta)^s(\beta_\varepsilon(y_\varepsilon(\lambda, f_1)) - \beta_\varepsilon(y_\varepsilon(\lambda, f_2)))$$
$$+\lambda \operatorname{div} D_\varepsilon(b_\varepsilon^*(y_\varepsilon(\lambda, f_1)) - b_\varepsilon^*(y_\varepsilon(\lambda, f_2))) = f_1 - f_2.$$

If we multiply the latter by $\mathcal{X}_\delta(\beta_\varepsilon(y_\varepsilon(\lambda, f_1)) - \beta_\varepsilon(y_\varepsilon(\lambda, f_2)))$ ($\in H^1$) and integrate over $\mathbb{R}^d$, we get

$$\int_{\mathbb{R}^d} (y_\varepsilon(\lambda, f_1) - y_\varepsilon(\lambda, f_2))\mathcal{X}_\delta(\beta_\varepsilon(y_\varepsilon(\lambda, f_1)) - \beta_\varepsilon(y_\varepsilon(\lambda, f_2)))dx$$
$$\leq \lambda \frac{1}{\delta} \int_{E_\varepsilon^\delta} |b_\varepsilon^*(y_\varepsilon(\lambda, f_1)) - b_\varepsilon^*(y_\varepsilon(\lambda, f_2))|\,|D_\varepsilon|\,|\nabla(\beta_\varepsilon(y_\varepsilon(\lambda, f_1)) - \beta_\varepsilon(y_\varepsilon(\lambda, f_2)))|dx$$
$$+|f_1 - f_2|_1,$$

because we have

$$\int_{\mathbb{R}^d} (\varepsilon I - \Delta)^s (\beta_\varepsilon(y_\varepsilon, f_1) - \beta_\varepsilon(y_\varepsilon, f_2)) \mathscr{X}_\delta(\beta_\varepsilon(y_\varepsilon, f_1) - \beta_\varepsilon(y_\varepsilon, f_2)) dx \geq 0.$$

This yields

$$|y_\varepsilon(\lambda, f_1) - y_\varepsilon(\lambda, f_2)|_1 \leq |f_1 - f_2|_1, \ \forall \lambda \in (0, \lambda_\varepsilon). \tag{2.333}$$

Arguing as in the previous cases, one gets the estimate

$$|J_\lambda^\varepsilon(f)|_1 + |J_\lambda^\varepsilon(f)|_\infty \leq c_1, \ \forall \varepsilon, \lambda > 0,$$

where $c_1 = c_1(|f|_1, |f|_\infty)$ is independent of ε and λ.

Now, fix $\lambda \in (0, \lambda_0)$ and $f \in L^1 \cap L^\infty$. For $\varepsilon \in (0, 1]$ set $y_\varepsilon := J_\lambda^\varepsilon(f)$. Then, since $\beta_\varepsilon(y_\varepsilon) \in H^1$, by (2.330) we get

$$\begin{aligned}(y_\varepsilon, (\varepsilon I - \Delta)^{-\frac{1}{2}} \beta_\varepsilon(y_\varepsilon))_2 + \lambda_{H^{-1}}\Big\langle(\varepsilon I - \Delta)^s \beta_\varepsilon(y_\varepsilon), (\varepsilon I - \Delta)^{-\frac{1}{2}} \beta_\varepsilon(y_\varepsilon)\Big\rangle_{H^1} \\ = -\lambda(\operatorname{div}(D_\varepsilon b_\varepsilon^*(y_\varepsilon)), (\varepsilon I - \Delta)^{-\frac{1}{2}} \beta_\varepsilon(y_\varepsilon))_2 + ((\varepsilon I - \Delta)^{-\frac{1}{2}} f, \beta_\varepsilon(y_\varepsilon))_2\end{aligned}$$

which, because $((\varepsilon I - \Delta)^{-\frac{1}{2}} y_\varepsilon, \beta_\varepsilon(y_\varepsilon))_2 \geq 0$ and $H^1 \subset H^\alpha$ for all $\alpha > 0$, implies that

$$\begin{aligned}\lambda|(\varepsilon I - \Delta)^{\frac{s}{2}} \beta_\varepsilon(y_\varepsilon)|_2^2 &\leq \lambda|D_\varepsilon|_\infty |b_\varepsilon|_\infty |y_\varepsilon|_\infty |\nabla(\varepsilon I - \Delta)^{-\frac{1}{2}} \beta_\varepsilon(y_\varepsilon)|_2 \\ &\leq \lambda|D|_\infty |b|_\infty |\beta_\varepsilon(y_\varepsilon)|_2 (1 + ||D| + \operatorname{div} D|^- |_\infty^{\frac{1}{2}}) |f|_\infty |f|_1.\end{aligned}$$

Since $|\beta_\varepsilon(r)| \leq (\operatorname{Lip}(\beta) + 1)|r|, \ r \in \mathbb{R}$, we obtain that

$$\sup_{\varepsilon \in (0,1]} |(\varepsilon I - \Delta)^{\frac{s}{2}} \beta_\varepsilon(y_\varepsilon)|_2^2 \leq C < \infty.$$

Since

$$|(-\Delta)^\alpha u|_2^2 \leq |(\varepsilon I - \Delta)^\alpha u|_2^2 \leq |(-\Delta)^\alpha u|_2^2 + \varepsilon^{2\alpha} |u|_2^2 \leq 2|(\varepsilon I - \Delta)^\alpha u|_2^2,$$

and as $\beta_\varepsilon(y_\varepsilon) \in H^1 \subset H^s$ with $|\beta_\varepsilon(r)| \leq (1 + \operatorname{Lip}(\beta))|r|$, we conclude that $\{\beta_\varepsilon(y_\varepsilon)\}$ is bounded in $H^s \subset L^2$ and, therefore (along a subsequence) as $\varepsilon \to 0$,

$$\begin{aligned}\beta_\varepsilon(y_\varepsilon) &\to z \text{ weakly in } H^s \text{ strongly in } L^2_{\text{loc}}, \\ (\varepsilon I - \Delta)^s \beta_\varepsilon(y_\varepsilon) &\to (-\Delta)^s z \text{ in } S', \\ y_\varepsilon &\to y \text{ weak-star in } L^\infty \text{ and weakly in } L^2.\end{aligned} \tag{2.334}$$

Then, we can let $\varepsilon \to 0$ in (2.330) to find that

$$y + \lambda(-\Delta)^s \beta(y) + \lambda \operatorname{div}(Db^*(y)) = f \text{ in } S'. \tag{2.335}$$

If we define $J_\lambda(f) := y \in D(A_0)$, it follows by (2.333) and Fatou's lemma that, for $f_1, f_2 \in L^1 \cap L^\infty$,

$$|J_\lambda(f_1) - J_\lambda(f_2)|_1 \le |f_1 - f_2|_1. \tag{2.336}$$

Hence J_λ extends continuously to all of L^1, still satisfying (2.336) for all $f_1, f_2 \in L^1$.

Moreover, arguing as in the proof of Theorem 2.14 below, it follows that (2.335) has at most one distributional solution $y \in L^\infty$.This implies that (2.334) holds for $\varepsilon \to 0$ (ont only on a subsequence) and that J_λ satisfies

$$J_{\lambda_2}(f) = J_{\lambda_1}\left(\frac{\lambda_1}{\lambda_2} f + \left(1 - \frac{\lambda_1}{\lambda_2}\right) J_{\lambda_2}(f)\right).$$

To prove that $\int_{\mathbb{R}^d} J_\lambda(f)dx = \int_{\mathbb{R}^d} f\,dx$, it suffices to note that

$$\int_{\mathbb{R}^d} A_0 y\,dx = 0.$$

Now define

$$\begin{aligned} D(A) &:= J_\lambda(L^1)\ (\subset D(A_0)), \\ A(y) &:= A_0(y),\ y \in D(A). \end{aligned} \tag{2.337}$$

Again it is easy to see that $J_\lambda(L^1)$ is independent of $\lambda \in (0, \lambda_0)$ and that

$$J_\lambda = (I + \lambda A)^{-1},\ \lambda \in (0, \lambda_0).$$

Therefore, by Lemma 2.15, the operator A defined by (2.337) is m-accretive in L^1 and $(I + \lambda A)^{-1} = J_\lambda$, $\lambda \in (0, \lambda_0)$. Moreover, if $\beta \in C^\infty(\mathbb{R})$, then $\overline{D(A)} = L^1$. Then, by the Crandall & Liggett theorem we have that the Cauchy problem (2.325) has, for each $u_0 \in \overline{D(A)}$, a unique mild solution $u = u(t, u_0) \in C([0, \infty); L^1)$ and $S(t)u_0 = u(t, u_0)$ is a C_0-semigroup of contractions on L^1. This yields

Theorem 2.14 *Assume that $\frac{1}{2} < s < 1$ and Hypotheses* (n)–(nnn) *hold. Then, there is a C_0-semigroup of contractions $S(t) : L^1 \to L^1$, $t \ge 0$, such that for each $u_0 \in \overline{D(A)}$, which is L^1 if $\beta \in C^\infty(\mathbb{R})$, $u(t, u_0) = S(t)u_0$ is a mild solution to* (2.324). *Moreover, if $u_0 \ge 0$, a.e. in $\mathbb{R}^d$,*

$$u(t, u_0) \ge 0,\ \ \textit{a.e. in } \mathbb{R}^d,\ \forall t \ge 0,$$

and

$$\int_{\mathbb{R}^d} u(t,u_0)(x)dx = \int_{\mathbb{R}^d} u_0(x)dx, \quad \forall t \geq 0.$$

Moreover, u is a distributional solution to (2.324) *on* $[0,\infty)\times\mathbb{R}^d$. *Finally, if* $u_0 \in L^1 \cap L^\infty$, *then all above assertions remain true, if we drop the assumption* $\beta \in \mathrm{Lip}(\mathbb{R})$ *from Hypothesis (n), and additionally we have that* $u \in L^\infty((0,T)\times\mathbb{R}^d)$, $T>0$.

We can prove the uniqueness of distributional solutions to (2.324), where $s \in \left(\frac{1}{2},1\right)$, under the following hypotheses:

(p) $\beta \in C^1(\mathbb{R}),\ \beta'(r) > 0,\ \forall r \in \mathbb{R},\ \beta(0)=0.$

(pp) $D \in L^\infty(\mathbb{R}^d;\mathbb{R}^d)$.

(ppp) $b \in C^1(\mathbb{R})$.

Theorem 2.15 *Under assumptions of Theorem* 2.14, *let* $d \geq 2$, $s \in \left(\frac{1}{2},1\right)$, $T>0$, *and let* $y_1, y_2 \in L^\infty((0,T)\times\mathbb{R}^d)$ *be two distributional solutions to* (2.324) *on* $(0,T)\times\mathbb{R}^d$ *(in the sense of* (2.327)) *such that* $y_1 - y_2 \in L^1((0,T)\times\mathbb{R}^d) \cap L^\infty(0,T;L^2)$ *and*

$$\lim_{t\to 0}\ \operatorname*{ess\,sup}_{s\in(0,t)} |(y_1(s)-y_2(s),\varphi)_2| = 0,\ \forall \varphi \in C_0^\infty(\mathbb{R}^d). \tag{2.338}$$

Then $y_1 \equiv y_2$. *If* $D \equiv 0$, *then Hypothesis* (p) *can be relaxed to*

(p)′ $\beta \in C^1(\mathbb{R}),\ \beta'(r) \geq 0,\ \forall r \in \mathbb{R},\ \beta(0)=0.$

The proof is similar to that of Theorem 2.6, so it will be outlined only (see [17] for details). Also, in this case, replacing β and b by β_N and b_N, respectively, may assume that (2.159)–(2.161) hold. We set

$$\begin{aligned}
\Phi_\varepsilon(y) &= (\varepsilon I + (-\Delta)^s)^{-1}y,\ \forall y \in L^2,\\
z = y_1 - y_2,\ w &= \beta(y_1)-\beta(y_2),\ b^*(y_i) \equiv b(y_i)y_i,\ i=1,2,\\
z_\varepsilon = z*\theta_\varepsilon,\ w_\varepsilon &= w*\theta_\varepsilon,\ \zeta_\varepsilon = (D(b^*(y_1)-b^*(y_2)))*\theta_\varepsilon,
\end{aligned}$$

and note that

$$(\Phi_\varepsilon(z_\varepsilon))_t = -(-\Delta)^s\Phi_\varepsilon(w_\varepsilon) - \mathrm{div}\Phi_\varepsilon(\zeta_\varepsilon) = 0 \text{ in } \mathcal{D}'(0,T;L^2).$$

We set $h_\varepsilon(t) = (\Phi_\varepsilon(z_\varepsilon(t)), z_\varepsilon(t))_2$ and get that

$$0 \le h_\varepsilon(t) \le h_\varepsilon(0+) + 2\varepsilon \int_0^t (\Phi_\varepsilon(z_\varepsilon(s)), w_\varepsilon(s))_2 ds - 2\alpha_3 \int_0^t |w_\varepsilon(s)|_2^2 ds$$
$$+2\alpha_1 |D|_\infty \int_0^t |\nabla \Phi_\varepsilon(z_\varepsilon(s))|_2 |w_\varepsilon(s)|_2 ds + 2\int_0^t |\gamma_\varepsilon(s)| ds, \ \forall t \in [0, T]. \tag{2.339}$$

We have

$$|\nabla \Phi_\varepsilon(z_\varepsilon(t))|_2^2 \le R^{2(1-s)} \int_{[|\xi| \le R]} \frac{|\mathscr{F}(z_\varepsilon(t))(\xi)|^2}{\varepsilon + |\xi|^{2s}} d\xi$$
$$+ \int_{[|\xi| \ge R]} |\mathscr{F}(z_\varepsilon(t))(\xi)|^2 |\xi|^{2(1-2s)} d\xi$$
$$\le R^{2(1-s)} h_\varepsilon(t) + R^{2(1-2s)} |z_\varepsilon(t)|_2^2, \ \forall t \in (0, T), \ R > 0, \tag{2.340}$$

because $2s \ge 1$. Taking into account (2.339), it remains to prove that

$$\lim_{\varepsilon \to 0} \varepsilon (\Phi_\varepsilon(z_\varepsilon(t)), w_\varepsilon(t))_2 = 0, \ \text{a.e. } t \in (0, T).$$

and to this end it suffices to show that

$$\lim_{\varepsilon \to 0} \varepsilon |\Phi_\varepsilon(z_\varepsilon(t))|_\infty = 0, \ \text{a.e. } t \in (0, T),$$

which follows by Lemma A.6 in [17]. Then, we get, for λ, $R > 0$, suitably chosen,

$$0 \le h_\varepsilon(t) \le \eta_\varepsilon(t) + C \int_0^t h_\varepsilon(s) ds, \quad \text{for } t \in [0, T],$$

where $C > 0$ is independent of ε and $\lim_{\varepsilon \to 0} \eta_\varepsilon(t) = 0$ for all $t \in [0, T]$. This implies that $h_\varepsilon(t) \to 0$ as $\varepsilon \to 0$ for every $t \in [0, T]$. Thus, $0 = \lim_{\varepsilon \to 0} z_\varepsilon(t) = z(t)$ in S' for a.e. $t \in (0, T)$, which implies $y_1 \equiv y_2$.

We also have the following linearized uniqueness theorem.

Theorem 2.16 *Under assumptions of Theorem* 2.15*, let* $T > 0$*,* $u \in L^\infty((0, T) \times \mathbb{R}^d)$ *and let* $y_1, y_2 \in L^\infty((0, T) \times \mathbb{R}^d)$ *with* $y_1 - y_2 \in L^1((0, T) \times \mathbb{R}^d) \cap L^\infty(0, T; L^2)$ *be two distributional solutions to the equation*

$$y_t + (-\Delta)^s \left(\frac{\beta(u)}{u} y \right) + \text{div}(yDb(u)) = 0 \ \textit{in} \ \mathscr{D}'((0, T) \times \mathbb{R}^d),$$
$$y(0) = u_0,$$

where u_0 is a measure of finite variation on $\mathbb{R}^d$ and $\frac{\beta(0)}{0} := 0$. If (2.338) *holds, then* $y_1 \equiv y_2$.

The proof is completely similar to that of Theorem 2.7, so it will be omitted.

As regards the existence of a weak solution to (2.328) we have by Theorem 2.14 and [89] (see also Remark 5.2 below)

Theorem 2.17 *Assume that Hypotheses* (n)–(nnn) *hold and let $u_0 \in L^1 \cap L^2$. Assume that either $u_0 \in \overline{D(A)}$ or that $\beta \in C^\infty(\mathbb{R})$ and let u be the solution of* (2.324) *from Theorem* 2.14*. Then, there exists a stochastic basis $\mathbb{B} := (\Omega, \mathcal{F}, (\mathcal{F}_t)_{t\geq 0}, \mathbb{P})$ and a d-dimensional isotropic $2s$-stable process L with Lévy measure $\frac{dz}{|z|^{d+2s}}$ as well as an $(\mathcal{F}_t)$-adapted càdlàg process (X_t) on Ω such that, for*

$$\mathcal{L}_{X_t}(x) := \frac{d(\mathbb{P} \circ X_t^{-1})}{dx}(x),\ t \geq 0, \tag{2.341}$$

we have

$$\begin{aligned} dX_t &= D(X_t)b(\mathcal{L}_{X_t}(X_t))dt + \left(\frac{\beta(\mathcal{L}_{X_t}(X_{t-}))}{\mathcal{L}_{X_t}(X_{t-})}\right)^{\frac{1}{2s}} dL_t, \\ \mathcal{L}_{X_0} &= u_0. \end{aligned} \tag{2.342}$$

Furthermore,

$$\mathcal{L}_{X_t} = u(t,\cdot),\ t \geq 0, \tag{2.343}$$

in particular, $((t,x) \mapsto \mathcal{L}_{X_t}(x)) \in L^\infty([0,T] \times \mathbb{R}^d)$ for every $T > 0$.

Proof By the well known formula that

$$(-\Delta)^s f(x) = -c_{d,s} P.V. - \int_{\mathbb{R}^d} (f(x+z) f(x)) \frac{dz}{|z|^{d+2s}}$$

with $c_{d,s} \in (0,\infty)$ (see [91, Section 13]), and since, as an easy calculation shows,

$$\int_A \frac{\beta(u(t,x))}{u(t,x)} \frac{dz}{|z|^{d+2s}} = \int_{\mathbb{R}^d} \mathbf{1}_A \left(\left(\frac{\beta(u(t,x))}{u(t,x)} \right)^{\frac{1}{2s}} z \right) \frac{dz}{|z|^{d+2s}}, \quad A \in B(\mathbb{R}^d \setminus \{0\}), \tag{2.344}$$

we have

$$\begin{aligned} &\frac{\beta(u(t,x))}{u(t,x)} (-\Delta)^s f(x) \\ &= -c_{d,s} P.V. \int_{\mathbb{R}^d} \left(f\left(x + \left(\frac{\beta(u(t,x))}{u(t,x)} \right)^{\frac{1}{2s}} z \right) - f(x) \right) \frac{dz}{|z|^{d+2s}}. \end{aligned} \tag{2.345}$$

As is easily checked, Hypotheses (i)–(iii) imply that condition (1.18) in [89] holds. Furthermore, for u as defined in the assertion

$$\mu_t(dx) := u(t,x)(dx),\ t \geq 0,$$

solves the Fokker–Planck equation (2.324) with $u_0(dx) := u_0(x)dx$. Hence, by [89, Theorem 1.5], (2.344), (2.345) and [66, Theorem 2.26, p. 157], there exists a stochastic basis $\mathbb{B}$ and $(X_t)_{t\geq 0}$ as in the assertion of the theorem, as well as a Poisson random measure N on $\mathbb{R}^d \times [0,\infty)$ with intensity $|z|^{-d-2s}dz\,dt$ on the stochastic basis $\mathbb{B}$ such that, for

$$L_t := \int_0^t \int_{|z|\leq 1} z\widetilde{N}(dz\,ds) + \int_0^t \int_{|z|>1} zN(dz\,ds), \tag{2.346}$$

(2.341), (2.342), and (2.343) hold. Here,

$$\widetilde{N}(dz\,dt) := N(dz\,dt) - |z|^{-d-2s}dz\,dt.$$

By Theorems 2.15 and 2.16, we obtain weak uniqueness for solutions to the McKean–Vlasov equation (2.328). Namely,

Theorem 2.18 *Under assumptions of Theorem* 2.17 *let* (X_t) *and* $(\widetilde{X}_t)$ *be two càdlàg processes on two (possibly different) stochastic bases* $\mathbb{B}, \widetilde{\mathbb{B}}$ *that are weak solutions to* (2.328)*. Assume that*

$$\left((t,x) \mapsto \mathcal{L}_{X_t}(x)\right), \left((t,x) \mapsto \mathcal{L}_{\widetilde{X}_t}(x)\right) \in L^\infty((0,T)\times\mathbb{R}^d).$$

Then X *and* $\widetilde{X}$ *have the same laws, i.e.,*

$$\mathbb{P}\circ X^{-1} = \widetilde{\mathbb{P}}\circ \widetilde{X}^{-1}.$$

Proof Clearly, by Dynkin's formula, both

$$\mu_t(dx) := \mathcal{L}_{X_t}(x)dx \text{ and } \widetilde{\mu}_t(dx) := \mathcal{L}_{\widetilde{X}_t}(x)dx$$

solve the NFPL (2.324) with the same initial condition $u_0(dx) := u_0(x)dx$, hence satisfy (2.338) with $y_1(t) := \mathcal{L}_{X_t}$ and $y_2 := \mathcal{L}_{\widetilde{X}_i}$.

Therefore, by Theorem 2.15,

$$\mathcal{L}_{X_t} = \mathcal{L}_{\widetilde{X}_t} \quad \text{for all } t \geq 0,$$

since $t \mapsto \mathcal{L}_{X_t}(x)dx$ and $t \mapsto \mathcal{L}_{\widetilde{X}_t}(x)dx$ are both narrowly continuous and are probability measures for all $t \geq 0$, so both are in $L^\infty(0,T;L^1\cap L^\infty) \subset L^\infty(0,T;L^2)$.

Now, consider the liner Fokker–Planck equation

$$\begin{aligned} v_t + (-\Delta)^s \frac{\beta(\mathscr{L}_{X_t})}{\mathscr{L}_{X_t}} v + \operatorname{div}(Db(\mathscr{L}_{X_t})v) = 0, \\ v(0, x) = u_0(x), \end{aligned} \tag{2.347}$$

again in the weak (distributional) sense analogous o (1.7). Then, by Theorem 2.16 we conclude that $\mathscr{L}_{X_t}$, $t \in [0, T]$, is the unique solution to (2.347) in $L^\infty(0, T; L^1 \cap L^\infty)$. Both $P \circ X^{-1}$ and $\widetilde{P} \circ \widetilde{X}^{-1}$ also solve the martingale problem with initial condition $u_0(dx) := u_0(x)dx$ for the linear Kolmogorov operator

$$K_{\mathscr{L}_{X_t}} := \frac{\beta(\mathscr{L}_{X_t})}{\mathscr{L}_{X_t}} (-\Delta)^s + b(\mathscr{L}_{X_t})D \cdot \nabla.$$

Since the above is true for all $u_0 \in L^1 \cap L^\infty$, and also holds when we consider (2.324) and (2.347) with start in any $s_0 > 0$ instead of zero, it follows by exactly the same arguments as in the proof of Lemma 2.12 in [98] that

$$P \circ X^{-1} = \widetilde{P} \circ \widetilde{X}^{-1}.$$

Remark 2.7 Let for $s \in [0, \infty)$ and $\zeta \in \mathscr{P}_0 := \{\zeta \equiv \zeta(x)dx \mid \zeta \in L^1 \cap L^\infty, \zeta \geq 0, |\zeta|_1 = 1\}$

$$P_{s,\zeta} := P \circ X^{-1}(s, \zeta),$$

where $(X_t(s, \zeta))_{t \geq s}$ on a stochastic basis $\mathbb{B}$ denotes the solution of (2.341) and (2.342) with initial condition ζ at s. Then, by Theorems 2.15–2.17, exactly the same way as Theorem 5.2 below, one proves that $P_{s,\zeta}$, $(s, \zeta) \in [0, \infty) \times \mathscr{P}_0$, form a nonlinear Markov process in the sense of Definition 5.1. We refer to Sect. 5.2 below, in particular, Remark 5.5, and to [88], in particular, Section 4.2, Example (iii).

2.9 A Splitting Formula for NFPE

Consider herein the nonlinear Fokker–Planck equation

$$\begin{aligned} &\rho_t - \Delta\beta(\rho) + \operatorname{div}(a(\rho)\rho) = 0 \text{ in } (0, \infty) \times \mathbb{R}^d, \\ &\rho(0, x) = \rho_0(x), \qquad\qquad x \in \mathbb{R}^d, \end{aligned} \tag{2.348}$$

where the functions $\beta : \mathbb{R} \to \mathbb{R}$ and $a : \mathbb{R} \to \mathbb{R}^d$ are assumed to satisfy the following hypotheses:

(i) $\beta \in C^1(\mathbb{R})$; $\beta'(r) > 0$, $\forall r \in \mathbb{R}$, $\beta(0) = 0$,

$$|\beta(r)| \leq \alpha_1 |r|, \ \forall r \in \mathbb{R}, \tag{2.349}$$

where $\alpha_1 > 0$.

(ii) *$a \in C^1(\mathbb{R};\mathbb{R}^d) \cap C_b(\mathbb{R};\mathbb{R}^d)$ and*

$$|a^*(r) - a^*(\bar{r})| \le \alpha_2 |\beta(r) - \beta(\bar{r})|, \ \forall r, \bar{r} \in \mathbb{R}, \tag{2.350}$$

where $\alpha_1 > 0$ and $a^*(r) \equiv a(r)r$.

We note that Eq. (2.348) is of the form (2.1) and by Theorem 2.1 it has a mild solution. However, under assumptions (i), (ii), it follows also uniqueness of mild solutions. Namely, if we denote by $A : D(A) \subset L^1 \to L^1$ the operator

$$\begin{aligned} A(y) &= -\Delta\beta(y) + \mathrm{div}(a(y)y) \text{ in } \mathcal{D}'(\mathbb{R}^d), \ y \in D(A), \\ D(A) &= \{y \in L^1; -\Delta\beta(y) + \mathrm{div}(a(y)y) \in L^1\}, \end{aligned} \tag{2.351}$$

then, arguing as in the proof of Lemma 2.6, it follows that A is m-accretive in L^1 (see Proposition 3.1 in [8]). Hence, the Cauchy problem

$$\begin{aligned} &\frac{d\rho}{dt} + A(\rho) = 0, \ \ t \ge 0, \\ &\rho(0) = \rho_0, \end{aligned} \tag{2.352}$$

has for each $\rho_0 \in \overline{D(A)}$ (the closure of $D(A)$ in L^1) a unique *mild solution* $\rho \in C([0,\infty); L^1)$, that is,

$$\rho(t) = \lim_{n\to\infty} \left(I + \frac{t}{n}A\right)^{-n} \rho_0 \ \text{ in } L^1, \ t \ge 0. \tag{2.353}$$

Formally, Eq. (2.352) can be written as

$$\begin{aligned} &\frac{d\rho}{dt} + A_1(\rho) + A_2(\rho) = 0 \ \text{ in } (0,\infty), \\ &\rho(0) = \rho_0, \end{aligned} \tag{2.354}$$

where $A_1 : D(A_1) \subset L^1 \to L^1$ is defined by

$$A_1(y) = -\Delta\beta(y), \ y \in D(A_1) = \{y \in L^1; \Delta\beta(y) \in L^1\}, \tag{2.355}$$

while $A_2 : D(A_2) \subset L^1 \to L^1$ is the (multivalued) operator defined by $f \in A_2(y)$, $y \in D(A_2)$ if

$$\mathrm{div}((a^*(y) - a^*(k))\mathrm{sign}(y-k)) \le f \ \text{ in } \mathcal{D}'(\mathbb{R}^d), \ \forall k \in \mathbb{R}. \tag{2.356}$$

It is well known (see, e.g., [4], p. 118) that the operator A_1 is m-accretive in L^1 and so it generates the C_0-semigroup of contractions, $S_1(t) : L^1 \to L^1$, given by

$$S_1(t)\rho_0 = \lim_{n\to\infty} \left(I + \frac{t}{n} A_1\right)^{-n} \rho_0 \text{ in } L^1, \ \forall t \geq 0, \tag{2.357}$$

which is by definition the *mild solution* to the porous media equation

$$\begin{aligned} &p_t - \Delta\beta(p) = 0 \text{ in } (0,\infty) \times \mathbb{R}^d, \\ &p(0) = p_0 \text{ in } \mathbb{R}^d. \end{aligned} \tag{2.358}$$

As shown by M.G. Crandall [42] (see, also, [5], p. 120), also the operator A_2 is m-accretive in L^1, and so it generates a continuous semigroup of contractions $S_{A_2}(t)$ on $\overline{D(A_2)} = L^1$. Hence, for each $q_0 \in L^1$, the Cauchy problem

$$\begin{aligned} &\frac{dq}{dt} + A_2(q) = 0, \ \ t \geq 0, \\ &q(0) = q_0, \end{aligned} \tag{2.359}$$

has a unique mild solution $q \in S_{A_2}(t)q_0 = \lim\limits_{n\to\infty} \left(I + \frac{t}{n} A_2\right)^{-n} q_0$ in L^1 which is just the *Kružkov entropy solution* to the conservation law equation

$$\begin{aligned} &q_t + \operatorname{div}(a^*(q)) = 0 \text{ in } (0,\infty) \times \mathbb{R}^d, \\ &q(0,x) = q_0(x), \ \ x \in \mathbb{R}^d, \end{aligned} \tag{2.360}$$

that is (see [42, 69]), $q \in C([0,\infty); L^1)$, $q(0,x) = q_0(x)$, $\forall x \in \mathbb{R}^d$, and

$$|q-k|_t + \operatorname{div}(\operatorname{sign}(q-k)(a^*(q) - a^*(k))) \leq 0 \text{ in } \mathcal{D}'((0,\infty) \times \mathbb{R}^d), \tag{2.361}$$

for all $k \in \mathbb{R}$.

It should be mentioned that, though formally we have $A_1 + A_2 \subset A$, one cannot say that $A_1 + A_2 = A$ (more precisely, $A_1 + A_2$ is not necessarily m-accretive) and so (2.354) is not equivalent with (2.352).

However, the problem we address herein is the validity of the Trotter product formula

$$S_A(t)\rho_0 = \lim_{n\to\infty} \left[S_{A_1}\left(\frac{t}{n}\right) S_{A_2}\left(\frac{t}{n}\right)\right]^n \rho_0, \ \text{ in } L^1, \ \forall \rho_0 \in L^1, \tag{2.362}$$

uniformly in t on compact intervals. Indeed, we have

Theorem 2.19 *Under Hypotheses* (i), (ii) *the Trotter product formula* (2.362) *holds for all* $\rho_0 \in L^1$.

Proof We note first that (2.362) can be equivalently expressed for each $T > 0$

$$S_A(t)\rho_0 = \lim_{h\to 0} \rho_h(t) \text{ in } L^1;\ \forall t \in [0, T], \tag{2.363}$$

where $\rho_h : [0, T] \to L^1$ is the step function

$$\rho_h(t) = \rho_h^j,\ \forall t \in [jh, (j+1)h),\ j = 0, 1, \dots, N_h = \left[\frac{T}{h}\right], \tag{2.364}$$

$$\begin{aligned} &(\rho_h^j)_t - \Delta\beta(\rho_h^j) = 0 \text{ in } (jh, (j+1)h) \times \mathbb{R}^d, \\ &\rho_h(jh, x) = q_h^j(h, x),\ x \in \mathbb{R}^d, \end{aligned} \tag{2.365}$$

$$\begin{aligned} &(q_h^j)_t + \operatorname{div}(a^*(q_h^j)) = 0 \text{ in } (0, h) \times \mathbb{R}^d, \\ &q_h^j(0, x) = \rho_h^j(jh - 0, x);\ q_h^0(0, x) = \rho_0(x),\ x \in \mathbb{R}^d. \end{aligned} \tag{2.366}$$

Herein, the solution ρ_h^j to the porous media equation (2.365) is considered in the mild sense, while q_h^j is the entropy solution to the conservation law equation (2.366). This algorithm is splitting the Fokker–Planck flow into two parts corresponding to an anomalous diffusion part and the conservation law flow.

As seen earlier, $S_{A_2}(t)$ is generated in L^1 by the m-accretive operator A_2 defined by (2.356).

For each $\varepsilon > 0$, consider the operator $A_2^\varepsilon : D(A_2^\varepsilon) \subset L^1 \to L^1$,

$$\begin{aligned} &A_2^\varepsilon(y) = -\varepsilon\Delta y + \operatorname{div}(a^*(y)),\ \forall y \in D(A_2^\varepsilon), \\ &D(A_2^\varepsilon) = \{y \in L^1 \cap H^1, -\varepsilon\Delta y + \operatorname{div}(a^*(y)) \in L^1\}. \end{aligned} \tag{2.367}$$

As proved in [42], the operator A_2^ε is m-accretive in L^1 and so it generates a continuous semigroup of contractions $S_2^\varepsilon(t) = \exp(-tA_2^\varepsilon)$ on L^1. We have

Lemma 2.16 *For all $\lambda \in (0, \lambda_0)$, where $\lambda_0 > 0$ and $f \in L^1$,*

$$\lim_{\varepsilon\to 0} (I + \lambda A_2^\varepsilon)^{-1} f = (I + \lambda A_2)^{-1} f \ \ in\ L^1. \tag{2.368}$$

Moreover, for all $y_0 \in L^1$,

$$\lim_{\varepsilon\to 0} S_2^\varepsilon(t)y_0 = S_{A_2}(t)y_0 \ \ in\ L^1,\ t \ge 0, \tag{2.369}$$

uniformly in t on compact intervals and

$$|S_{A_2}(t)y_0 - S_{A_2}(t)\bar{y}_0|_1 \le |y_0 - \bar{y}_0|_1,\ \forall t \ge 0,\ \forall y_0, \bar{y}_0 \in L^1. \tag{2.370}$$

Proof First, we note that (2.369) follows by (2.368) via the Trotter–Kato theorem. On the other hand, it was proved in [42] (see also [5], p. 125) that (2.368) holds in L^1_{loc}, so it suffices to check that

$$\lim_{N\to\infty}\int_{[|x|\ge N]}|(I+\lambda A^\varepsilon_2)^{-1}f|dx=0,\ \ \forall\, f\in L^1, \tag{2.371}$$

uniformly in ε. Taking into account that

$$|(I+\lambda A^\varepsilon_2)^{-1}f_1-(I+\lambda A^\varepsilon_2)^{-1}f_2|_1\le|f_1-f_2|_1,\ \forall f_1,f_2\in L^1,$$

it suffices to prove (2.371) for $f\in C^\infty_0(\mathbb{R}^d)$.

To this purpose, we set $y_\varepsilon=(I+\lambda A^\varepsilon_2)^{-1}f$, that is,

$$y_\varepsilon-\lambda\varepsilon\Delta y_\varepsilon+\lambda\operatorname{div}(a^*(y_\varepsilon))=f\ \text{ in }\mathbb{R}^d. \tag{2.372}$$

It is easily seen that $y_\varepsilon\in H^1\cap L^1$, $|y_\varepsilon|_1\le|f|_1$ and

$$|y_\varepsilon|^2_2+\lambda\varepsilon|\nabla y_\varepsilon|^2_2\le C|f|^2_2,\ \forall\lambda\in(0,\lambda_0). \tag{2.373}$$

Consider now the function

$$\Phi(x)\equiv(1+|x|^2)^{\frac12},\ \ \varphi_\theta=\Phi\exp(-\theta\Phi),\ \theta>0. \tag{2.374}$$

If we multiply (2.372) by $\mathscr{X}_\delta(y_\varepsilon)\varphi_\theta$ and integrate on $\mathbb{R}^d$, we get as in the proof of Lemma 2.9

$$\begin{aligned}
\int_{\mathbb{R}^d}y_\varepsilon\varphi_\theta\mathscr{X}_\delta(y_\varepsilon)dx&=-\lambda\varepsilon\int_{\mathbb{R}^d}\nabla y_\varepsilon\cdot\nabla(\varphi_\theta\mathscr{X}_\delta(y_\varepsilon))dx\\
&+\lambda\int_{\mathbb{R}^d}a^*(y_\varepsilon)\cdot(\mathscr{X}_\delta(y_\varepsilon)\nabla\varphi_\theta+\varphi_\theta\mathscr{X}'_\delta(y_\varepsilon)\nabla y_\varepsilon)dx+\int_{\mathbb{R}^d}\varphi_\theta f\mathscr{X}_\delta(y_\varepsilon)dx\\
&\le-\lambda\varepsilon\int_{\mathbb{R}^d}(\nabla y_\varepsilon\cdot\nabla\varphi_\theta)\mathscr{X}_\delta(y_\varepsilon)dx+\lambda\int_{\mathbb{R}^d}|a^*(y_\varepsilon)|\,|\nabla\varphi_\theta|dx\\
&-\lambda\varepsilon\int_{\mathbb{R}^d}(\nabla\varphi_\theta\cdot\nabla y_\varepsilon)\mathscr{X}_\delta(y_\varepsilon)dx\\
&+\lambda\int_{\mathbb{R}^d}|a^*(y_\varepsilon)|\,|\varphi_\theta|\mathscr{X}'_\delta(y_\varepsilon)|\nabla y_\varepsilon|dx+\int_{\mathbb{R}^d}|\varphi_\theta|\,|f|dx.
\end{aligned}$$

As seen earlier, we have

$$\lim_{\delta\to0}\int_{\mathbb{R}^d}|a^*(y_\varepsilon)|\varphi_\theta|\mathscr{X}'_\delta(y_\varepsilon)|\,|\nabla y_\varepsilon|dx\le|a|_\infty\lim_{\delta\to0}\frac1\delta\int_{[|y_\varepsilon|\le\delta]}|y_\varepsilon|\,|\varphi_\theta|\,|\nabla y_\varepsilon|dx=0.$$

Then, keeping in mind that $\nabla y_\varepsilon \cdot \mathscr{X}_\delta(y_\varepsilon) = \nabla j_\delta(y_\varepsilon)$, $j_\delta(r) = \int_0^r \mathscr{X}_\delta(s)ds$, letting $\delta \to 0$ and taking into account (2.373), we get that

$$\begin{aligned}\int_{\mathbb{R}^d} \varphi_\theta(x)|y_\varepsilon(x)|dx &\le \int_{\mathbb{R}^d} \varphi_\theta(x)|f(x)|dx \\ &\quad +\lambda \int_{\mathbb{R}^d} |a^*(y_\varepsilon)|\,|\nabla\varphi_\theta|dx + \lambda\varepsilon|y_\varepsilon|_1|\Delta\varphi_\theta|_\infty \\ &\le \int_{\mathbb{R}^d} \varphi_\theta(x)|f(x)|dx + \lambda|a|_\infty|y_\varepsilon|_1|\nabla\varphi_\theta|_\infty + \lambda\varepsilon|y_\varepsilon|_1|\Delta\varphi_\theta|_\infty.\end{aligned} \tag{2.375}$$

On the other hand, we have

$$\begin{aligned}\nabla\varphi_\theta &= (\nabla\Phi - \theta\Phi\nabla\Phi)\exp(-\theta\Phi) \\ \Delta\varphi_\theta &= (\Delta\Phi - 2\theta|\nabla\Phi|^2 - \theta\Phi\Delta\Phi + \theta^2\Phi|\nabla\Phi|^2)\exp(-\theta\Phi).\end{aligned} \tag{2.376}$$

Then, letting $\theta \to 0$ in (2.375) and recalling that $|y_\varepsilon|_1 \le |f|_1$, we get by (2.376) that

$$\begin{aligned}&\int_{\mathbb{R}^d} \Phi(x)|y_\varepsilon(x)|dx \\ &\quad\le \int_{\mathbb{R}^d} \Phi(x)|f(x)|dx + \lambda|a|_\infty|y_\varepsilon|_1|\nabla\Phi|_\infty + \lambda\varepsilon|f|_1|\Delta\Phi|_\infty \\ &\quad< \int_{\mathbb{R}^d} \Phi(x)|f(x)|dx + \lambda|a|_\infty|f|_1|\nabla\Phi|_\infty + \lambda\varepsilon|f|_1|\Delta\Phi|_\infty,\ \forall\varepsilon > 0.\end{aligned} \tag{2.377}$$

This yields, for some C independent of ε,

$$\int_{[\|x\|\ge N]} |y_\varepsilon(x)|dx \le C(\inf\{\Phi(x);\ |x| \ge N\})^{-1},\ \forall N > 0,$$

which implies (2.371), as claimed.

We note that also in the case of the semigroup $S_{A_2}(t)$ we have

$$S_{A_1}(t)y_0 = \lim_{\varepsilon\to 0} S_1^\varepsilon(t)y_0 \text{ in } L^1,\ \forall t \ge 0, \tag{2.378}$$

where $S_1^\varepsilon(t) = \exp(-tA_1^\varepsilon)$, $A_1^\varepsilon(y) = -\varepsilon\Delta y - \Delta\beta(y)$, $D(A_1^\varepsilon) = L^1 \cap H^1$. By (2.378), we have that

$$|S_{A_1}(t)y_0 - S_{A_1}(t)\bar{y}_0|_1 \le |y_0 - \bar{y}_0|_1,\ \forall t \ge 0,\ y_0, \bar{y}_0 \in L^1, \tag{2.379}$$

as claimed. By (2.369) and (2.378), it also follows that

$$S_{A_i}(t)(\mathscr{P}) \subset \mathscr{P}, \ \ \forall t \geq 0, \ i = 1, 2, \tag{2.380}$$

$$|S_{A_i}(t)y_0|_p \leq |y_0|_p, \ \ \forall t \geq 0, \ y_0 \in L^p, \ 1 \leq p \leq \infty, \tag{2.381}$$

for $i = 1, 2$.

Proof of Theorem 2.19 (Continued) We consider the family of continuous operators $F_\nu : L^1 \to L^1$,

$$F_\nu(y) = S_{A_1}(\nu)S_{A_2}(\nu)y, \ \ y \in L^1, \ \nu > 0. \tag{2.382}$$

For each $\nu > 0$, the operator $\nu^{-1}(I - F_\nu)$ is m-accretive in L^1 and so, by virtue of the Chernoff theorem for nonlinear semigroups of contractions in Banach spaces (see [33]), formula (2.362) follows by the lemma below. □

Lemma 2.17 *For all $\lambda > 0$ and $f \in L^1$, we have*

$$\lim_{\nu \to 0} \left(I + \lambda \frac{I - F_\nu}{\nu} \right)^{-1} f = (I + \lambda A)^{-1} f \ \text{ in } L^1. \tag{2.383}$$

Proof of Lemma 2.17 Let $f \in L^1$ and let $y_\nu \in L^1$ be defined by

$$y_\nu = \left(I + \lambda \frac{I - F_\nu}{\nu} \right)^{-1} f. \tag{2.384}$$

Equivalently,

$$(\lambda + \nu)y_\nu - \lambda S_{A_1}(\nu)S_{A_2}(\nu)y_\nu = \nu f. \tag{2.385}$$

We set $z_\nu(t) = S_{A_1}(t)S_{A_2}(\nu)y_\nu$ and note that by (2.385) and (2.358) it follows that

$$\begin{aligned} &(z_\nu)_t - \Delta\beta(z_\nu) = 0 \text{ in } (0, \nu) \times \mathbb{R}^d, \\ &z_\nu(0, x) = q_\nu(\nu, x), \ \ x \in \mathbb{R}^d, \end{aligned} \tag{2.386}$$

$$\begin{aligned} &(q_\nu)_t + \operatorname{div}(a^*(q_\nu)) = 0 \ \text{ in } (0, \nu) \times \mathbb{R}^d, \\ &q_\nu(0, x) = y_\nu(x), \ \ x \in \mathbb{R}^d, \end{aligned} \tag{2.387}$$

(Here $z_\nu = z_\nu(t, x)$ is the mild solution to the porous media equation (2.386) while $q_\nu = q_\nu(t, x)$ is the entropy solution to the conservation law equation (2.360).) Then, we may rewrite (2.385) as

$$(\lambda + \nu)y_\nu - \lambda z_\nu(\nu) = \nu f \ \text{ in } \mathbb{R}^d, \tag{2.388}$$

and so (2.383) reduces to

$$\lim_{\nu\to 0} y_\nu = y = (I+\lambda A)^{-1} f \ \text{ in } L^1. \tag{2.389}$$

Assume first that $f \in L^1 \cap L^2$. Then, by (2.388) and (2.381) we see that

$$|y_\nu|_i \le \frac{\lambda}{\lambda+\nu}|z_\nu(\nu)|_i + \frac{\nu}{\lambda+u}|f|_i \le \frac{\lambda}{\lambda+\nu}|y_\nu|_i + \frac{\nu}{\lambda+\nu}|f|_i,\ i=1,2.$$

This yields

$$\begin{aligned}&|y_\nu|_i \le |f|_i,\ \ \forall \nu>0,\ i=1,2.\\ &|z_\nu(t)|_i \le |y_\nu|_i \le |f|_i,\ i=1,2;\ t\ge 0.\end{aligned} \tag{2.390}$$

To prove (2.389), we shall show first that $\{y_\nu\}$ is relatively compact in $L^1_{\rm loc}$. To this end, we set

$$\begin{aligned}&f^k(x)\equiv f(x+k)-f(x),\ f_k(x)\equiv f(x+k),\ y^k_\nu(x)\equiv y_\nu(x+k)-y_\nu(x),\\ &q^k_\nu(x)\equiv q_\nu(x+k)-q_\nu(x),\ z^k_\nu(t,x)\equiv z_\nu(t,x+k)-z_\nu(t,x),\ \forall k\in\mathbb{R}^d,\ x\in\mathbb{R}^d.\end{aligned}$$

By (2.386) and (2.387) we have

$$\begin{aligned}&(z^k_\nu)_t - \Delta(\beta(z_\nu))^k = 0 \ \text{ in } (0,\nu)\times\mathbb{R}^d,\\ &z^k_\nu(0,\cdot) = q^k_\nu(\nu,\cdot) \qquad \text{in } \mathbb{R}^d,\end{aligned} \tag{2.391}$$

$$\begin{aligned}&(q^k_\nu)_t + {\rm div}(a^*(q_\nu))^k = 0 \ \text{ in } (0,\nu)\times\mathbb{R}^d,\\ &q^k_\nu(0,\cdot) = y^k_\nu \qquad \text{in } \mathbb{R}^d.\end{aligned} \tag{2.392}$$

Therefore,

$$\begin{aligned}z^k_\nu(t,x) &= S_{A_1}(t)(g_k)(x) - S_{A_1}(t)(g)(x),\\ q^k_\nu(t,x) &= S_{A_2}(t)(h_k)(x) - S_{A_2}(t)(h)(x),\end{aligned}$$

where $g_k(x)\equiv q_\nu(x+k)$, $g(x)\equiv q_\nu(x)$, $h_k\equiv y_\nu(x+k)$, $h(x)=y_\nu(x)$. Then, again by (2.370) and (2.379), we get

$$|z^k_\nu(t)|_1 \le |z^k_\nu(0)|_1 = |q^k_\nu(\nu)|_1,\ \forall t\in(0,\nu), \tag{2.393}$$

$$|q^k_\nu(t)|_1 \le |y^k_\nu|_1,\ \ \forall t\in(0,\nu). \tag{2.394}$$

This yields

$$|y_\nu^k|_1 \le \frac{\lambda}{\lambda+\nu}\,|z_\nu^k(\nu)|_1 + \frac{\nu}{\lambda+\nu}|f^k|_1 \le \frac{\lambda}{\lambda+\nu}|y_\nu^k|_1 + \frac{\nu}{\lambda+\nu}|f^k|_1,$$

and, therefore,

$$|y_\nu^k|_1 \le |f^k|_1 \to 0 \text{ as } \nu \to 0.$$

Then, by the Riesz–Kolmogorov theorem, it follows that the set $\{y_\nu\}_{\nu>0}$ is relatively compact in $L^1_{\rm loc}$ and, by (2.393) and (2.394), a similar conclusion for $\{z_\nu\}_{\nu>0}$ and $\{q_\nu\}_{\nu>0}$. Hence, on a subsequence $\{\nu'\} \to 0$,

$$\lim_{\nu'\to 0} y_{\nu'} = y \text{ in } L^1_{\rm loc}. \tag{2.395}$$

Let us prove now that (2.395) holds in L^1. To this purpose, we consider the function Φ defined above and get (see (2.377))

$$\int_{\mathbb{R}^d} |y_\lambda(x)|\Phi(x)|dx \le \frac{\lambda}{\lambda+\nu}\int_{\mathbb{R}^d}|z_\nu(\nu,x)|\Phi(x)dx + \frac{\nu}{\lambda+\nu}\int_{\mathbb{R}^d}|f(x)|\Phi(x)dx. \tag{2.396}$$

On the other hand, by (2.391) and (2.392), we have

$$\int_{\mathbb{R}^d} |z_\nu(t,x)|\Phi(x)dx \le \int_{\mathbb{R}^d}|z_\nu(0,x)|\Phi(x)dx + \mu_1 t|z_\nu(0)|_1,\ t\in[0,\nu], \tag{2.397}$$

$$\int_{\mathbb{R}^d} |q_\nu(t,x)|\Phi(x)dx \le \int_{\mathbb{R}^d}|q_\nu(0,x)|\Phi(x)dx + \mu_2 t|q_\nu(0)|_1,\ t\in[0,\nu], \tag{2.398}$$

where $\mu_i > 0$, $i = 1, 2$, are independent of t and $z_\nu(0)$, $q_\nu(0)$. Here is the argument. Taking into account that $S_{A_i}(t)z_0 = \lim_{n\to\infty}\left(I + \frac{t}{n}A_i\right)^{-n} z_0$, $i = 1, 2$, in L^1, for (2.397) and (2.398) it suffices to prove that

$$\int_{\mathbb{R}^d} |(I+\lambda A_1)^{-1}f|\Phi\,dx \le \int_{\mathbb{R}^d}|f|\Phi\,dx + \mu_1\lambda|f|_1,\ \forall f\in L^1,\ \lambda>0, \tag{2.399}$$

$$\int_{\mathbb{R}^d} |(I+\lambda A_2)^{-1}f|\Phi\,dx \le \int_{\mathbb{R}^d}|f|\Phi\,dx + \mu_2\lambda|f|_1,\ \forall\lambda>0,\ f\in L^1. \tag{2.400}$$

To this end, we set $z = (I + \lambda A_1)^{-1} f$, that is, $z - \lambda \Delta\beta(z) = f$ in $\mathscr{D}'(\mathbb{R}^d)$ and assume first that $f \in L^1 \cap L^\infty$. Then, $z \in L^1 \cap L^\infty, \nabla\beta(z) \in L^2$. Multiplying the latter by $\mathscr{X}_\delta(\beta(z))\varphi_\theta$, $\varphi_\theta = \Phi \exp(-\theta\Phi)$, and integrating on $\mathbb{R}^d$, we get

$$\begin{aligned}
\int_{\mathbb{R}^d} z\mathscr{X}_\delta(\beta(z))\varphi_\theta\, dx &= \int_{\mathbb{R}^d} f\mathscr{X}_\delta(\beta(z))\varphi_\theta\, dx - \lambda \int_{\mathbb{R}^d} (\nabla\beta(z)\cdot\nabla\varphi_\theta)\mathscr{X}_\delta(\beta(z))dx \\
&\quad -\lambda \int_{\mathbb{R}^d} (\nabla\beta(z)\cdot\nabla\mathscr{X}_\delta(\beta(z)))\varphi_\theta\, dx \\
&\le \int_{\mathbb{R}^d} f\mathscr{X}_\delta(\beta(z))\varphi_\theta\, dx - \lambda \int_{\mathbb{R}^d} \nabla j_\delta(\beta(z))\cdot\nabla\varphi_\theta\, dx \\
&= \int_{\mathbb{R}^d} f\mathscr{X}_\delta(\beta(z))\varphi_\theta\, dx + \lambda \int_{\mathbb{R}^d} j_\delta(\beta(z))\Delta\varphi_\theta\, dx,
\end{aligned}$$

where $j_\delta(r) = \int_0^r \mathscr{X}_\delta(s)ds$. □

Now, taking into account (2.376) and hypotheses (2.349), it follows for $\theta \to 0$,

$$\int_{\mathbb{R}^d} z\mathscr{X}_\delta(\beta(z))\Phi\, dx \le \int_{\mathbb{R}^d} |f|\Phi\, dx + \alpha_1\lambda|f|_1|\Delta\Phi|_\infty,\ \forall\lambda > 0,$$

because $|z|_1 \le |f|_1$. Then, for $\delta \to 0$ we get (2.399) for $\mu_1 = \alpha_1|\Delta\Phi|_\infty$, $f \in L^1 \cap L^\infty$. By density this extends to all $f \in L^1$. Similarly, it follows by (2.377) that

$$\overline{\lim_{\varepsilon\to 0}} \int_{\mathbb{R}^d} |(I + \lambda A_2^\varepsilon)^{-1} f|\Phi\, dx \le \int_{\mathbb{R}^d} |f|\Phi\, dx + \lambda|a|_\infty|\nabla\Phi|_\infty|f|_1,\ \forall\varepsilon > 0.$$

Taking into account (2.368), it follows via Fatou's lemma that (2.400) holds. In particular, by (2.398) and (2.387) it follows that

$$\int_{\mathbb{R}^d} |q_\nu(\nu, x)|\Phi(x)dx \le \int_{\mathbb{R}^d} |y_\nu(x)|\Phi(x)dx + \mu_2\nu|y_\nu|_1, \tag{2.401}$$

and so, by (2.386) and (2.397)

$$\begin{aligned}
\int_{\mathbb{R}^d} |z_\nu(\nu, x)|\Phi(x)dx &\le \int_{\mathbb{R}^d} |q_\nu(\nu, x)|\Phi(x)dx + \mu_1\nu|q_\nu(\nu)|_1 \\
&\le \int_{\mathbb{R}^d} |y_\nu(x)|\Phi(x)dx + (\mu_1 + \mu_2)\nu|y_\nu|_1.
\end{aligned}$$

Substituting in (2.396), we get

$$\begin{aligned}
\int_{\mathbb{R}^d} |y_\nu(x)|\Phi(x)dx &\le \int_{\mathbb{R}^d} |f(x)|\Phi(x)dx + \lambda(\mu_1 + \mu_2)|y_\nu|_1 \\
&\le \int_{\mathbb{R}^d} |f(x)|\Phi(x)dx + \lambda(\mu_1 + \mu_2)|y_\nu|_1,\ \ \forall\nu > 0.
\end{aligned}$$

Then, as seen above, this implies by (2.395) $y_{\nu_n} \to y$ strongly in L^1 on a subsequence $\{\nu_n\} \to 0$ and so (2.389) follows. It also follows by (2.387) that

$$q_{\nu_n}(t) \to S_{A_1}(t)y \text{ in } L^1, \ \forall\, t \geq 0. \tag{2.402}$$

Let us prove now that such a function y, that is,

$$y = \lim_{n\to\infty} y_{\nu_n} \text{ in } L^1 \tag{2.403}$$

is a solution to the equation $(I + \lambda A)y = f$. To this end, we rewrite (2.388) as

$$y_\nu + \frac{\lambda}{\nu}(y_\nu - q_\nu(\nu)) - \frac{\lambda}{\nu}(z_\nu(\nu) - q_\nu(\nu)) = f,$$

where (z_ν, q_ν) is the solution to (2.386) and (2.387). This yields

$$\begin{aligned}\int_{\mathbb{R}^d} y_\nu\varphi\, dx + \frac{\lambda}{\nu}\int_{\mathbb{R}^d}(y_\nu - q_\nu(\nu))\varphi\, dx - \frac{\lambda}{\nu}\int_{\mathbb{R}^d}(z_\nu(\nu) - q_\nu(\nu))\varphi\, dx \\ = \int_{\mathbb{R}^d} f\,\varphi\, dx, \ \ \forall\, \varphi \in C_0^\infty(\mathbb{R}^d).\end{aligned} \tag{2.404}$$

By (2.387), we have

$$\begin{aligned}\frac{1}{\nu}\int_{\mathbb{R}^d}(y_\nu - q_\nu(\nu))\varphi\, dx + \frac{1}{\nu}\int_0^\nu\int_{\mathbb{R}^d} a^*(q_\nu(t))\cdot\nabla\varphi\, dt\, dx = 0, \\ \forall\varphi \in C_0^\infty(\mathbb{R}^d),\end{aligned} \tag{2.405}$$

and so, for $\nu_n \to 0$, we get by (2.402)

$$\lim_{n\to\infty}\frac{1}{\nu_n}\int_{\mathbb{R}^d}(y_{\nu_n} - q_{\nu_n}(\nu_n))\varphi\, dx = -\int_{\mathbb{R}^d} a^*(y)\cdot\nabla\varphi\, dx, \ \forall\,\varphi \in C_0^\infty(\mathbb{R}^d). \tag{2.406}$$

(To get rigorously (2.406), we use the fact that q_ν is the mild solution to (2.387), and so it is limit in $L^\infty(0, T; L^1)$ of the associated finite difference scheme or, by (2.369), of the solution q_ν^ε to the approximating equation

$$\frac{dq_\nu^\varepsilon}{dt} + A_2^\varepsilon(q_\nu^\varepsilon) = 0, \ t \in (0, \nu).)$$

Similarly, taking into account (2.378), it follows by (2.386) and (2.402) that, as $n \to \infty$,

$$\begin{aligned}\frac{1}{\nu_n}\int_{\mathbb{R}^d}(z_{\nu_n}(\nu) - q_{\nu_n}(\nu))\varphi\, dx = \frac{1}{\nu_n}\int_0^{\nu_n}\int_{\mathbb{R}^d}\beta(z_{\nu_n}(t))\Delta\varphi\, dx\, dt \to \int_{\mathbb{R}^d}\beta(y)\Delta\varphi\, dx \\ \forall\varphi \in C_0^\infty(\mathbb{R}^d).\end{aligned}$$

Then, by (2.404) and (2.406), it follows that

$$\int_{\mathbb{R}^d} (y\varphi - \lambda\beta(y)\Delta\varphi - \lambda a^*(y)\cdot\nabla\varphi)dx = \int_{\mathbb{R}^d} f\varphi\, dx,\ \forall\varphi\in C_0^\infty(\mathbb{R}^d),$$

which, by (2.351), means that $y + \lambda A(y) = f$.

Since, as seen in Lemma 2.5, the solution $y \in L^1$ to this equation is unique, we may infer that the limit y in (2.403) is independent of $\{\nu_n\}$ and so (2.383) follows. This completes the proof of Lemma 2.17 and, therefore, of Theorem 2.17.

Comments to Chap. 2

The main results of Sect. 2.1 were given in an appropriate form in the authors work [11]. For related results, see also [13, 15]. The results of Sects. 2.2, 2.5 and 2.6, 2.8 were established in the authors works [12, 16, 17, 21]. An alternative way to treat NFPE (2.80) is to treat it as a Cauchy problem in the space H^{-1} by taking A_0 of the form (2.90) with $D(A_0) = H^{-1}$. In this case, A_0 generates a continuous semigroup of quasi-contractions in H^{-1} which is everywhere differentiable from the right. In the special case $b \equiv 0$ and $\beta(r) \equiv \alpha r H(r - r_c)$, the probabilistic representation of solution was given in [22] and [26]. (See also [24, 46, 72, 73, 82, 86].) The results of Sect. 2.9 were established in [8]. Under appropriate conditions, $u(t) = S(t)u_0$ can be represented as a gradient flow on the differential manifold $\mathscr{P}$ (see [88] and [81] in the case $D \equiv 0$). The existence and uniqueness results presented in this chapter partially extend to nonlinear generalized Fokker–Planck equations of the form

$$u_t - \Delta\beta(u) + \mathrm{div}(K(u)u) = 0,$$

where K is a nonlinear operator in L^1. Very often $K(u) = K_0 \neq u$, where K_0 is a singular kernel derived from Riesz potentials (see, e.g., [40, 63, 79, 80]) or K is the Biot–Savart operator as happens for $2D$ vorticity Navier–Stokes equations (see [23]). However, the L^1-nonlinear semigroup approach is not applicable in this case because, if K is not a Nemytskii operator, the operator $u \to \mathrm{div}(K(u)u)$ is not accretive in L^1. Fokker–Planck equations of the form

$$\begin{aligned}
&u_t - \Delta\beta(u) + \mathrm{div}(Db(u)u) = 0 \ \text{ in } (0,\infty)\times\mathcal{O},\\
&u(0,x) = u_0(x),\ x\in\mathbb{R}^d,\\
&\frac{\partial}{\partial\mathbf{n}}\beta(u) - (D\cdot\mathbf{n})b(u)u = 0 \ \text{ on } (0,\infty)\times\partial\mathcal{O},
\end{aligned}$$

on an open domain $\mathcal{O} \subset \mathbb{R}^d$ with smooth boundary $\partial\mathcal{O}$, can be treated in a similar way, that is, as a Cauchy problem in $L^1(\mathcal{O})$ by defining $A_0 : D(A_0) \subset L^1(\mathcal{O}) \to L^1(\mathcal{O})$ as

$$
\begin{aligned}
A_0(u) &= -\Delta\beta(u) + \mathrm{div}(Db(u)u) \\
D(A_0) &= \{u \in L^1(\mathcal{O});\ \beta(u) \in W^{1,1}(\mathcal{O});\ \frac{\partial}{\partial \mathbf{n}}\beta(u) - (D \cdot \mathbf{n})b(u)u = 0 \text{ on } \partial\mathcal{O})\}.
\end{aligned}
$$

Under suitable conditions on β, D, b, it turns out as above that A_0 has an m-accretive realization A in $L^1(\mathcal{O})$ and derive so the existence of a mild solution. (On these lines see [7, 9]). However, a detailed analysis of this equation remains to be done. We note that it is equivalent with a McKean–Vlasov SDE of variational type with reflecting barriers.

Chapter 3
Time Dependent Fokker–Planck Equations

We shall study in this chapter the continuous semiflow associated with time dependent nonlinear Fokker–Planck equations, viewed as nonlinear evlution equations in appropriate Hilbert spaces. Such an equation arises in the statistical mechanics in the context of nonequilibrium thermodynamics and in the study of periodic stochastic processes.

3.1 Time Varying Fokker–Planck Flows

Consider herein the equation

$$\begin{aligned} &u_t(t,x) - \Delta(a(t,x,u(t,x))u(t,x)) + \mathrm{div}(b(t,x,u(t,x))u(t,x)) = 0 \\ &\qquad\qquad\qquad\qquad \text{on } (0,T)\times\mathbb{R}^d, \\ &u(0,\cdot) = u_0, \end{aligned} \tag{3.1}$$

where $d \geq 1$ and $0 < T < \infty$. In the following, for simplicity, we shall use the notations

$$\beta(t,x,r) \equiv a(t,x,r)r,\ b^*(t,x,r) \equiv b(t,x,r)r,\ (t,x,r) \in [0,T]\times\mathbb{R}^d\times\mathbb{R}. \tag{3.2}$$

Hypotheses

(i) $a \in C^1([0,T]\times\mathbb{R}^d\times\mathbb{R})$, a, a_r are bounded and, for all $t, s \in [0,T]$, $x \in \mathbb{R}^d$, $r, \bar{r} \in \mathbb{R}$,

$$(\beta(t,x,r) - \beta(t,x,\bar{r}))(r-\bar{r}) \geq \nu|r-\bar{r}|^2, \tag{3.3}$$

V. Barbu, M. Röckner, *Nonlinear Fokker-Planck Flows and their Probabilistic Counterparts*, Lecture Notes in Mathematics 2353,
https://doi.org/10.1007/978-3-031-61734-8_3

$$|\beta_r(t,x,r)-\beta_r(s,x,r)|\le h(x)|t-s|\beta_r(t,x,r), \tag{3.4}$$

$$|\beta(t,x,r)-\beta(s,x,r)|+|\beta_x(t,x,r)-\beta_x(s,x,r)|\le h(x)|t-s|(1+|r|), \tag{3.5}$$

$$|\beta_t(t,x,r)|+|\beta_x(t,x,r)|\le h(x)|r|,\ \forall(t,x,r)\in[0,T]\times\mathbb{R}^d\times\mathbb{R}. \tag{3.6}$$

(ii) $r\to b(t,x,r)$ is C^1, $\forall(t,x)\in[0,T]\times\mathbb{R}^d$, b, rb_r are bounded, and $\forall t,s\in[0,T]$, $x\in\mathbb{R}^d$, $r\in\mathbb{R}$,

$$|b^*(t,x,r)-b^*(s,x,r)|\le h(x)|t-s|(1+|b^*(t,x,r)|), \tag{3.7}$$

$$|b^*(t,x,r)|\le h(x)|r|, \tag{3.8}$$

where $\nu>0$, $h\in L^\infty(\mathbb{R}^d)\cap L^2(\mathbb{R}^d)$, $h\ge 0$. Here $a_r=\frac{\partial a}{\partial r}$, $b_r=\frac{\partial b}{\partial r}$, $\beta_x=\nabla_x\beta$, $\beta_r=\frac{\partial}{\partial r}\beta$ and div, Δ, ∇ are taken with respect to the spatial variable $x=(x_j)_{j=1}^d$. Here, we note that by (3.3) we have $\beta_r\ge\nu>0$. As in the autonomous case studied earlier, it is of great interest to prove the existence of a Schwartz distributional solution u to (3.1) which is weakly t-continuous, nonnegative for nonnegative initial data, that is,

$$u(t,x)\ge 0,\ \forall t\in[0,T] \text{ and a.e. } x\in\mathbb{R}^d, \tag{3.9}$$

$$\lim_{t\to t_0}\int_{\mathbb{R}^d}u(t,x)\psi(x)dx=\int_{\mathbb{R}^d}u(t_0,x)\psi(x)dx,\ \forall t_0\in[0,T],\ \psi\in C_b(\mathbb{R}^d), \tag{3.10}$$

$$\int_{\mathbb{R}^d}u(t,x)dx=1,\ \forall t\in[0,T], \tag{3.11}$$

for $u_0\in\mathscr{P}$. If u is a solution to (3.1) satisfying (3.9)–(3.11), then one can find, as in the autonomous case, a probabilistic representation of u as the time marginal law of the solution $X=X(t)$, $t\in[0,T]$, to the McKean-Vlasov stochastic differential equation (of Nemytskii-type)

$$\begin{aligned}dX(t)&=b\left(t,X(t),\frac{d\mathscr{L}_{X(t)}}{dx}(X(t))\right)+\sigma\left(t,X(t),\frac{d\mathscr{L}_{X(t)}}{dx}(X(t))\right)dW(t),\\X(0)&=\xi_0,\end{aligned} \tag{3.12}$$

where $\sigma=\sqrt{2a}$, $\mathscr{L}_{X(t)}$ is the law of $X(t)$, $W=\{W_j\}_{j=1}^d$ is an $(\mathscr{F}_t)$-Brownian motion on a filtered probability space $(\Omega,\mathscr{F},(\mathscr{F}_t),\mathbb{P})$ and $\xi_0:\Omega\to\mathbb{R}^d$ is $\mathscr{F}_0$-measurable such that $\mathbb{P}\circ\xi_0^{-1}(dx)=u_0(x)dx$. Namely, there is a *probabilistically weak solution* to (3.12) such that

$$u(t,x)dx=\mathbb{P}\circ(X(t))^{-1}(dx),\ \forall t\ge 0. \tag{3.13}$$

In Chap. 2 we have proved the existence of a weak solution u for (3.1) in the autonomous case $\beta \equiv \beta(x,r)$, $b \equiv b(x,u)$. In this case, the generation theorem for nonlinear semigroups of contractions in L^1 was the essential instrument to prove the well posedness of (3.1) but, as it is well known, the nonlinear semigroup theory does not extend completely to the Cauchy problem for differential equations with time-dependent accretive operators, and so here we shall use a different approach based on the general existence theory of the time-dependent Cauchy problem in Banach spaces.

In the following we shall use the same notations as those used in Chap. 1.

Namely, for $1 \le p \le \infty$, denote by the space $L^p(\mathbb{R}^d)$ with norm denoted $|\cdot|_p$. By $L^p_{\rm loc}(\mathbb{R}^d) = L^p_{\rm loc}$ we denote the corresponding local space on $\mathbb{R}^d$. The scalar product in L^2 is denoted by $(\cdot,\cdot)_2$. By $H^{-1}(\mathbb{R}^d)$, also written H^{-1}, we denote the dual space of $H^1(\mathbb{R}^d)$, also denoted H^1, with the scalar product

$$\langle u,v\rangle_{-1,\varepsilon} = \int_{\mathbb{R}^d} (\varepsilon I - \Delta)^{-1} u v\,dx,\ \forall u,v \in H^{-1},$$

and the corresponding norm $|\cdot|_{-1,\varepsilon}$. Here $(\varepsilon I - \Delta)^{-1}u = y \in H^2(\mathbb{R}^d)$ is defined for $\varepsilon > 0$ and $u \in L^2(\mathbb{R}^d)$ by the equation

$$\varepsilon y - \Delta y = u \text{ in } \mathscr{D}'(\mathbb{R}^d).$$

$H^{-2}(\mathbb{R}^d)$ is the dual space of $H^2(\mathbb{R}^d)$ with the norm $|u|_{-2} = |(\varepsilon I - \Delta)^{-1}u|_2$, $u \in H^{-2}(\mathbb{R}^d)$. By $W^{1,p}([0,T];H^{-1})$, $1 \le p \le \infty$ we denote the space of absolutely continuous functions $u : [0,T] \to H^{-1}$ such that $\frac{du}{dt} \in L^p(0,T;H^{-1})$. Here $\frac{du}{dt}$ is the derivative of u in the sense of H^{-1}-valued distributions on $(0,T)$ or, equivalently, a.e. on $(0,T)$.

Denote by $C([0,T];L^2)$ the space of continuous L^2-valued functions and by $C_w([0,T];L^2)$ the space of weakly continuous functions $u : [0,T] \to L^2$. Finally, denote by $C_b(\mathbb{R}^d)$ and $C_b([0,T] \times \mathbb{R}^d)$ the space of all continuous and bounded functions on $\mathbb{R}^d$ and $[0,T] \times \mathbb{R}^d$, respectively.

Theorem 3.1, which is the main existence result for Eq. (3.1), amounts to saying that, under the above hypotheses, there is a continuous semi-flow $u_0 \to U(t,s)u_0 = u(t,s,u_0)$ associated with Eq. (3.1).

Theorem 3.1 *Assume that Hypotheses* (i), (ii) *hold. Let* $0 < T < \infty$ *be arbitrary, but fixed. Let* $D_0 := \{u \in L^2;\ \beta(0,\cdot,u) \in H^1\}$. *Then, for each* $u_0 \in D_0$, *there is a unique strong solution* $u = u(t,u_0)$ *to* (3.1) *satisfying*

$$u \in C([0,T];L^2) \cap W^{1,2}([0,T];H^{-1}), \tag{3.14}$$

$$u,\beta(\cdot,u) \in L^2(0,T;H^1). \tag{3.15}$$

$$|u(t,u_0) - u(t,\bar u_0)|_2^2 \le C|u_0 - \bar u_0|_2^2,\ \forall t \in [0,T],\ u_0,\bar u_0 \in D_0. \tag{3.16}$$

If $u_0 \in L^1 \cap L^2$, *then* $u \in L^\infty(0,T;L^1)$. *Moreover, it satisfies the weak continuity condition* (3.10) *and* $|u(t)|_1 \le |u_0|_1$, *for all* $t \in [0,T]$.

In addition, if $u_0 \in L^1 \cap D_0$ *is such that*

$$u_0 \ge 0, \quad or \quad \int_{\mathbb{R}^d} u_0(x)dx = 1, \tag{3.17}$$

then u *satisfies also conditions* (3.9) *and* (3.11).

Finally, assume that (3.8) *is replaced by the stronger condition*

$$|b^*(t,x,r) - b^*(t,x,\bar{r})| \le h(x)|r-\bar{r}|, \ \forall t \in [0,T], \ x \in \mathbb{R}^d, \ r, \bar{r} \in \mathbb{R}. \tag{3.18}$$

Then, for all $u_0, \bar{u}_0 \in L^1 \cap D_0$, *the corresponding solutions* $u(t,u_0)$, $u(t,\bar{u}_0)$ *to* (3.1) *satisfy*

$$|u(t,u_0) - u(t,\bar{u}_0)|_1 \le |u_0 - \bar{u}_0|_1, \ \forall t \in [0,T]. \tag{3.19}$$

Here, as above, $\beta(t,x,r) \equiv a(t,x,r)r$.

By *strong solution* to (3.1) we mean a function u satisfying (3.14), (3.15), and (3.1) in H^{-1}, a.e. in $t \in (0,T)$ with Δ and div taken in the sense of Schwartz distributions on $\mathbb{R}^d$. (We note that, by (3.14) and (3.15), we have $\frac{\partial u}{\partial t} \in L^1(0,T;H^{-1})$, $\Delta\beta(\cdot,u) \in L^2(0,T;H^{-1})$, $\mathrm{div}(b(\cdot,u)u) \in L^2(0,T;H^{-1})$.) In other words, $u \in L^2(0,T;L^2)$ is a solution to (3.1) in the sense of H^{-1}-valued vectorial distributions on $(0,T)$. It is easily seen that the solution u given by Theorem 3.1 is also a solution in the sense of Schwartz distributions on $(0,T)\times\mathbb{R}^d$, that is,

$$\begin{aligned}\int_0^T \int_{\mathbb{R}^d} &(u(t,x)\cdot\varphi_t(t,x) + a(t,x,u(t,x))u(t,x)\Delta\varphi(t,x) \\ &+u(t,x)b(t,x,u(t,x))\cdot\nabla\varphi(t,x))dt\,dx + \int_{\mathbb{R}^d}\varphi(0,x)u_0(x)dx = 0, \\ &\forall \varphi \in C_0^2([0,T)\times\mathbb{R}^d),\end{aligned} \tag{3.20}$$

where $C_0^2((0,T]\times\mathbb{R}^d)$ is the space of all twice differentiable functions with compact support in $(0,T]\times\mathbb{R}^d$. In particular, by (3.19) the L^1-stability of the solution $u = u(t,u_0)$ follows and also that $t \to u(t,u_0)$ is a contraction flow in the space L^1. More precisely, by (3.19), it follows via a density argument that, in fact, for every $u_0 \in L^1$, there is a distributional solution u in the sense of (3.20) satisfying (3.19). It follows by [1, Lemma 8.1.2] that $t \mapsto u(t,x)dx$ has a weakly continuous dt-version on $[0,T]$, which thus satisfies (3.10).

If $\overline{D}_0$ is the closure of D_0 in L^2, it follows by (3.16) that $u = u(t,u_0)$ extends as solution to (3.1) on all of $u_0 \in \overline{D}_0$, the closure of D_0 in L^2. Then the operator

$$U(t,s) = u(t,u(s,u_0)), \ 0 \le s \le t \le T, \ u_0 \in \overline{D}_0,$$

is a time-evolution operator generated by the time dependent Fokker–Planck equation (3.1), that is,

$$U(t,t) = I,\ \forall t \in [0,T],\ U(t_2,s) = U(t_2,t_1)U(t_1,s),\ s < t_1 < t_1.$$

As noted earlier, by virtue of Theorem 3.1, we have (see Chap. 5)

Corollary 3.1 *Assume that* (i), (ii) *and* $u_0 \in L^1 \cap D_0$ *such that* (3.18) *holds. Then, there exists a (probabilistically) weak solution* X *to the McKean–Vlasov SDE* (3.12), *with* $\sigma = \sqrt{2a}$, *where* a *is as in* (i), *such that the strong solution* u *to* (3.1) *given by Theorem* 3.1 *is its time marginal law density. In particular, we have the representation formula* (3.13) *for* u. *The same is true for all* $u_0 \in L^1$ *if* (3.18) *holds.*

We shall prove Theorem 3.1 in several steps and the first one is an existence result for an approximating equation corresponding to (3.1).

Proposition 3.1 *Assume that* (i), (ii) *hold and that* $u_0 \in D_0$. *Then, for each* $\varepsilon > 0$, *the equation*

$$\begin{aligned} &u_t - \Delta\beta(t,x,u) + \varepsilon\beta(t,x,u) + \mathrm{div}(b(t,x,u)u) = 0,\ t \in (0,T),\ x \in \mathbb{R}^d,\\ &u(0,x) = u_0(x), \end{aligned} \tag{3.21}$$

has a unique strong solution $u = u_\varepsilon \in W^{1,\infty}([0,T];H^{-1}) \cap L^2(0,T;H^1)$, *such that* $\beta(u_\varepsilon) \in L^2(0,T;H^1)$. *In particular,* $u_\varepsilon \in C([0,T];L^2)$.

Then the first part of Theorem 3.1 will follow from Proposition 3.1 by letting $\varepsilon \to 0$ in Eq. (3.21). As regards Proposition 3.1, it will be derived from an abstract existence result for the Cauchy problem in the space H^{-1}

$$\begin{aligned} &\frac{du}{dt} + A_\varepsilon(t)u = 0,\ t \in (0,T),\\ &u(0) = u_0, \end{aligned} \tag{3.22}$$

where $A_\varepsilon(t) : D(A_\varepsilon(t)) \subset H^{-1} \to H^{-1}$ is for each $t \in [0,T]$ a quasi-m-accretive operator in H^{-1} to be defined below.

Proof of Proposition 3.1 We define for $t \in [0,T]$ and $\varepsilon > 0$ the operator $A_\varepsilon(t) : D(A_\varepsilon(t)) \subset H^{-1} \to H^{-1}$

$$\begin{aligned} (A_\varepsilon(t)y)(x) =& -\Delta\beta(t,x,y(x)) + \varepsilon\beta(t,x,y(x))\\ &+ \mathrm{div}(b(t,x,y(x))y(x)),\ \forall y \in D(A(t)),\ x \in \mathbb{R}^d,\\ D(A_\varepsilon(t)) =& \{y \in L^2;\ \beta(t,\cdot,y) \in H^1\},\ t \in [0,T], \end{aligned} \tag{3.23}$$

where the differential operators Δ and div are taken in the sense of the Schwartz distribution space $\mathscr{D}'(\mathbb{R}^d)$. □

By (i) it follows immediately that

$$H^1 \cap L^\infty \subset D(A_\varepsilon(t)) \subset H^1,$$

hence $D(A_\varepsilon(t))$ is dense in L^2.

By Hypotheses (i), (ii), it is easily seen that, setting $D_t := D(A_\varepsilon(t))$,

$$A_\varepsilon(t)(D_t) \subset H^{-1}, \ \forall t \in [0, T].$$

Since below we fix $\varepsilon > 0$, for simplicity we write $\langle \cdot, \cdot \rangle_{-1}$ and $| \cdot |_{-1}$ instead of $\langle \cdot, \cdot \rangle_{-1,\varepsilon}$, $| \cdot |_{-1,\varepsilon}$, respectively. We also have

Lemma 3.1 *Let $t \in [0, T]$ and $\lambda_0 := \frac{2\nu}{|b_r^*|_\infty^2}$. Then, the operator $A_\varepsilon(t)$ is quasi-m-accretive in H^{-1}, that is, $(I + \lambda A_\varepsilon(t))^{-1} : H^{-1} \to H^{-1}$ is single-valued and*

$$\|(I + \lambda A_\varepsilon(t))^{-1}\|_{\mathrm{Lip}_\varepsilon(H^{-1})} \le \left(1 - \frac{\lambda}{\lambda_0}\right)^{-1}, \ \forall \lambda \in (0, \lambda_0), \tag{3.24}$$

where $\| \cdot \|_{\mathrm{Lip}_\varepsilon}$ means Lipschitz norm w.r.t. the norm $| \cdot |_{-1,\varepsilon}$ on H^{-1}.

In particular, $(A_\varepsilon(t), D_t)$ is closed for every $t \in [0, T]$.

Moreover, $D_t = D_0$, for every $t \in [0, T]$, and for all $\lambda \in (0, \lambda_0)$ we have

$$\begin{aligned} &|(I + \lambda A_\varepsilon(t))^{-1}u - (I + \lambda A_\varepsilon(s))^{-1}u|_{-1} \\ &\le \lambda |t - s| L(|u|_{-1})(1 + |A_\varepsilon(t)u|_{-1}), \ \forall u \in D_0, \ s, t \in [0, T], \end{aligned} \tag{3.25}$$

where $L : [0, \infty) \to [0, \infty)$ is a monotone nondecreasing function.

Proof Let v be arbitrary in H^{-1} and consider the equation

$$u + \lambda A_\varepsilon(t)u = v. \tag{3.26}$$

In the following, we shall simply write

$$\beta(t, x, u) = \beta(t, u), b(t, x, u) = b(t, u), b^*(t, x, u) = b^*(t, u).$$

We seek a function $u \in L^2$ such that

$$u - \lambda(\Delta - \varepsilon I)(\beta(t, u)) + \lambda \mathrm{div}(b^*(t, u)) = v \text{ in } \mathscr{D}'(\mathbb{R}^d). \tag{3.27}$$

Equivalently,

$$(\varepsilon I - \Delta)^{-1}u + \lambda\beta(t, u) + \lambda(\varepsilon I - \Delta)^{-1}\mathrm{div}(b^*(t, u)) = (\varepsilon I - \Delta)^{-1}v. \tag{3.28}$$

Let $u, \bar{u} \in L^2$. Then

$$|u-\bar{u}|_{-1}^2 = \varepsilon|(\varepsilon I - \Delta)^{-1}(u-\bar{u})|_2^2 + |\nabla(\varepsilon I - \Delta)^{-1}(u-\bar{u})|_2^2.$$

Hence, by (L2) we have

$$\begin{aligned}
&\left\langle (\varepsilon I - \Delta)^{-1}\mathrm{div}(b^*(t,u) - b^*(t,\bar{u})), u-\bar{u}\right\rangle_2 \\
&\quad \geq -|b_r^*|_\infty |\nabla(\varepsilon I - \Delta)^{-1}(u-\bar{u})|_2 |u-\bar{u}|_2 \\
&\quad \geq -|b_r^*|_\infty |u-\bar{u}|_{-1}|u-\bar{u}|_2 \geq -\frac{|b_r^*|_\infty^2}{2\nu}\,|u-\bar{u}|_{-1}^2 - \frac{\nu}{2}\,|u-\bar{u}|_2^2,
\end{aligned} \tag{3.29}$$

and, by (3.3),

$$\langle \beta(t,u) - \beta(t,\bar{u}), u-\bar{u}\rangle_2 \geq \nu|u-\bar{u}|_2^2.$$

This implies that the operator $F : L^2 \to L^2$,

$$Fu = (\varepsilon I - \Delta)^{-1}u + \lambda\beta(t,u) + \lambda(\varepsilon I - \Delta)^{-1}\mathrm{div}(b^*(t,u)),\ u \in L^2,$$

satisfies the condition

$$\langle Fu - F\bar{u}, u-\bar{u}\rangle_2 \geq \frac{\nu\lambda}{2}\,|u-\bar{u}|_2^2 + \left(1 - \frac{\lambda|b_r^*|_\infty^2}{2\nu}\right)|u-\bar{u}|_{-1}^2,\ \forall u, \bar{u} \in L^2. \tag{3.30}$$

Hence, for $\lambda \in (0, \lambda_0)$, the operator $F : L^2 \to L^2$ is accretive, continuous and coercive. Therefore, it is surjective (see, Theorems 6.3 and 6.4). So, (3.28) has a solution $u \in L^2$ such that $\beta(t,u) \in H^1$. By (3.30), we also have

$$|F^{-1}(v) - F^{-1}(\bar{v})|_2 \leq \frac{2}{\nu\lambda}\,|v-\bar{v}|_2,\ \forall v, \bar{v} \in L^2,\ \text{for } 0 < \lambda \leq \lambda_0.$$

Moreover, by (3.28)–(3.30), we see that, for $v, \bar{v} \in H^{-1}$ and the corresponding solutions $u, \bar{u} \in L^2$ to (3.24), we have

$$\frac{\nu\lambda}{2}\,|u-\bar{u}|_2^2 + \left(1 - \frac{\lambda}{\lambda_0}\right)|u-\bar{u}|_{-1}^2 \leq |u-\bar{u}|_{-1}|v-\bar{v}|_{-1},$$

and, therefore,

$$|(I+\lambda A_\varepsilon(t))^{-1}v - (I+\lambda A_\varepsilon(t))^{-1}\bar{v}|_{-1} \leq \left(1 - \frac{\lambda}{\lambda_0}\right)^{-1}|v-\bar{v}|_{-1},\ \forall v, \bar{v} \in H^{-1},$$

for $0 < \lambda < \lambda_0$. This implies (3.24), as claimed.

Let us show now that $D(A_\varepsilon(t))$ is independent of t. Indeed, by (3.4) we have

$$\beta_r(s,x,r) \le (h(x)|t-s|+1)\beta_r(t,x,r),\ \forall\, t,s \in [0,T],\ x \in \mathbb{R}^d,\ r \in \mathbb{R}. \tag{3.31}$$

Let $y \in L^2$. Then, by (3.5), $y \in D(A_\varepsilon(t))$ if and only if $\beta_r(t,\cdot,y)\nabla y \in L^2$. Hence, by (3.31) we also have

$$\beta_r(s,,\cdot,y)\nabla y \in L^2,\quad \forall\, s \in [0,T],$$

that is, $y \in D(A_\varepsilon(s))$, $\forall\, s \in [0,T]$, as claimed.

Therefore, $D(A_\varepsilon(t))$ is independent of t, and

$$D(A_\varepsilon(t)) = D_0 = \{u_0 \in L^2;\ \beta(0,u_0) \in H^1\}.$$

We note that condition (3.25) holds if

$$\begin{aligned} |A_\varepsilon(t)u - A_\varepsilon(s)u|_{-1} \le C|t-s|L(|u|_{-1})(1+|A_\varepsilon(t)u|_{-1}),\\ \forall u \in D_0,\ s,t \in [0,T], \end{aligned} \tag{3.32}$$

where $L : [0,\infty) \to [0,\infty)$ is a nondecreasing monotone function. To prove (3.32), we note that

$$\begin{aligned} |A_\varepsilon(t)u - A_\varepsilon(s)u|_{-1}^2 &= \Big\langle \beta(t,u)-\beta(s,u)\\ &\quad +(\varepsilon I-\Delta)^{-1}\operatorname{div}(b^*(t,u)-b^*(s,u)),\ \varepsilon(\beta(u(t,u))-\beta(s,u))\\ &\quad -\Delta(\beta(t,u)-\beta(s,u))+\operatorname{div}(b^*(t,u)-b^*(s,u))\Big\rangle_2\\ &= |\nabla(\beta(t,u)-\beta(s,u))|_2^2 + |\operatorname{div}(b^*(t,u)-b^*(s,u))|_{-1}^2\\ &\quad +\varepsilon|\beta(t,u)-\beta(s,u)|_2^2 + 2\,\langle\beta(t,u)-\beta(s,u),\operatorname{div}(b^*(t,u)-b^*(s,u))\rangle_2,\\ &\qquad \forall\, s,t \in [0,T],\ u \in D_0. \end{aligned} \tag{3.33}$$

We also have

$$|\operatorname{div} f|_{-1} \le C|f|_2,\ \forall\, f \in L^2,$$

and so, by (3.7),

$$\begin{aligned} |\operatorname{div}(b^*(t,u)-b^*(s,u))|_{-1} \le C|b^*(t,u)-b^*(s,u)|_2 \le C|t-s|(1+|b^*(t,u)|_2),\\ \forall (s,t) \in (0,T), \end{aligned}$$

and, by (3.3)–(3.6),

$$\begin{aligned} &|\nabla(\beta(t,u)-\beta(s,u))|_2 + |\beta(t,u)-\beta(s,u)|_2\\ &\quad \le C|t-s|(1+|\beta(t,u)|_2+|\nabla\beta(t,u)|_2),\ \forall s,t \in [0,T]. \end{aligned}$$

Then, by (3.33) we see that

$$\begin{aligned}|A_\varepsilon(t)u - A_\varepsilon(s)u|_{-1} \le C|t-s|(1+|\beta(t,u)|_2 \\ +|\nabla\beta(t,u)|_2 + |b^*(t,u)|_2),\ \forall\, t,s\in[0,T],\ u\in D_0.\end{aligned} \tag{3.34}$$

On the other hand, we have by (3.6)

$$\begin{aligned}|A_\varepsilon(t)u|^2_{-1} &= |\nabla\beta(t,u)|_2^2 + |\operatorname{div} b^*(t,u)|^2_{-1} \\ &\quad +2\,\langle\beta(t,u), \operatorname{div} b^*(t,u)\rangle_2 + \varepsilon|\beta(t,u)|_2^2 \\ &\ge \frac{1}{2}\,|\nabla\beta(t,u)|_2^2 + \varepsilon|\beta(t,u)|_2^2 + |b^*(t,u)|_2^2 - C|b^*(t,u)|_2^2 \\ &\ge \frac{1}{2}\,|\nabla\beta(t,u)|_2^2 + \varepsilon|\beta(t,u)|_2^2 + |b^*(t,u)|_2^2 - C|u|_2^2,\ \forall\, u\in D_0.\end{aligned} \tag{3.35}$$

We recall also the interpolation inequality

$$|u|_2^2 \le C|u|_{H^1}|u|_{-1},\quad \forall\, u\in H^1.$$

(Here, we have denoted by C several positive constants depending on ε, but independent of u.) Noting that, by (3.3) and (3.6),

$$\begin{aligned}|\nabla\beta(t,u)|_2^2 &= |\beta_x(t,u) + \beta_r(t,u)\nabla u|_2^2 \\ &\ge \frac{1}{2}\,\nu^2|\nabla u|_2^2 - |h|^2_\infty|u|_2^2 \\ &\ge \frac{1}{4}\,\nu^2|\nabla u|_2^2 - C|u|^2_{-1},\end{aligned} \tag{3.36}$$

(3.33) yields

$$|u|_1^2 + |A_\varepsilon(t,u)|^2_{-1} \ge C(|\beta(t,u)|_2^2 + |\nabla\beta(t,u)|_2^2 + |b^*(t,u)|_2^2,\ \forall u\in D_0,$$

and so, by (3.34), we get

$$|A_\varepsilon(t)u - A_\varepsilon(s)u|_{-1} \le C|t-s|(|A_\varepsilon(t)u|_{-1} + |u|_{-1} + 1),\ \forall (t,s)\in[0,T],\ u\in D_0.$$

This yields

$$|A_\varepsilon(t)u - A_\varepsilon(s)u|_{-1} \le C|t-s|L(|u|_{-1})(1+|A_\varepsilon(t)u|_{-1}),\ \forall\, t,s\in[0,T],\ u\in D_0,$$

where $L:[0,\infty)\to[0,\infty)$ is defined by

$$L(r) = \max\{1, r+1\},\quad \forall r\ge 0.$$

Hence (3.32) follows.

Proof of Proposition 3.1 By Lemma 3.1 and Theorem 6.10, it follows that, for each $u_0 \in D_0$, there is a unique strong solution $u = u_\varepsilon \in C([0,T]; H^{-1})$ to Eq. (3.22), given by the exponential formula

$$u_\varepsilon(t) = \lim_{n\to\infty} \prod_{k=1}^{n} \left(I + \frac{t-s}{n} A_\varepsilon\left(s + k\frac{t-s}{n}\right)\right)^{-1} u_0, \ \forall t \in [0,T], \tag{3.37}$$

and (3.37) is uniform in t, where the limit is taken in the strong topology of H^{-1}. As a matter of fact, (3.37) is just the finite difference scheme corresponding to the Cauchy problem (3.22). Namely,

$$\begin{aligned}&\mu^{-1}(u^\varepsilon_{i+1} - u^\varepsilon_i) + A_\varepsilon((i+1)\mu)u^\varepsilon_{i+1} = 0,\ i = 0,1,\dots,N = \left[\tfrac{T}{\mu}\right]\\ &u^\varepsilon_0 = u_0\end{aligned} \tag{3.38}$$

$$\lim_{\mu\to 0} u^\varepsilon_\mu(t) = u_\varepsilon(t) \text{ strongly in } H^{-1},\ \forall t \in [0,T], \tag{3.39}$$

where for $\mu > 0$ the step function $u^\varepsilon_\mu : [0,T] \to H^{-1}$ is defined by

$$u^\varepsilon_\mu(t) = u^\varepsilon_i,\ \forall t \in [i\mu, (i+1)\mu). \tag{3.40}$$

Moreover, since $u_0 \in D_0$, then we have the regularity properties $u_\varepsilon(t) \in D_0$ for every $t \in [0,T]$ and

$$\frac{du_\varepsilon}{dt} \in L^\infty(0,T;H^{-1}),\ A_\varepsilon u_\varepsilon \in L^\infty(0,T;H^{-1}), \tag{3.41}$$

$$u_\varepsilon(t), \beta(t, u_\varepsilon(t)) \in H^1,\ \ \forall t \in [0,T], \tag{3.42}$$

where (3.41) follows from (3.32) by similar arguments as in the proof of [4, Theorem 4.19, p. 182].

By (L2) and (3.21), it follows that

$$\begin{aligned}\frac{1}{2}\frac{d}{dt}|u_\varepsilon(t)|^2_{-1} + \langle\beta(u_\varepsilon(t)), u_\varepsilon(t)\rangle_2 &= \langle \mathrm{div}(b(t,u_\varepsilon)u_\varepsilon(t), u_\varepsilon(t))\rangle_{-1}\\ &\le |u_\varepsilon(t)|_{-1}|\mathrm{div}(b(t,u_\varepsilon(t))u_\varepsilon(t))|_{-1} \le C|u_\varepsilon(t)|_{-1}|b(t,u_\varepsilon(t))u_\varepsilon(t)|_2\\ &\le C_1|u_\varepsilon(t)|_{-1}|u_\varepsilon(t)|_2,\ \text{ a.e. } t \in (0,T).\end{aligned}$$

By (3.3), this yields

$$\frac{d}{dt}|u_\varepsilon(t)|^2_{-1} + \nu|u_\varepsilon(t)|^2_2 \le C_2|u_\varepsilon(t)|^2_{-1},\ \text{ a.e. } t \in (0,T),$$

and, therefore, by Gronwall's lemma

$$|u_\varepsilon(t)|_{-1}^2 + \int_0^T |u_\varepsilon(t)|_2^2 dt \le C_3 |u_0|_{-1}^2, \ \forall t \in [0, T]. \tag{3.43}$$

Next, we have

$$\beta(u_\varepsilon(t)) = (\varepsilon I - \Delta)^{-1}(A_\varepsilon(t)u_\varepsilon(t)) + (\varepsilon I - \Delta)^{-1}\mathrm{div}(b(t, u_\varepsilon(t))u_\varepsilon(t)), \quad \text{a.e. } t \in (0, T),$$

and since, by (3.41), $A_\varepsilon u_\varepsilon \in L^2(0, T; H^{-1})$, we have

$$(\varepsilon I - \Delta)^{-1} A_\varepsilon u_\varepsilon \in L^2(0, T; H^1),$$

while

$$|(\varepsilon I - \Delta)^{-1}\mathrm{div}(b(t, u_\varepsilon(t))u_\varepsilon(t))|_{H^1} \le C|b(t, u_\varepsilon(t))u_\varepsilon(t)|_2 \le C_1|u_\varepsilon(t)|_2, \quad \text{a.e. } t \in (0, T).$$

Then, by (3.43), we infer that

$$\beta(u_\varepsilon), u_\varepsilon \in L^2(0, T; H^1). \tag{3.44}$$

This completes the proof of Proposition 3.1. □

Proof of Theorem 3.1 In the following we shall omit x in the notations $\beta(t, x, u)$ and $b(t, x, u)$. For $\varepsilon > 0$, we consider the solution u_ε to (3.21) and get first a few apriori estimates. In this proof, we consider H^{-1} with its usual inner product $\langle u, v\rangle_{-1} = \langle (I - \Delta)^{-1}u, v\rangle_{L^2}$. We note first the estimate

$$\begin{aligned} &|u_\varepsilon(t)|_2^2 + \int_0^t \int_{\mathbb{R}^d} (|\nabla u_\varepsilon(s, x)|^2 + |\nabla\beta(s, x, u_\varepsilon(s, x))|^2 ds\, dx \\ &+\varepsilon \int_0^t \int_{\mathbb{R}^d} |\beta(s, x, u_\varepsilon)|^2 ds\, dx \le C_T |u_0|_2^2, \ \forall t \in [0, T]. \end{aligned} \tag{3.45}$$

□

To show (3.45), we prove first, for some $C\ (= C(|a|_\infty, |b|_\infty, \nu, T) \in (0, \infty))$,

$$|u_\varepsilon(t)|_2^2 + \int_0^t |\nabla u_\varepsilon(s)|_2^2 ds + \varepsilon \int_0^t |\beta(s, u_\varepsilon(s))|_2^2 ds \le C|u_0|_2^2, \ t \in [0, T], \tag{3.46}$$

which, by (3.41) and [4, Theorem 1.19, p. 25] in turn implies

$$u_\varepsilon \in C([0, T]; L^2). \tag{3.47}$$

To this end, we note that by (3.41), (3.44) in (3.21) (with u replaced by u_ε) we can take L^2-inner product with u_ε and integrate over $(0, t)$ to obtain for every $t \in [0, T]$

$$\begin{aligned}\frac{1}{2}\,|u_\varepsilon(t)|_2^2 + \int_0^t \langle \nabla u_\varepsilon(s), \nabla \beta(s, u_\varepsilon(s))\rangle_2\, ds + \varepsilon \int_0^t \langle u_\varepsilon(s), \beta(s, u_\varepsilon(s))\rangle_2\, ds \\ = \frac{1}{2}\,|u_0|_2^2 + \int_0^t \big\langle \nabla u_\varepsilon(s), b^*(s, u_\varepsilon(s))\big\rangle_2\, ds,\end{aligned}$$

where the first integral on the left hand side by (3.3) and (3.6) is bigger than

$$\int_0^t \left(\frac{3\nu}{4}\,|\nabla u_\varepsilon(s)|_2^2 - \frac{1}{\nu}\,|h|_\infty^2 |u_\varepsilon(s)|_2^2\right) ds$$

and the integral on the right hand side by (L2) is dominated by

$$\int_0^t \left(\frac{\nu}{4}\,|\nabla u_\varepsilon(s)|_2^2 + \frac{1}{\nu}\,|b|_\infty^2 |u_\varepsilon(s)|_2^2\right) ds.$$

Hence, since by (i) $|\beta(s, u_\varepsilon(s))| \le |a|_\infty |u_\varepsilon(s)|$, by Gronwall's inequality we obtain (3.46). Furthermore, we need that, for a.e. $t \in [0, T]$,

$$\begin{aligned}\frac{d}{dt}\int_{\mathbb{R}^d} j(t, x, u_\varepsilon(t, x))dx = {}_{H^1}\Big\langle \beta(t, u_\varepsilon(t)), \tfrac{du_\varepsilon(t)}{dt}\Big\rangle_{H^{-1}} \\ + \int_{\mathbb{R}^d} j_t(t, x, u_\varepsilon(t, x))dx,\end{aligned} \tag{3.48}$$

where $j(t, x, r) = \int_0^r \beta(t, x, \bar r)d\bar r$, $r \in \mathbb{R}$, $x \in \mathbb{R}^d$, $t \in [0, T]$.

To prove (3.48), we first note that due to (i), for all $(t, x, r) \in [0, T] \times \mathbb{R}^d \times \mathbb{R}$,

$$\nu|r|^2 \le j(t, x, r) \le |a|_\infty |r|^2,\ |j_t(t, x, r)| \le \frac{1}{2}\,h(x)|r|^2. \tag{3.49}$$

Fix $t \in [0, T]$ and define the convex lower semi-continuous function $\varphi^t : H^{-1} \to [0, \infty]$ by

$$\varphi^t(u) := \begin{cases} \displaystyle\int_{\mathbb{R}^d} j(t, x, u)dx & \text{if } u \in L^2, \\ +\infty & \text{if } u \in H^{-1} \setminus L^2. \end{cases}$$

Then, it is elementary to check that for its subdifferential $\partial\varphi^t$ on H^{-1} we have

$$(I - \Delta)\beta(t, u) = \partial\varphi^t(u),\ \forall u \in L^2.$$

Furthermore, (3.44) and (3.4)–(3.6) imply that $\beta(t, u_\varepsilon) \in L^2(0, T; H^1)$. Hence we conclude by Barbu [4, Lemma 4.4, p. 158] that

$$\frac{d}{ds}\varphi^t(u_\varepsilon(s)) = {}_{H^1}\left\langle \beta(t, u_\varepsilon(t)), \frac{du_\varepsilon(s)}{ds}\right\rangle_{H^{-1}} \quad \text{for } ds\text{-a.e. } s \in [0, T]. \tag{3.50}$$

Applying (3.50) with $s = t$, we find for dt-a.e. $t \in [0, T]$

$$\begin{aligned}
&\frac{d}{dt}\int_{\mathbb{R}^d} j(t, x, u_\varepsilon(t, x))dx \qquad (3.51)\\
&= \lim_{\delta\to 0}\Big[\frac{1}{\delta}\int_{\mathbb{R}^d}(j(t+\delta, x, u_\varepsilon(t+\delta, x)) - j(t, x, u_\varepsilon(t+\delta, x)))dx\\
&\qquad + \frac{1}{\delta}(\varphi^t(u_\varepsilon(t+\delta)) - \varphi^t(u_\varepsilon(t)))\Big]\\
&= \lim_{\delta\to 0}\int_{\mathbb{R}^d}\int_0^1 j_t(t+s\delta, x, u_\varepsilon(t+\delta, x))ds\,dx + {}_{H^1}\left\langle \beta(t, u_\varepsilon(t)), \frac{du_\varepsilon(t)}{dt}\right\rangle_{H^{-1}}.
\end{aligned}$$

Since, as $\delta \to 0$,

$$j_t(t+s\delta, x, u_\varepsilon(t+\delta, x)) \to j_t(t, x, u_\varepsilon(t, x)) \text{ in } dt \otimes dx \text{ measure},$$

and by (3.49) and (3.6) it follows that

$$j_t(t+s\delta, x, u_\varepsilon(t+\delta, x)) \le \frac{1}{2}h(x)|u_\varepsilon(t+\delta, x)|^2$$

for all $\delta < \delta_0$ and all $s \in [0, 1]$, $(t, x) \in [0, T] \times \mathbb{R}^d$, where the latter term by (3.47) converges in L^1 as $\delta \to 0$, (3.51) implies (3.48).

To prove (3.45), we integrate (3.48) over $(0, t)$. By (3.21) and (3.49), we find

$$\begin{aligned}
0 &\le \int_{\mathbb{R}^d} j(t, x, u_\varepsilon(t, x))dx\\
&\le \int_{\mathbb{R}^d} j(0, x, u_0(x))\,dx + \frac{1}{2}|h|_\infty\int_0^t |u_\varepsilon(s)|_2^2 ds\\
&\quad - \int_0^t |\nabla\beta(s, u_\varepsilon(s))|_2^2 - \varepsilon\int_0^t |\beta(s, u_\varepsilon(s))|_2^2 ds\\
&\quad + \int_0^t \langle b^*(s, u_\varepsilon(s)), \nabla\beta(s, u_\varepsilon(s))\rangle_2\,dx,
\end{aligned}$$

which by (i), (ii) and (3.46), (3.49) implies that

$$0 \le \frac{1}{2}(|a|_\infty + (|h|_\infty + |b|_\infty^2)T)|u_0|_2^2 - \frac{1}{2}\int_0^t |\nabla\beta(s, u_\varepsilon(s))|_2^2 ds.$$

Therefore, by (3.46), inequality (3.45) follows. Hence, along a subsequence, again denoted $\{\varepsilon\} \to 0$, we have

$$\begin{aligned} u_\varepsilon &\to u \quad \text{weak-star in } L^\infty(0,T;L^2),\ \text{weakly in } L^2(0,T;H^1) \\ \beta(u_\varepsilon) &\to \eta \quad \text{weakly in } L^2(0,T;H^1) \\ \frac{du_\varepsilon}{dt} &\to \frac{du}{dt} \quad \text{weakly in } L^2(0,T;H^{-1}) \\ b(u_\varepsilon)u_\varepsilon &\to \zeta \quad \text{weakly in } L^2(0,T;L^2) \end{aligned}$$

$$\begin{aligned} \frac{du}{dt} - \Delta\eta + \operatorname{div}\zeta &= 0 \text{ in } \mathscr{D}'((0,T)\times\mathbb{R}^d), \\ u(0,x) &= u_0(x), \quad x \in \mathbb{R}^d. \end{aligned} \tag{3.52}$$

Moreover, since H^1 is compactly embedded in $L^2_{\text{loc}}(\mathbb{R}^d)$, by the Aubin–Lions compactness theorem (see, e.g., [4], p. 26), it follows that, for $\varepsilon \to 0$, we have

$$u_\varepsilon \longrightarrow u \text{ strongly in } L^2(0,T;L^2_{\text{loc}}(\mathbb{R}^d)), \tag{3.53}$$

hence, selecting another subsequence, if necessary, $u_\varepsilon \to u$, a.e. in $(0,T)\times\mathbb{R}^d$, and so

$$\beta(u_\varepsilon) \longrightarrow \beta(u), \text{ a.e. in } (0,T)\times\mathbb{R}^d.$$

Hence $\eta = \beta(u)$ and $\zeta = b(u)u$, a.e. on $(0,T)\times\mathbb{R}^d$ and thus u solves (3.1). Letting $\varepsilon \to 0$ in (3.45) and taking into account (3.21), we obtain the estimates

$$\begin{aligned} |u(t)|^2_{L^\infty(0,T;L^2)} + \int_0^T \int_{\mathbb{R}^d} (|\nabla u(t,x)|^2 + |\nabla\beta(t,x,u(t,x))|^2)dt\,dx &\le C|u_0|_2^2, \\ \left|\frac{du}{dt}\right|_{L^2(0,T;H^{-1})} &\le C. \end{aligned}$$

Hence $u \in L^\infty(0,T;L^2) \cap L^2(0,T;H^1) \cap W^{1,2}([0,T];H^{-1})$ and $\beta(u) \in L^2(0,T;H^1)$, in particular, $u \in C([0,T];L^2)$.

This implies that u satisfies (3.14) and (3.15), and so it is a strong solution to (3.1). The uniqueness proof implied by (3.16), which is based on the monotonicity of $r \to \beta(\cdot,\cdot,r)$ and the fact that b^* is Lipschitz by (ii), is immediate and so it will be omitted.

Assume now that $u_0 \in L^1 \cap D_0$. To prove that u satisfies (3.9)–(3.11), we multiply equation (3.1) by $\mathscr{X}_\delta(u(t,x))$, where $\delta > 0$ and $\mathscr{X}_\delta$ is the function (2.41). We note that, for dt-a.e. $t \in (0,T)$, since $u(t) \in H^1$, it follows that $\mathscr{X}_\delta(u(t)) \in H^1$.

(Here, $\mathscr{X}_\delta$ is the function (2.41).) If we apply ${}_{H^{-1}}\langle \cdot, \mathscr{X}_\delta(u(t,\cdot))\rangle_{H^1}$ to (3.1) and integrate over $(0, t)$, we get

$$\begin{aligned}
&\int_{\mathbb{R}^d} j_\delta(u(t,x))dx + \int_0^t \int_{\mathbb{R}^d} \nabla\beta(s,x,u(s,x)) \cdot \nabla \mathscr{X}_\delta(u(s,x))dx\,ds \\
&\quad = \int_0^t \int_{\mathbb{R}^d} b^*(s,u(s,x)) \cdot \nabla \mathscr{X}_\delta(u(s,x))dx\,ds \\
&\qquad + \int_{\mathbb{R}^d} j_\delta(u_0(x))dx, \ \forall t \in [0,T],
\end{aligned}$$

$$j_\delta(r) = \int_0^r \mathscr{X}_\delta(s)ds, \ \forall r \in \mathbb{R}. \tag{3.54}$$

Taking into account that by (3.3) β_u, $\mathscr{X}_\delta' \geq 0$, a.e. on $(0,T) \times \mathbb{R}^d$ and $\mathbb{R}$, respectively, we have by (3.6) that, for dt-a.e. $t \in (0,T)$,

$$\begin{aligned}
&\int_{\mathbb{R}^d} \nabla\beta(t,x,u(t,x)) \cdot \nabla\mathscr{X}_\delta(u(t,x))dx \\
&\quad = \int_{\mathbb{R}^d} (\beta_u(t,x,u(t,x))\nabla u(t,x) + \beta_x(t,x,u(t,x)))\nabla\mathscr{X}_\delta(u(t,x))dx \\
&\quad \geq \int_{\mathbb{R}^d} \beta_x(t,x,u(t,x)) \cdot \nabla\mathscr{X}_\delta(u(t,x))dx \\
&\quad \geq -\frac{1}{\delta}\int_{[|u(t,x)|\leq\delta]} h(x)|u(t,x)| \cdot |\nabla u(t,x)|dx \\
&\quad \geq -|h|_2 \left(\int_{[|u(t,x)|\leq\delta]} |\nabla u(t,x)|^2 dx\right)^{\frac{1}{2}} = -\eta_\delta(t).
\end{aligned} \tag{3.55}$$

Since $\nabla u(t,x) = 0$, a.e. on $[x \in \mathbb{R}^d;\ u(t,x) = 0]$, it follows that, for dt-a.e. $t \in (0,T)$, $\eta_\delta(t) \to 0$ as $\delta \to 0$. We have, therefore,

$$\liminf_{\delta\to 0} \int_0^t \int_{\mathbb{R}^d} \nabla\beta(s,x,u(s,x)) \cdot \nabla\mathscr{X}_\delta(u(s,x))dx\,ds \geq 0,$$

and, similarly, it follows by (L2), part (3.8), that

$$\begin{aligned}
&\lim_{\delta\to 0} \left|\int_0^t \int_{\mathbb{R}^d} b^*(s,u(s,x)) \cdot \nabla\mathscr{X}_\delta(u(s,x))\right| dx\,ds \\
&\quad \leq \lim_{\delta\to 0} \int_0^t \int_{[|u(s,x)|\leq\delta]} h(x)|\nabla u(s,x)|dx\,ds = 0,
\end{aligned} \tag{3.56}$$

because $h \in L^2$. This yields, since $0 \leq j_\delta(r) \leq |r|, r \in \mathbb{R}$,

$$\liminf_{\delta \to 0} \int_{\mathbb{R}^d} j_\delta(u(t,x))dx \leq \int_{\mathbb{R}^d} |u_0(x)|dx$$

and, since $\lim\limits_{\delta \to 0} j_\delta(r) = |r|, r \in \mathbb{R}$, we infer by Fatou's lemma that

$$\int_{\mathbb{R}^d} |u(t,x)|dx \leq \int_{\mathbb{R}^d} |u_0(x)|dx, \ \forall t \in [0,T], \tag{3.57}$$

and, therefore, $u \in L^\infty(0,T;L^1)$ with $|u(t)|_1 \leq |u_0|_1$, for all $t \in [0,T]$.

If $u_0 \in D_0$ and $u_0 \geq 0$, a.e. on $\mathbb{R}^d$, then applying ${}_{H^{-1}}\langle \cdot, u^-(t)\rangle_{H^1}$ to (3.1) and integrating over $(0,T)$, we see that

$$\int_{\mathbb{R}^d} |u^-(t,x)|^2 dx = 0, \ \forall t \in [0,T],$$

and so $u \geq 0$, a.e. in $(0,T) \times \mathbb{R}^d$ and (3.5) is proved.

Finally, integrating (3.1) over $\mathbb{R}^d$, we see that (3.11) holds. (More precisely, one multiplies (3.1) with $\psi_n \in C_0^2(\mathbb{R}^d)$, $n \in \mathbb{N}$, with uniformly bounded first and second order derivatives such that $0 \leq \psi_n \nearrow 1$, integrates over $\mathbb{R}^d$ and lets $n \to \infty$.)

It remains to prove the weak-continuity condition (3.10). So, let $u_0 \in L^1 \cap D_0$ with $u_0 \geq 0$ and $|u_0|_1 = 1$. Let u be the solution to (3.1) with the initial condition u_0 and let $t_n, t \in [0,T]$, such that $t_n \to t$ as $n \to \infty$. Then, the probability measures $\mu_n(dx) := u(t_n,x)dx$ converge vaguely to the probability measure $\mu(dx) := u(t,x)dx$, because $u \in C([0,T];L^2)$. Since, for probability measures, vague and weak convergence are equivalent, (3.6) follows.

Assume now (3.18) and let us prove (3.19). To this end, it is convenient to use the finite difference scheme (3.38) and (3.39). Namely,

$$\begin{aligned} &\frac{1}{\mu}(u^\varepsilon_{i+1} - u^\varepsilon_i) - \Delta\beta((i+1)\mu, u^\varepsilon_{i+1}) + \varepsilon\beta((i+1)\mu, u^\varepsilon_{i+1}) \\ &\qquad + \mathrm{div}(b^*((i+1)\mu, u^\varepsilon_{i+1})) = 0, \ i = 0,1,\dots,N \text{ in } \mathbb{R}^d, \\ &u^\varepsilon_0 = u_0, \end{aligned} \tag{3.58}$$

for the solution $u^\varepsilon = u^\varepsilon(t,u_0)$ to (3.21) and, similarly, for $\bar{u}^\varepsilon = u^\varepsilon(t,\bar{u}_0)$, where $u_0, \bar{u}_0 \in L^1 \cap D_0$. Here again we supress the x-dependence in the notation. We have

$$\int_{\mathbb{R}^d} |u^\varepsilon_{i+1}|dx \leq \int_{\mathbb{R}^d} |u_0|dx, \ \ \forall i = 0,1,\dots,N. \tag{3.59}$$

We postpone for the time being the proof of (3.59) and we set $u_{i+1} = u^\varepsilon_{i+1}$ and $\bar{u}_{i+1} = \bar{u}^\varepsilon_{i+1}$ (corresponding to $\bar{u}_0$). Also we shall write

$$\beta_i(u_i) = \beta(i\mu, x, u_i(x)), \ \ \beta_i(\bar{u}_i) = \beta(i\mu, x, \bar{u}_i(x)),$$

and recall also that $b^*(t, x, u)$ is simply denoted $b^*(t, u)$. We get

$$\begin{aligned}&\frac{1}{\mu}(u_{i+1}-\bar{u}_{i+1})-\Delta(\beta_{i+1}(u_{i+1})-\beta_{i+1}(\bar{u}_{i+1}))+\varepsilon(\beta_{i+1}(u_{i+1})-\beta_{i+1}(\bar{u}_{i+1}))\\ &\quad+\operatorname{div}(b^*((i+1)\mu, u_{i+1})-b^*((i+1)\mu, \bar{u}_{i+1})) = \frac{1}{\mu}(u_i-\bar{u}_i), \ i = 0, 1, \dots, N.\end{aligned} \tag{3.60}$$

Then, multiplying by $\mathscr{X}_\delta(\beta_{i+1}(u_{i+1}) - \beta_{i+1}(\bar{u}_{i+1}))$ and integrating over $\mathbb{R}^d$, we get

$$\begin{aligned}&\frac{1}{\mu}\int_{\mathbb{R}^d}(u_{i+1}-\bar{u}_{i+1})\mathscr{X}_\delta(\beta_{i+1}(u_{i+1})-\beta_{i+1}(\bar{u}_{i+1}))dx\\ &\quad+\varepsilon\int_{\mathbb{R}^d}(\beta_{i+1}(u_{i+1})-\beta_{i+1}(\bar{u}_{i+1}))\mathscr{X}_\delta(\beta_{i+1}(u_{i+1})-\beta_{i+1}(\bar{u}_{i+1}))dx\\ &\quad+\int_{\mathbb{R}^d}|\nabla(\beta_{i+1}(u_{i+1})-\beta_{i+1}(\bar{u}_{i+1}))|^2\mathscr{X}_\delta'(\beta_{i+1}(u_{i+1})-\beta_{i+1}(\bar{u}_{i+1}))dx\\ &\quad-\int_{\mathbb{R}^d}(b^*((i+1)\mu, u_{i+1})-b^*((i+1)\mu, \bar{u}_{i+1}))\cdot\\ &\quad\cdot\nabla(\beta_{i+1}(u_{i+1})-\beta_{i+1}(\bar{u}_{i+1}))\mathscr{X}_\delta'(\beta_{i+1}(u_{i+1})-\beta_{i+1}(\bar{u}_{i+1}))dx\\ &\quad=\frac{1}{\mu}\int_{\mathbb{R}^d}(u_i-\bar{u}_i)\mathscr{X}_\delta(\beta_{i+1}(u_{i+1})-\beta_{i+1}(\bar{u}_{i+1}))dx.\end{aligned} \tag{3.61}$$

We set

$$v^\delta_{i+1} = \mathscr{X}_\delta(\beta_{i+1}(u_{i+1}) - \beta_{i+1}(\bar{u}_{i+1})) - \mathscr{X}_\delta(u_{i+1} - \bar{u}_{i+1})$$

and note that

$$\sup_{\delta\in(0,1)}\|v^\delta_{i+1}\|_\infty \le 2 \ \text{ and } \ \lim_{\delta\to 0} v^\delta_{i+1} = 0, \ \text{a.e. on } \mathbb{R}^d, \tag{3.62}$$

$$\begin{aligned}&((u_{i+1}-\bar{u}_{i+1})-(u_i-\bar{u}_i))\mathscr{X}_\delta(u_{i+1}-\bar{u}_{i+1})\\ &\quad\ge j_\delta(u_{i+1}-\bar{u}_{i+1})-j_\delta(u_i-\bar{u}_i), \ \forall i,\end{aligned}$$

where j_δ is defined by (3.54). Then, by (3.61), this

$$\begin{aligned}\int_{\mathbb{R}^d} j_\delta(u_{i+1}-\bar{u}_{i+1})dx &\le \int_{\mathbb{R}^d} j_\delta(u_0-\bar{u}_0)dx\\ &-\sum_{j=0}^{i}\int_{\mathbb{R}^d} v^\delta_{j+1}[(u_{j+1}-\bar{u}_{j+1})-(u_j-\bar{u}_j)]dx\\ &+\sum_{j=0}^{i}\mu\int_{\mathbb{R}^d}(b^*((j+1)\mu,u_{j+1})-b^*((j+1)\mu,\bar{u}_{j+1}))\cdot\\ &\cdot\nabla(\beta_{j+1}(u_{j+1})-\beta_{j+1}(\bar{u}_{j+1}))\mathscr{X}'_\delta(\beta_{j+1}(u_{j+1})-\beta_{j+1}(\bar{u}_{j+1}))dx,\end{aligned} \tag{3.63}$$

for all $i \le N$.

On the other hand, we have by (3.3) and (3.18)

$$\begin{aligned}&\lim_{\delta\to 0}\int_{\mathbb{R}^d}|b^*((j+1)\mu,u_{j+1})-b^*((j+1)\mu,\bar{u}_{j+1})|\\ &\quad\cdot|\nabla(\beta_{j+1}(u_{j+1})-\beta_{j+1}(\bar{u}_{j+1}))|\mathscr{X}'_\delta(\beta_{j+1}(u_{j+1})-\beta_{j+1}(\bar{u}_{j+1}))dx\\ &\le\lim_{\delta\to 0}\frac{1}{\delta}\int_{\left[|u_{j+1}-\bar{u}_{j+1}|\le\frac{\delta}{\nu}\right]}h(x)|u_{j+1}-\bar{u}_{j+1}|\cdot|\nabla(\beta_{j+1}(u_{j+1})\\ &\quad-\beta_{j+1}(\bar{u}_{j+1}))|dx=0,\end{aligned}$$

because $h, \nabla\beta_{j+1}(u_{j+1}), \nabla\beta_{j+1}(\bar{u}_{j+1}) \in L^2$ and $\nabla(\beta_{j+1}(u_{j+1})-\beta_{j+1}(\bar{u}_{j+1})) = 0$ on $[|u_{j+1}-\bar{u}_{j+1}|=0]$. Then, by (3.62) and (3.63), we get

$$\lim_{\delta\to 0}\int_{\mathbb{R}^d} j_s(u_{i+1}-\bar{u}_{j+1})dx \le |u_0-\bar{u}_0|_1,\ \forall i=0,1,\dots N-1,$$

and, therefore,

$$|u_{i+1}-\bar{u}_{i+1}|_1 \le |u_0-\bar{u}_0|_1,\ \forall i=0,1,\dots,N.$$

Then, by (3.40), we obtain that

$$|u^\varepsilon_h(t)-\bar{u}^\varepsilon_h(t)|_1 \le |u_0-\bar{u}_0|_1,\ \forall u_0,\bar{u}_0\in L^1\cap L^2,$$

and so, by (3.39), it follows that

$$|u_\varepsilon(t,u_0)-u_\varepsilon(t,\bar{u}_0)|_1 \le |u_0-\bar{u}_0|_1,\ \forall u_0,\bar{u}_0\in L^1\cap L^2,\ \forall t\in(0,T]. \tag{3.64}$$

Moreover, by (3.53) it follows that, for $\varepsilon \to 0$, $u_\varepsilon(\cdot, u_0) \to u(\cdot, u_0)$ in $L^2(0, T; B_R)$ for all $B_R = \{x \in \mathbb{R}^d; \|x| \le R\}$ and, therefore, (3.64) yields

$$\|u(t, u_0) - u(t, \bar{u}_0)\|_{L^1(B_R)} \le |u_0 - \bar{u}_0|_1, \quad \text{a.e. } t \in (0, T),$$

for all $R > 0$ and so (3.19) follows.

Proof of (3.59) Taking into account that

$$\mathcal{X}_\delta(u_{i+1})(u_{i+1} - u_i) \ge j_\delta(u_{i+1}) - j_\delta(u_i), \ \forall i = 0, 1, \dots, N, \ \text{a.e. in } \mathbb{R}^d,$$

and arguing as in (3.61)–(3.63), we see that

$$\int_{\mathbb{R}^d} j_\delta(u_{i+1})dx \le \int_{\mathbb{R}^d} |u_0|dx + \mu \sum_{j=0}^{i} \int_{\mathbb{R}^d} b^*((j+1)\mu, u_{j+1}) \cdot \nabla u_{j+1} \mathcal{X}'_\delta(u_{j+1})dx$$
$$\forall i = 0, 1, \dots, N,$$

which, as above, by (3.8) for $\delta \to 0$ yields (3.59), as claimed. □

Finally, let us prove the application to the McKean–Vlasov SDE (3.12):

Proof of Corollary 3.1 Let $u_0 \in L^1 \cap D_0$, $u_0 \ge 0$ and $\int_{\mathbb{R}^d} u_0 dx = 1$. Let u be the strong solution from Theorem 3.1. Then $\mu_t(dx) := u(t, x)dx$, $t \in [0, T]$, are probability measures on $\mathbb{R}^d$, weakly continuous in t, which solve (3.20) and

$$\int_0^T \int_{\mathbb{R}^d} (|a(t, x, u(t, x))| + |b(t, x, u(t, x))|)u(t, x)dx\, dt < \infty,$$

since both a and b are bounded. Now the assertion follows by Theorem 5.1 below. □

3.2 Fokker–Planck Periodic Flows

Theorem 3.1 may be used to study the existence of almost-periodic or periodic flows generated by the nonlinear Fokker–Planck equation. Herein, we shall study the later case. Namely, we consider here the NFPE (2.9) with the μ-periodic drift $D(t, \cdot)$, namely,

$$u_t(t, x) - \Delta\beta(t, x)) + \operatorname{div}(D(t, x)b(u(t, x))u(t, x)) = 0, \ t \ge 0, \ x \in \mathbb{R}^d, \tag{3.65}$$

under the following hypotheses

(j) $\beta \in C^2(\mathbb{R}),\ \beta(0)=0,\ \beta'' \in C_b(\mathbb{R}),\ 0<\gamma \le \beta'(r) \le \gamma_1 < \infty,\ \forall r \in \mathbb{R}$.
(jj) $b \in C_b(\mathbb{R}) \cap C^1(\mathbb{R}),\ b(r) \ge b_0 > 0,\ \forall r \in \mathbb{R}$.
(jjj) $D \in C^1([0,\infty)\times\mathbb{R}^d) \cap C_b(0,\mu;\mathbb{R}^d) \cap L^2(0,\mu;L^2)$, $\mathrm{div}_x D \ge 0$ *on* $(0,\infty)\times\mathbb{R}^d$, $D(t+\mu,x) \equiv D(t,x),\ \forall (t,x) \in [0,\infty)\times\mathbb{R}^d$.
(jv) $D(t,x) \equiv -\nabla_x \Psi(t,x)$, *where* $\Psi \in C^1([0,\mu]\times\mathbb{R}^d)$, $\Psi_t, \nabla_x\Psi \in L^2((0,\mu)\times\mathbb{R}^d)$, $\Psi(0,x) \equiv \Psi(\mu,x)$; $\Psi \ge 1$, $\lim\limits_{|x|\to\infty} \Psi(t,x) = +\infty$, *uniformly in* $t \in [0,\mu)$.

It should be noted that Hypotheses (j)–(jj) agree with Hypotheses (i), (ii), and so Theorem 3.1 is true in the present situation.

Theorem 3.2 *Under Hypotheses* (j)–(jv), *there is at least one strong solution to Eq.* (3.65) *which satisfies* (3.17) *and is* μ*-periodic in t, that is,*

$$u(t+\mu,x) \equiv u(t,x),\ \forall t \in \mathbb{R},\ x \in \mathbb{R}^d. \tag{3.66}$$

Proof For given $M_1, M_2 > 0$, we consider the closed convex set of L^1,

$$K = \left\{ u \in L^1 \cap L^2;\ |u|_2 \le M_1,\ u \ge 0,\ \text{a.e.in } \mathbb{R}^d, \right.$$
$$\left. \int_{\mathbb{R}^d} u(x)dx = 1,\ \int_{\mathbb{R}^d} \Psi(0,x)u_0(x)dx \le M_2 \right\},$$

and define on K the operator (the Poincaré map)

$$\Gamma(u_0) = u(\mu,u_0),\ u_0 \in K,$$

where $u(t,u_0)$ is the strong solution to (3.55) given by Theorem 3.1. We are going to prove, via the Schauder theorem in L^1, that Γ has a fixed point on K and conclude so the proof of Theorem 3.2. To this end, we prove that Γ is continuous in L^1, $\Gamma(K) \subset K$ and that $\Gamma(K)$ is relatively compact in L^1.

These intermediate steps are given in Lemmas 3.2 and 3.3 which follow.

Lemma 3.2 $\Gamma : L^1 \to L^1$ *is continuous and* $\Gamma(K) \subset K$.

Proof We note that

$$|\Gamma(u_0) - \Gamma(\bar{u}_0)|_1 \le |u_0 - \bar{u}_0|_1,\ \forall u_0, \bar{u}_0 \in K, \tag{3.67}$$

because, by Theorem 3.1, part (3.19), we have $|u(t) - \bar{u}(t)|_1 \le |u_0 - \bar{u}_0|_1,\ \forall t \ge 0$. Moreover, if $u_0 \in K$, then, as seen earlier, we have

$$|\Gamma(u_0)|_1 \le |u_0|_1, \tag{3.68}$$

$$\Gamma(u_0) \ge 0,\ \text{ a.e. in } \mathbb{R}^d,\ \ \textstyle\int_{\mathbb{R}^d} \Gamma(u_0(x))dx = 1. \tag{3.69}$$

We also have

$$|\Gamma(u_0)|_2 \le |u_0|_2 \le M_2,\ \forall u_0 \in L^1 \cap L^2. \tag{3.70}$$

Indeed, if we multiply equation (3.65) by $u(t)$ and integrate on $(0,t)\times\mathbb{R}^d$, we get

$$\begin{aligned}
&\frac{1}{2}\,|u(t)|_2^2 + \gamma \int_0^t |\nabla u(s)|_2^2 ds \\
&\quad \le \frac{1}{2}\,|u_0|_2^2 + \int_0^t \int_{\mathbb{R}^d} D(s,x)\cdot\nabla u(s,x) b(u(s,x))dsdx \\
&\quad = \frac{1}{2}\,|u_0|_2^2 + \int_0^t \int_{\mathbb{R}^d} D(s,x)\cdot\nabla H(u(s,x))dsdx \\
&\quad = \frac{1}{2}\,|u_0|_2^2 - \int_0^t \int_{\mathbb{R}^d} \mathrm{div}_x D(s) H(u(s,x))dsdx \le \frac{1}{2}\,|u_0|_2^2,\ \forall t \ge 0.
\end{aligned} \tag{3.71}$$

(Here, $H(r) = \int_0^r b^*(s)ds \ge 0,\ \forall r \in \mathbb{R}$.) then, (3.70) follows.

Let us prove now that

$$\int_{\mathbb{R}^d} \Psi(\mu,x)u(\mu,x)dx \le M_2,\ \forall u_0 \in K. \tag{3.72}$$

To this end, we note that, by Hypothesis (jv), we have

$$\begin{aligned}
&\frac{d}{dt}\int_{\mathbb{R}^d} \Psi(t,x)u(t,x)dx \\
&\quad = \int_{\mathbb{R}^d} \Psi_t(t,x)u(t,x)dx + \int_{\mathbb{R}^d} \Psi(t,x)(\Delta\beta(u(t,x)) \\
&\qquad - \mathrm{div}_x(D(t,x)b^*(u(t,x)))dx \\
&\quad = \int_{\mathbb{R}^d} \Psi_t(t,x)u(t,x)dx + \int_{\mathbb{R}^d} \nabla_x\Psi(t,x)\cdot\nabla\beta(u(t,x))dx \\
&\qquad - \int_{\mathbb{R}^d} D(t,x)\cdot\nabla_x\Psi(t,x)b^*(u(t,x))dx \\
&\quad \le |\Psi_t(t)|_2|u(t)|_2 + \gamma_1|\nabla_x\Psi(t)|_2|\nabla u(t)|_2 \\
&\qquad + |b|_\infty|u(t)|_2|D(t)|_\infty|\nabla_x\Psi(t)|_2.
\end{aligned} \tag{3.73}$$

Then, by (3.71) and (jv), we get

$$\int_{\mathbb{R}^d} \Psi(t,x)u(t,x)dx \le \mu|u_0|_2\left(\int_0^\mu |\Psi_t(t)|_2 + |b|_\infty\right)$$
$$+|\nabla_x\Psi|_{L^2((0,\mu)\times\mathbb{R}^d)}(\gamma_1|u_0|_\infty + \sqrt{\mu}|D|_\infty) \le C,\ \forall t \in [0,\mu].$$

Hence, since $t \to \Psi(t,\cdot)$ is μ-periodic, we have

$$\int_{\mathbb{R}^d} \Psi(t,x)u(t,x)dx \le M_3,\ \forall t \in [0,\mu],\ u_0 \in K,$$

as claimed.

Lemma 3.3 *The set $\Gamma(K)$ is relatively compact in $L^1(\mathbb{R})$.*

Proof By (3.71), we have

$$|u(t)|_2^2 + \int_0^t |\nabla u(s)|_2^2 ds \le C(|u_0|_2^2 + 1),\ \forall t \in [0,\mu).$$

Hence,

$$\int_0^\mu |u(t)|_{H^1}^2 dt \le C(|u_0|_2^2 + 1) \le C(M_1^2 + 1),$$

where C is independent of $u_0 \in K$.

This implies that, for each $k \in \mathbb{N}$,

$$m\{t \in [0,\mu];\ |u(t)|_{H^1} \ge k\} \le C_1/k^2,$$

where C_1 is independent of k. Therefore, there is a sequence $\{t_n\} \in (0,\mu)$, $t_n \to \mu$, such that for each t_m the set $\{u(t_n,u_0);\ u_0 \in K\}$ is compact in $L^1_{\rm loc}$. On the other hand, by Eq. (3.65), we see that

$$|u(t+h,u_0) - u(t,u_0)|_1 \le C(|u(h,u_0) - u_0|_1 + h),\ \forall t,h \in [0,\mu],$$

while

$$|u(h,u_0) - u_0|_1 \le |\Delta\beta(u_0) - {\rm div}(D(h)b^*(u_0))|_1 \le Ch(|u_0|_{H^2} + 1).$$

Since, for each t_n, $\{u(t_n,u_0);\ u_0 \in K\}$ is compact in $L^1_{\rm loc}$, this implies that, for each $N > 0$, the set $\{u(\mu,u_0);\ u_0 \in K;\ |u_0|_{H^2} \le N\}$ is compact in $L^1_{\rm loc}$. Since the latter set is dense (in L^1-topology) in K, we infer that $\{u(\mu,u_0);\ u_0 \in K\}$ is relatively compact in $L^1_{\rm loc}$.

Finally, recalling that

$$\int_{\mathbb{R}^d} \Psi(\mu, x)u(\mu, x)dx \le M_3,\ \forall u_0 \in K,$$

it follows by (jv) and by the compactness of $\{u(\mu, u_0);\ u_0 \in K\}$ in L^1_{loc} that the latter is relatively compact in L^1. Hence, the set $\Gamma(K)$ is relatively compact in L^1, as claimed. This completes the proof.

Comments to Chap. 3

The main result, Theorem 3.1 was given in an appropriate form in [18]. It should be noted that the regularity conditions (3.4) and (3.5) were imposed herein in order to represent NFPE (3.1) as a smooth evolution equation in the space H^{-1}. An alternative way, we did not pursue herein, is to treat (3.1) as a time-dependent nonlinear Cauchy problem of monotone type in a pair of dual spaces (V, V^*), where $V = L^2$ and V^* is the dual of V in the pairing defined by the pivot space H^{-1}. In such a case, the measurability of functions $t \to a(t, \cdot)$, $t \to b(t, \cdot)$ is sufficient but the resulting solution u is less regular.

Chapter 4
Convergence to Equilibrium of Nonlinear Fokker–Planck Flows

The convergence to an equilibrium state of solutions is a central problem in the theory of Fokker–Planck equations for its implications in statistical mechanics. As a matter of fact, the solution $u = u(t, x)$ to NFPE is a transient state which links the initial nonequilibrium state u_0 with a final equilibrium state and the convergence to equilibrium state is the *H-theorem* as initially called it by Stefan Boltzmann. In this chapter, we shall prove with the semigroup techniques in $L^1(\mathbb{R}^d)$ the H-theorem for the NFP flow $S(t)$ in the special case of nondegenerate diffusion term β and also the behaviour of $S(t)u_0$ for $t \to \infty$ in the degenerate case and, in particular, for power-law nonlinear diffusions.

4.1 The H-Theorem for the Nonlinear Fokker–Planck Equations

In classical statistical mechanics, the Boltzmann H-theorem amounts to saying that the entropy functional

$$S(f) = -H(f) = \int f(x, t, v) \log f(x, v) dx dv$$

is nondecreasing in time and the state $f(t, x, v)$, which is the density of particles with position in x and velocity v, is approaching an equilibrium state as $t \to \infty$. The state $f = f(t, x, v)$ is the solution to the classical Boltzmann's equation which describes the dynamics of an isolated ideal gas. Here we shall study this problem for the nonlinear Fokker–Planck equation (2.2), with nondegenerate diffusion term

V. Barbu, M. Röckner, *Nonlinear Fokker-Planck Flows and their Probabilistic Counterparts*, Lecture Notes in Mathematics 2353,
https://doi.org/10.1007/978-3-031-61734-8_4

β and the drift $D = -\nabla\Phi$, where $\Phi = \Phi(x)$ is a given potential which goes to $+\infty$ as $|x| \to \infty$. Namely, consider the equation

$$\begin{aligned} u_t - \Delta\beta(u) + \operatorname{div}(Db(u)u) &= 0 \text{ in } (0,\infty) \times \mathbb{R}^d, \\ u(0,x) &= u_0(x),\ x \in \mathbb{R}^d, \end{aligned} \tag{4.1}$$

under the following hypotheses on the functions $\beta : \mathbb{R} \to \mathbb{R}$, $D : \mathbb{R}^d \to \mathbb{R}^d$ and $b : \mathbb{R} \to \mathbb{R}$, where $1 \le d < \infty$.

(K1) $\beta \in C^1(\mathbb{R}),\ \beta(0) = 0,\ \gamma \le \beta'(r) \le \gamma_1,\ \forall r \in \mathbb{R}$, for $0 < \gamma < \gamma_1 < \infty$.
(K2) $b \in C_b(\mathbb{R}) \cap C^1(\mathbb{R})$.
(K3) $D \in L^\infty(\mathbb{R}^d;\mathbb{R}^d) \cap W^{1,1}_{\mathrm{loc}}(\mathbb{R}^d;\mathbb{R}^d)$ and $\operatorname{div} D \in (L^2 + L^\infty)$.
(K4) $D = -\nabla\Phi$, *where* $\Phi \in C(\mathbb{R}^d) \cap W^{2,1}_{\mathrm{loc}}(\mathbb{R}^d),\ \Phi \ge 1,\ \lim\limits_{|x|\to\infty} \Phi(x) = +\infty$ *and there exists* $m \in [2,\infty)$ *such that* $\Phi^{-m} \in L^1$.
(K5) $b(r) \ge b_0 > 0$ for $r \ge 0$.

(Here, $L^p = L^p(\mathbb{R}^d)$, $1 \le p \le \infty$.)

A typical example is $\Phi(x) = C(1 + |x|^2)^\alpha$, $x \in \mathbb{R}^d$, with $\alpha \in \left(0, \frac{1}{2}\right]$, for which we even have that $\operatorname{div} D \in L^\infty$. Under Hypotheses (K1)–(K3), it follows by Theorem 2.8 that there is a C_0-semigroup of contraction $S(t) : L^1 \to L^1$ such that, for $u_0 \in L^1$, $u(t) = S(t)u_0$ is a mild solution to (4.1). (Under Hypotheses (i)–(iv) of Theorem 2.4, this solution is unique in the class of mild solutions to (4.1).) In the following, by solution to (4.1) we mean such a function.

If (K1)–(K4) hold, we shall prove here the convergence of the solutions to equilibrium in L^1 if, in addition, the following condition holds

(K6) $\gamma_1\Delta\Phi(x) - b_0|\nabla\Phi(x)|^2 \le 0$, for a.e. $x \in \mathbb{R}^d$.

An example of such a function Φ for $d \ge 2$ is

$$\Phi(x) = \begin{cases} |x|^2 \log|x| + \mu & \text{for } |x| \le \delta, \\ \varphi(|x|) + \eta|x| + \mu & \text{for } |x| > \delta, \end{cases} \tag{4.2}$$

$\delta = \exp\left(-\frac{d+2}{2d}\right)$, and

$$\varphi(r) = \delta^2 \log\delta - \eta\delta + \int_\delta^r h(s)ds, \tag{4.3}$$

for $r \ge \delta$, where $\mu, \eta > 0$ are sufficiently large and h is suitably chosen ([19]).

In the special case where $b \equiv b_0 > 0$ and $\beta' = \gamma_1$, condition (K6) reduces to

$$\gamma_1 \operatorname{div}(D(x)) + b_0|D(x)|^2 \ge 0,\ x \in \mathbb{R}^d.$$

One of our motivations is to apply our asymptotic results *to find an invariant (probability) measure for the nonlinear distorted Brownian motion on* $\mathbb{R}^d$. Condition (K6) requires a certain balance between the strength of the (in general nonlinear) diffusion coefficient β' and the strength of the nonlinear drift coefficient b in terms of the *potential* Φ. Without this additional condition, there is in general no equilibrium on $L^1(\mathbb{R}^d)$ for Eq. (4.1). Just consider the linear case $\beta = id$ and $D \equiv 0$, so the case where (4.1) is the heat equation. Hence, as in the linear case, we need a big enough *negative* drift. Condition (K6) is, however, not optimal, because for the Fokker-Planck equation associated to the classical Ornstein–Uhlenbeck process on $\mathbb{R}^d$, it does not hold, though the standard Gaussian measure is its equilibrium measure.

We would like to mention here another special case of (4.1), namely that of a power-law diffusion and a linear drift term, that is, with $\beta(u) = u^m$, $m > 1$, $b \equiv const.$ and $D(x) = x$, which is not covered by our results, but was analyzed in [38]. In this case, the equilibrium is given through an explicit formula and the decay rate in L^1-distance is calculated. So, the approach is completely different from ours which is to prove the so-called H-theorem (see below) to show convergence of solutions to a unique equilibrium of (1.1) in L^1 as $t \to \infty$. Such a case will be also studied in Sect. 4.2.

The Construction of the Fokker–Planck Semigroup

We shall recall here for every reference the construction of the semigroup $S(t): L^1 \to L^1$ given in Sect. 2.2 under weaker hypotheses on Eq. (4.1). Namely, consider in the space $L^1 = L^1(\mathbb{R}^d)$ the operator $A_0 : D(A_0) \subset L^1 \to L^1$, defined by

$$\begin{aligned} A_0(u) &= -\Delta\beta(u) + \operatorname{div}(Db(u)u), \ \forall u \in D(A_0), \\ D(A_0) &= \{u \in L^1;\ -\Delta\beta(u) + \operatorname{div}(Db(u)u) \in L^1\}. \end{aligned} \tag{4.4}$$

Lemma 4.1 *Assume that Hypotheses* (K1)–(K4) *hold. Then, there is* $\lambda_0 > 0$ *such that*

$$R(I + \lambda A_0) = L^1, \ \forall \lambda \in (0, \lambda_0), \tag{4.5}$$

and there is an operator $J_\lambda : L^1 \to L^1$ *such that* $J_\lambda(0) = 0$ *and*

$$J_{\lambda_2}(f) = J_{\lambda_1}\left(\frac{\lambda_1}{\lambda_2} f + \left(1 - \frac{\lambda_1}{\lambda_2}\right) J_{\lambda_2}(f)\right), \ \forall \lambda_1, \lambda_2 \in (0, \lambda_0), \tag{4.6}$$

$$(I + \lambda A_0) J_\lambda(f) = f, \ \forall f \in L^1, \tag{4.7}$$

$$|J_\lambda(f_1) - J_\lambda(f_2)|_1 \le |f_1 - f_2|_1, \ f_1, f_2 \in L^1, \tag{4.8}$$

$$\overline{D(A)} = L^1, \tag{4.9}$$

$$\int_{\mathbb{R}^d} J_\lambda(f) dx = \int_{\mathbb{R}^d} f(x) dx, \ \forall f \in L^1, \tag{4.10}$$

$$J_\lambda(f) \ge 0, \ \text{a.e. in } \mathbb{R}^d \text{ if } f \ge 0, \ \text{a.e. in } \mathbb{R}^d. \tag{4.11}$$

The proof of Lemma 4.1 which is essentially the same as that of Proposition 2.3 (see also Lemma 2.8) will be outlined later on for reader's convenience. Now, we use it to define as in Sect. 2.2 the operator $A : D(A) \subset L^1 \to L^1$,

$$A(u) = A_0(u), \forall u \in D(A) = J_\nu(L^1), \tag{4.12}$$

where $\nu \in (0, \lambda_0)$ is arbitrary. Hence, $D(A) \subset D(A_0)$ and by (4.6), it follows that $D(A)$ is independent of ν.

By (4.5)–(4.9), (4.16), it follows that A *is m-accretive in* L^1 *and*

$$(I + \lambda A)^{-1}(u) = J_\lambda(u), \forall u \in L^1, \lambda > 0. \tag{4.13}$$

We note that A is an accretive section of A_0 and if $(I+\lambda A_0)^{-1}$ is single valued, then $A = A_0$. As shown in Theorem 2.4, this happens for instance if, besides (K1)–(K3), the following conditions hold

$$\operatorname{div} D \in L^k_{\text{loc}},\ k > \frac{d}{2},\ |rb'(r) + b(r)| \le \alpha\beta'(r),\quad \forall r \in \mathbb{R};\ \alpha > 0. \tag{4.14}$$

(For instance, the latter holds if the function $r \to b'(r)r$ is in L^∞.)

Consider now the Cauchy problem associated with A, that is,

$$\begin{aligned} &\frac{du}{dt} + A(u) = 0,\ t \ge 0, \\ &u(0) = u_0, \end{aligned} \tag{4.15}$$

which, as in the previous case, has a *unique mild solution* to $u(t)$, that is, if

$$u(t) = \lim_{h \to 0} u_h(t) \text{ in } L^1, \tag{4.16}$$

uniformly on compacts of $[0, \infty)$, where $u_h^1 = u_0$, and

$$u_h(t) = u_h^i,\ t \in [ih, (i+1)h),\ i = 0, 1, \dots, \tag{4.17}$$

$$u_h^i + hA(u_h^i) = u_h^{i-1},\ i = 0, \dots. \tag{4.18}$$

We have, therefore,

Proposition 4.1 *The operator* A *generates a* C_0*-semigroup* $S(t) : L^1 \to L^1$*, that is,*

$$S(t)u_0 = \lim_{n \to \infty} \left(I + \frac{t}{n}A\right)^{-n} u_0,\ \forall t \ge 0, \tag{4.19}$$

uniformly on bounded intervals of $[0, \infty)$ *in the strong topology in* L^1.

$$|S(t)u_0 - S(t)\bar{u}_0|_1 \le |u_0 - \bar{u}_0|_1, \ \forall u_0, \bar{u}_0 \in L^1, \tag{4.20}$$

$$\int_{\mathbb{R}^d} S(t)u_0\, dx = \int_{\mathbb{R}^d} u_0(x)dx, \ \forall t \ge 0, \tag{4.21}$$

$$S(t)u_0(x) \ge 0, \ a.e. \ on \ (0, \infty) \times \mathbb{R}^d \ \ if \ u_0 \ge 0, \ a.e. \ in \ \mathbb{R}^d. \tag{4.22}$$

Moreover, for each $u_0 \in L^1$, $u(t) = S(t)u_0$ *is a mild solution to* (4.1).

We consider the following subspace of L^1

$$\mathscr{M} = \left\{ u \in L^1; \int_{\mathbb{R}^d} \Phi(x)|u(x)|dx < \infty \right\} \tag{4.23}$$

with the norm

$$\|u\| = \int_{\mathbb{R}^d} \Phi(x)|u(x)|dx, \ \forall u \in \mathscr{M}. \tag{4.24}$$

We also set $\mathscr{M}_+ = \{u_0 \in \mathscr{M};\ u_0 \ge 0, \text{ a.e. on } \mathbb{R}^d\}$.

It turns out that the semigroup $S(t)$ leaves invariant $\mathscr{M}$. More precisely, as we shall prove later on, we have

Proposition 4.2 *Assume that Hypotheses* (K1)–(K4) *hold and that* $\operatorname{div} D \in L^\infty$. *Then*

$$\|S(t)u_0\| \le \|u_0\| + \gamma t |u_0|_1, \ \forall u_0 \in \mathscr{M}, \tag{4.25}$$

where $\gamma = \gamma_1(m+1)|\Delta\Phi|_\infty + |b|_\infty(1+m)^2|D|^2_\infty$.

Note that by (4.21) we have $S(t)(\mathscr{P}) \subset \mathscr{P}$, $\forall t \ge 0$, where

$$\mathscr{P} = \left\{ u \in L^1;\ u \ge 0, \int_{\mathbb{R}^d} u(x)dx = 1 \right\}. \tag{4.26}$$

Now, we come back to the proof of Lemma 4.1, which, as mentioned earlier, is essentially identical with Proposition 2.3. Namely, we define, for each $\varepsilon > 0$, the operator $A_\varepsilon : D(A_\varepsilon) \subset L^1 \to L^1$,

$$A_\varepsilon(u) = (\varepsilon I - \Delta)\beta(u) + \operatorname{div}(D_\varepsilon b^*_\varepsilon(u)), \tag{4.27}$$

$$D(A_\varepsilon) = \{u \in L^1,\ (\varepsilon I - \Delta)\beta(u) + \operatorname{div}(D_\varepsilon b^*_\varepsilon(u)) \in L^1\}. \tag{4.28}$$

Here,

$$b_\varepsilon \equiv b * \varphi_\varepsilon, \quad b_\varepsilon^*(r) \equiv \frac{b_\varepsilon(r) r}{1+\varepsilon|r|}, \; r \in \mathbb{R}, \tag{4.29}$$

where $\varphi_\varepsilon(r) \equiv \frac{1}{\varepsilon}\,\varphi\left(\frac{r}{\varepsilon}\right)$, $\varphi \in C_0^\infty(\mathbb{R})$, $\varphi \geq 0$. Moreover,

$$D_\varepsilon = -\nabla \Phi_\varepsilon, \quad \Phi_\varepsilon(x) \equiv \frac{\Phi(x)}{(1+\varepsilon\Phi(x))^k}.$$

Then $\Phi_\varepsilon \in L^2$, since $m \geq 2$, and

$$D_\varepsilon = D(1+\varepsilon\Phi)^{-k} - k\varepsilon\Phi D(1+\varepsilon\Phi)^{-(k+1)} \tag{4.30}$$

and, therefore, by Hypothesis (K4),

$$\begin{aligned} &D_\varepsilon \in (L^\infty \cap L^1)(\mathbb{R}^d;\mathbb{R}^d) \\ &|D_\varepsilon(x)| \leq (1+k)|D(x)|, \quad \lim_{\varepsilon\to 0} D_\varepsilon(x) = D(x), \quad \text{for a.e. } x \in \mathbb{R}^d, \\ &\varepsilon^k |D_\varepsilon| \leq (1+k)|D|_\infty \Phi^{-k}, \; \forall \varepsilon > 0. \end{aligned} \tag{4.31}$$

We also note that b_ε^*, b_ε are bounded and Lipschitz and that, for $\varepsilon \to 0$,

$$b_\varepsilon^*(r) \to b(r) r \;\; \text{uniformly on compacts.} \tag{4.32}$$

Lemma 4.2 *Assume that Hypotheses* (K1)–(K4) *hold. Then, for each* $\varepsilon > 0$, A_ε *is m-accretive in* L^1 *and* $J_\lambda^\varepsilon = (I + \lambda A_\varepsilon)^{-1}$ *satisfies* (4.6)–(4.8). *Namely,*

$$J_{\lambda_2}^\varepsilon(f) = J_{\lambda_1}^\varepsilon\left(\frac{\lambda_1}{\lambda_2} f + \left(1 - \frac{\lambda_1}{\lambda_2}\right) J_\lambda^\varepsilon(f)\right), \; \forall \lambda_1, \lambda_2 \in (0,\infty), \tag{4.33}$$

$$(I + \lambda A_\varepsilon) J_\lambda^\varepsilon(f) = f, \; \forall f \in L^1, \tag{4.34}$$

$$|J_\lambda^\varepsilon(f_1) - J_\lambda^\varepsilon(f_2)|_1 \leq |f_1 - f_2|_1, \; \forall f_1, f_2 \in L^1, \tag{4.35}$$

$$J_\lambda^\varepsilon(f) \geq 0, \text{ a.e. in } \mathbb{R}^d \text{ if } f \geq 0, \text{ a.e. in } \mathbb{R}^d, \; \forall \lambda \in (0,\infty), \tag{4.36}$$

$$\int_{\mathbb{R}^d} J_\lambda^\varepsilon(f) dx = \int_{\mathbb{R}^d} f\, dx, \; \forall \lambda > 0, \; \forall f \in L^1. \tag{4.37}$$

Moreover, for all $\lambda \in (0, \lambda_0)$,

$$\lim_{\varepsilon \to 0} J_\lambda^\varepsilon(f) = J_\lambda(f) \;\; \text{in } L^1, \; \forall f \in L^1, \tag{4.38}$$

where J_λ *satisfies* (4.10) *and* (4.11).

Proof Taking into account that, by Hypotheses (K1)–(K3) and by (4.29)–(4.32), Hypotheses (i)–(iv) in Sect. 2.2 are satisfied, it follows by Lemma 2.7 that A_ε is m-accretive in L^1 and that (4.33)–(4.37) hold. It remains to prove (4.38). We note first that for $f \in L^1 \cap L^2$, by virtue of (K1)–(K4) it follows by the equation

$$u_\varepsilon + \lambda A_\varepsilon(u) = f$$

or, equivalently,

$$u_\varepsilon + \lambda(\varepsilon I - \Delta)\beta(u_\varepsilon) + \lambda \operatorname{div}(D_\varepsilon b_\varepsilon^*(u_\varepsilon)) = f \text{ in } \mathscr{D}'(\mathbb{R}^d), \tag{4.39}$$

that $u_\varepsilon, \beta(u_\varepsilon) \in H^1(\mathbb{R}^d)$ and

$$\gamma |u_\varepsilon|_2^2 + \lambda |\nabla \beta(u_\varepsilon)|_2^2 + \varepsilon\lambda |\beta(u_\varepsilon)|_2^2 \le C_{\lambda_1}, \ \forall \lambda \in (0, \lambda_1), \tag{4.40}$$

where λ_1 is independent of ε. We also have

$$|u_\varepsilon|_1 \le |f|_1, \ \forall \varepsilon > 0. \tag{4.41}$$

Let $f \in L^1 \cap L^2$. By (4.40) it follows that $\{u_\varepsilon\}$ and $\{\beta(u_\varepsilon)\}$ are bounded in $H^1 = H^1(\mathbb{R}^d)$ and so, by (4.40) $\{u_\varepsilon\}$ is bounded in L^1. Hence, along a subsequence $\{\varepsilon'\} \subset \{\varepsilon\}$, for simplicity again denoted $\{\varepsilon\} \to 0$, we have

$$\begin{aligned} u_\varepsilon &\to u && \text{weakly in } H^1, \text{ strongly in } L^2_{\text{loc}}, \\ \beta(u_\varepsilon) &\to \beta(u) && \text{weakly in } H^1 \text{ and strongly in } L^2_{\text{loc}}, \\ \Delta\beta(u_\varepsilon) &\to \Delta\beta(u) && \text{weakly in } H^{-1}, \end{aligned} \tag{4.42}$$

and, by Hypothesis (K2) and (4.32),

$$b_\varepsilon^*(u_\varepsilon) \longrightarrow b(u)u \text{ strongly in } L^2_{\text{loc}}. \tag{4.43}$$

This yields

$$D_\varepsilon b_\varepsilon^*(u_\varepsilon) \to Db(u)u \text{ strongly in } L^2_{\text{loc}}. \tag{4.44}$$

Passing to the limit in (4.39), we obtain

$$u - \lambda\Delta\beta(u) + \lambda \operatorname{div}(Db(u)u) = f \text{ in } \mathscr{D}'(\mathbb{R}^d), \tag{4.45}$$

where $u = u(\lambda, f) \in H^1$. Taking into account Hypotheses (K1)–(K3), it is easily seen that Eq. (4.45) has a unique solution $u \in H^1$. Hence, for $f \in L^1 \cap L^2$, the convergence $u_\varepsilon \to u$ in (4.42) is not for a subsequence of $\{\varepsilon\}$ only, but for $\{\varepsilon\} \to 0$.

By (4.35) and (4.42), it follows that

$$|u(\lambda, f_1) - u(\lambda, f_2)|_1 \le |f_2 - f_2|_1, \ \forall f_1, f_2 \in L^2 \cap L^1, \tag{4.46}$$

and hence

$$u(\lambda, f) + \lambda A_0(u_\lambda), f = f, \ \forall f \in L^1 \cap L^2. \tag{4.47}$$

Now, let $f \in L^1$ and $f_n \in L^1 \cap L^2$, $n \in \mathbb{N}$, such that $f_n \to f$ in L^1. Then, by (4.46), $u(\lambda, f_n) \to u = u(\lambda, f)$ in L^1 and, therefore, since each $u(\lambda, f_n)$ satisfies (4.47), we conclude that $u(\lambda, f) \in D(A_0)$ and that u also satisfies (4.47), and so (4.5) follows for all $\lambda \in (0, \lambda_0)$. This extends to all $\lambda > 0$.

We define $J_\lambda : L^1 \to L^1$ as $J_\lambda(f) = u(\lambda, f)$, then (4.8) follows by (4.46). Moreover, letting $\varepsilon \to 0$ in (4.33)–(4.36), it follows that J_λ satisfies (4.6)–(4.8) and (4.10)–(4.11), as claimed.

By (4.42), we have for $0 < \lambda < \lambda_0$

$$u_\varepsilon \to u = u(\lambda, f) = J_\lambda(f) \text{ in } L^1_{\text{loc}}. \tag{4.48}$$

(In fact, as seen earlier, this follows for $f \in L^1 \cap L^2$, but taking into account (4.35) and (4.46) it extends by density to all $f \in L^1$.)

To prove that (4.38), meaning that (4.48) holds in L^1, we can invoke Lemma 2.10, but this also follows by the next lemma, which has an intrinsic interest to be used later and where we use Hypothesis (K4) for the first time.

Lemma 4.3 *Assume that Hypotheses* (K1)–(K4) *hold and let* $u_0 \in \mathcal{M} \cap L^2$. *Then, we have*

$$\sup_{\varepsilon \in (0,1)} \int_{\mathbb{R}^d} |u_\varepsilon| \Phi \, dx < \infty. \tag{4.49}$$

Assume that div $D \in L^\infty$. *Then, for all* $\lambda \in (0, \lambda_0)$,

$$\|(I + \lambda A_\varepsilon)^{-1} u_0\| \le \|u_0\| + \rho_\varepsilon \lambda |u_0|_1, \tag{4.50}$$

where $\rho_\varepsilon = \gamma_1(m+1)|\Delta\Phi|_\infty + \gamma_1 m(m+3)\varepsilon|D|^2_\infty + |b|_\infty(1+k)^2|D|^2_\infty$.

Proof By approximation we may restrict to the case $u_0 \in \mathcal{M} \cap L^2$.

If we multiply equation (4.39) by $\varphi_\nu \mathscr{X}_\delta(\beta(u_\varepsilon))$, where $u_\varepsilon = (I + \lambda A_\varepsilon)^{-1}u_0$, $\varphi_\nu(x) = \Phi_\varepsilon(x)\exp(-\nu\Phi_\varepsilon(x))$ and integrate over $\mathbb{R}^d$, we get, since $\mathscr{X}'_\delta \geq 0$,

$$
\begin{aligned}
\int_{\mathbb{R}^d} u_\varepsilon \mathscr{X}_\delta(\beta(u_\varepsilon))\varphi_\nu\, dx &\leq -\lambda \int_{\mathbb{R}^d} \nabla\beta(u_\varepsilon)\cdot\nabla(\mathscr{X}_\delta(\beta(u_\varepsilon))\varphi_\nu)dx \\
&\quad -\lambda\varepsilon \int_{\mathbb{R}^d} \beta(u_\varepsilon)\mathscr{X}_\delta(\beta(u_\varepsilon)\varphi_\nu\, dx \\
&\quad +\lambda \int_{\mathbb{R}^d} D_\varepsilon b^*_\varepsilon(u_\varepsilon)\cdot\nabla(\mathscr{X}_\delta(\beta(u_\varepsilon))\varphi_\nu)dx + \int_{\mathbb{R}^d} |u_0|\varphi_\nu dx \\
&\leq -\lambda \int_{\mathbb{R}^d} \nabla\beta(u_\varepsilon)\cdot\nabla\varphi_\nu \mathscr{X}_\delta(\beta(u_\varepsilon))dx \\
&\quad +\lambda \int_{\mathbb{R}^d} D_\varepsilon b^*_\varepsilon(u_\varepsilon)\cdot\nabla\beta(u_\varepsilon)\mathscr{X}'_\delta(\beta(u_\varepsilon))\varphi_\nu dx \\
&\quad +\lambda \int_{\mathbb{R}^d} (D_\varepsilon\cdot\nabla\varphi_\nu)b^*_\varepsilon(u_\varepsilon)\mathscr{X}_\delta(\beta(u_\varepsilon))dx + \int_{\mathbb{R}^d} |u_0|\varphi_\nu dx.
\end{aligned}
\tag{4.51}
$$

Letting $\delta \to 0$, we get as above

$$
\begin{aligned}
\int_{\mathbb{R}^d} |u_\varepsilon|\varphi_\nu dx &\leq -\lambda \int_{\mathbb{R}^d} \nabla|\beta(u_\varepsilon)|\cdot\nabla\varphi_\nu dx \\
&\quad +\overline{\lim_{\delta\to 0}}\, \frac{\lambda}{\delta} \int_{[|\beta(u_\varepsilon)|\leq\delta]} |D_\varepsilon|\,|b^*_\varepsilon(u_\varepsilon)|\,|\nabla\beta(u_\varepsilon)|\varphi_\nu dx \\
&\quad +\lambda \int_{\mathbb{R}^d} \operatorname{sign} u_\varepsilon b^*_\varepsilon(u_\varepsilon)D_\varepsilon\cdot\nabla\varphi_\nu\, dx + \int_{\mathbb{R}^d} |u_0|\varphi_\nu dx \\
&\leq \lambda \int_{\mathbb{R}^d} (|\beta(u_\varepsilon)|\Delta\varphi_\nu + |b^*_\varepsilon(u_\varepsilon)|\,|\nabla\Phi_\varepsilon\cdot\nabla\varphi_\nu|)dx + \int_{\mathbb{R}^d} |u_0|\varphi_\nu dx,
\end{aligned}
\tag{4.52}
$$

because $|b^*(u_\varepsilon)| \leq C|u_\varepsilon| \leq \frac{C}{\gamma}|\beta(u_\varepsilon)|$, a.e. in $\mathbb{R}^d$, and so

$$
\frac{1}{\delta}\int_{[|\beta(u_\varepsilon)|\leq\delta]} |D_\varepsilon|\,|b^*(u_\varepsilon)|\,|\nabla\beta(u_\varepsilon)|\varphi_\nu dx \leq \frac{C}{\gamma}|D_\varepsilon|_2\left(\int_{[|\beta(u_\varepsilon)|\leq\delta]} |\nabla\beta(u_\varepsilon)|^2 dx\right)^{\frac{1}{2}}
$$

$$
\lim_{\delta\to 0}\int_{[|v|\leq\delta]} |\nabla v|^2 dx = 0, \quad \forall v \in H^1(\mathbb{R}^d).
$$

We have

$$
\nabla\varphi_\nu(x) = (1-\nu\Phi_\varepsilon)\nabla\Phi_\varepsilon\exp(-\nu\Phi_\varepsilon), \tag{4.53}
$$

$$
\Delta\varphi_\nu(x) = ((1-\nu\Phi_\varepsilon)\Delta\Phi_\varepsilon - 2\nu|\nabla\Phi_\varepsilon|^2 + \nu^2\Phi_\varepsilon|\nabla\Phi_\varepsilon|^2)\exp(-\nu\Phi_\varepsilon), \tag{4.54}
$$

$$\Delta\Phi_\varepsilon = -\operatorname{div} D_\varepsilon = (1 - m\varepsilon\Phi(1+\varepsilon\Phi)^{-1})(1+\varepsilon\Phi)^{-k}\Delta\Phi \qquad (4.55)$$
$$+m\varepsilon((m+1)\varepsilon\Phi(1+\varepsilon\Phi)^{-1} - 2)(1+\varepsilon\Phi)^{-(m+1)}|D|^2.$$

Then, letting $\nu \to 0$, since $\beta(u_\varepsilon)$, $\varepsilon \in (0,1)$, is bounded in $L^1 \cap L^2$, we get by (4.52)–(4.55) and Hypothesis (K3) that

$$\sup_{\varepsilon\in(0,1)} \int_{\mathbb{R}^d} |u_\varepsilon|\Phi\, dx < \infty,$$

and (4.49) follows. If $\operatorname{div} D \in L^\infty$, we additionally get from (4.52) that

$$\|u_\varepsilon\| \le \|u_0\| + \lambda\gamma_1|\Delta\Phi_\varepsilon|_\infty|u_0|_1 + \lambda|b|_\infty|u_0|_1|\nabla\Phi_\varepsilon|_2^2,\ \forall\varepsilon > 0.$$

By (4.55), we have

$$|\Delta\Phi_\varepsilon(x)| \le (m+1)|\Delta\Phi(x)| + m(m+3)\varepsilon|D|^2(x) \text{ for a.e. } x \in \mathbb{R}^d, \qquad (4.56)$$

and this, together with (4.31), yields (4.50), as claimed.

Proof of (4.38) By (4.49) and Hypothesis (K4), it follows that, if $f \in \mathcal{M} \cap L^2$, then we have, for all $\lambda \in (0,\lambda_0)$ and $\varepsilon \in (0,1)$, $N > 0$,

$$\int_{\{\Phi\ge N\}} |(I+\lambda A_\varepsilon)^{-1} f|dx \le \frac{1}{N}\|(I+\lambda A_\varepsilon)^{-1} f\| \le \frac{C}{N}.$$

Recalling (4.38) and that $\{\Phi \le N\}$ is compact, the latter implies that, if $f \in \mathcal{M} \cap L^2$, then $\lim_{\varepsilon\to 0}|u_\varepsilon - u|_1 = 0$, i.e.,

$$\lim_{\varepsilon\to 0}(I+\lambda A_\varepsilon)^{-1} f = (I+\lambda A)^{-1} f \text{ in } L^1,\ \forall f \in \mathcal{M} \cap L^2. \qquad (4.57)$$

Since $L^2 \cap \mathcal{M}$ is dense in L^1 and $(I+\lambda A_\varepsilon)^{-1}$, $\varepsilon > 0$, are equicontinuous, (4.38) follows. □

Proof of (4.9) Let $f \in C_0^\infty(\mathbb{R}^d)$ and $u_\lambda = J_\lambda(f) \in D(A)$, $\lambda > 0$. Since $D(A) \subset D(A_0)$, we have

$$u_\lambda + \lambda A_0(u_\lambda) = f, \qquad (4.58)$$

where $u_\lambda \in L^1 \cap L^\infty$, $|u_\lambda|_1 \le |f|_1$, and, by Proposition 2.3,

$$\sup_{\lambda\in(0,\lambda_0)} |u_\lambda|_\infty = C_\infty < \infty. \qquad (4.59)$$

By (4.40), we also have

$$\sup_{\lambda \in (0,\lambda_0)} |u_\lambda|_2^2 = C_2 < \infty, \tag{4.60}$$

for some $\lambda_0 > 0$. Taking into account (4.59) and that $b^*(r) \equiv b(r)r$ is locally Lipschitz, it follows as in the proof of Lemma 2.8 that

$$\sup_{\lambda \in (0,\lambda_0)} \int_{\mathbb{R}^d} |u_\lambda(x)|\Phi(x)dx < \infty. \tag{4.61}$$

Next, by (4.58) we see that since $A_0(u_\lambda) \in L^2$, we have

$$(A_0(u_\lambda), u_\lambda)_2 + \lambda |A_0(u_\lambda)|_2^2 = (A_0(u_\lambda), f)_2 \leq |A_0(u_\lambda)|_2 |f|_2.$$

This yields

$$(\nabla \beta(u_\lambda), \nabla u_\lambda)_2 \leq (D, b^*(u_\lambda)\nabla u_\lambda)_2 + (\nabla \beta(u_\lambda), \nabla f)_2 - (D, b^*(u_\lambda)\nabla f)_2$$

and so, by Hypotheses (K1)–(K2) we get, for $\delta > 0$,

$$\gamma |\nabla u_\lambda|_2^2 \leq \delta(1+\gamma_1^2)|\nabla u_\lambda|^2 + \frac{1}{\delta}(|D|_\infty^2 |b|_\infty^2 |u_\lambda|_2^2 + |\nabla f|_2^2) + |D|_\infty |b|_\infty |u_\lambda|_2 |\nabla f|_2.$$

This yields

$$|\nabla u_\lambda|_2^2 \leq K_\delta(\gamma - \delta(1+\gamma_1^2))^{-1}. \tag{4.62}$$

By (4.58)–(4.62), it follows that

$$\lambda A_0(u_\lambda) \to 0 \text{ in } H^{-1} \text{ as } \lambda \to 0$$

and, therefore, $u_\lambda \to f$ in H^{-1} as $\lambda \to 0$ and so, by (4.62), we have on a subsequence $\{\lambda\} \to 0$

$$u_\lambda \to f \text{ in } L^2_{\text{loc}} \subset L^1_{\text{loc}}.$$

Then, by (4.61), we infer that for $\lambda \to 0$, $u_\lambda \to f$ in L^1 and so $f \in \overline{D(A)}$. Hence, $C_0^\infty(\mathbb{R}^d) \subset \overline{D(A)}$ and so (4.9) follows. We note that, similarly, it follows that

$$\overline{D(A_\varepsilon)} = L^1, \ \forall \varepsilon > 0.$$

This completes the proof of Lemma 4.2. □

Proof of Proposition 4.2 By (4.50) in Lemma 4.3, we have, for $\lambda \in (0, \lambda_0)$, and $\delta > 0$,

$$\|(I + \lambda A)^{-1}u_0\| \le \|u_0\| + \rho\lambda|u_0|_1, \ \forall u_0 \in \mathcal{M}.$$

This yields

$$\|(I + \lambda A)^{-n}u_0\| \le \|u_0\| + n\lambda\rho|u_0|_1, \ \forall n \in \mathbb{N},$$

and so, by (4.19), we get

$$\|S(t)u_0\| \le \|u_0\| + \rho t|u_0|_1, \ \forall t \ge 0, \ u_0 \in \mathcal{M}, \tag{4.63}$$

as claimed. □

As seen above, the operator A which generates the semigroup $S(t)$ arising in Propositions 4.1 and 4.2 is constructed via the family $\{J_\lambda\}_{\lambda>0}$ by the approximating scheme (4.27), (4.28) and (4.38), (4.39). If $S_\varepsilon(t) = \exp(-tA_\varepsilon)$ is the semigroup generated by the operator A_ε, that is,

$$\frac{d}{dt} S_\varepsilon(t)u_0 + A_\varepsilon(S_\varepsilon(t)u_0) = 0, \ t \ge 0, \tag{4.64}$$

then by (4.38) it follows via the Trotter–Kato theorem for nonlinear semigroup of contractions (see Theorem 6.9) that, for all $u_0 \in \overline{D(A)} = L^1$,

$$S(t)u_0 = \lim_{\varepsilon \to 0} S_\varepsilon(t)u_0 \text{ in } L^1(\mathbb{R}^d) \tag{4.65}$$

uniformly on compact intervals of $[0, \infty)$. Then Propositions 4.1 and 4.2 can be completed as follows (see also Theorem 2.10).

Proposition 4.3 *Under Hypotheses* (K1)–(K4) *there is a* C_0*-semigroup* $S(t)$: $L^1 \to L^1$ *satisfying* (4.19)–(4.22) *and* (4.64). *Moreover, if* $\operatorname{div} D \in L^\infty$, *then* (4.25) *holds. If, in addition,* (4.14) *holds, then the semigroup* $S(t)$ *and the mild solution* u *to* (4.1) *are unique.*

Everywhere in the following we shall work with this semigroup-flow for NFPE (4.1) which was defined, as mentioned earlier, a viscosity-mild solution $u \in C([0, \infty); L^1)$ to (4.1). Of course, also this solution is dependent of approximating scheme chosen to solve the equation $u + \lambda A_0(u) = f$ in $\mathcal{D}'(\mathbb{R}^d)$, but it has the advantage to limit of a family $\{u_\varepsilon(t) = S_\varepsilon(t)u_\varepsilon\}$ of more regular functions.

The H-Theorem for NFPE (4.1)
Let $S(t)$ be the continuous semigroup of contractions arising in Proposition 4.1. A lower semicontinuous function $V : L^1 \to (-\infty, \infty]$ is said to be a *Lyapunov function* for $S(t)$ (equivalently, for Eq. (1.1) or (2.6)) if

$$V(S(t)u_0) \leq V(S(s)u_0), \quad \text{for } 0 \leq s \leq t < \infty, \ u_0 \in L^1.$$

In the following, we shall restrict the semigroup $S(t)$ to the probability density set $\mathscr{P}$ (see (4.25)) and, for each $u_0 \in \mathscr{P}$, consider the ω-limit set

$$\omega(u_0) = \{w = \lim S(t_n)u_0 \text{ in } L^1_{\text{loc}} \text{ for some } \{t_n\} \to \infty\}.$$

Our aim here is to construct a Lyapunov function for $S(t)$, to prove that $\omega(u_0) \neq \emptyset$ and also that every $u_\infty \in \omega(u_0)$ is an equilibrium state of Eq. (4.1), that is, $Au_\infty = 0$. To this end, we shall assume that, besides (K1)–(K4), Hypothesis (K5) also holds.

Consider the function $\eta \in C(\mathbb{R})$,

$$\eta(r) = -\int_0^r d\tau \int_\tau^1 \frac{\beta'(s)}{sb(s)}\, ds, \ \forall r \geq 0, \tag{4.66}$$

and define the function $V : D(V) = \mathscr{M}_+ = \{u \in \mathscr{M}; u \geq 0, \text{ a.e. on } \mathbb{R}^d\} \to \mathbb{R}$

$$V(u) = \int_{\mathbb{R}^d} \eta(u(x))dx + \int_{\mathbb{R}^d} \Phi(x)u(x)dx = -\widetilde{S}[u] + F[u]. \tag{4.67}$$

Since, by (K1), (K4) and (K5),

$$\frac{\gamma}{r|b|_\infty} \leq \frac{\beta'(r)}{rb(r)} \leq \frac{\gamma_1}{rb_0}, \ \forall r > 0, \tag{4.68}$$

we have

$$\begin{aligned} &\frac{\gamma_1}{b_0} z_{[0,1]}(r)r(\log r - 1) + \frac{\gamma}{|b|_\infty} \mathbf{1}_{(1,\infty)}(r)r(\log r - 1) \leq \eta(r) \\ &\leq \frac{\gamma}{|b|_\infty} \mathbf{1}_{[0,1]}(r)r(\log r - 1) + \frac{\gamma_1}{b_0} \mathbf{1}_{(1,\infty)}(r)r(\log r - 1). \end{aligned} \tag{4.69}$$

We also have that $\eta \in C([0,\infty))$, $\eta \in C^2((0,\infty))$, $\eta'' \geq 0$. Since Φ is Lipschitz, hence of at most linear growth, $F[u]$ is well-defined and finite if $u \in \mathscr{M}$. Furthermore, arguing as in [67], p. 16, one proves that $(u \log u)^- \in L^1$ if $u \in D(V)$. Hence $\widetilde{S}[u]$ is well-defined and $-\widetilde{S}[u] \in (-\infty, \infty]$ because of (4.69) and thus $V(u) \in (-\infty, \infty]$ for all $u \in D(V)$. We define $V = \infty$ on $L^1 \setminus D(V)$. Then, obviously, $V : L^1 \to (-\infty, \infty]$ is convex and L^1_{loc}-lower semicontinuous on balls in $\mathscr{M}$, as easily follows (4.69) from (4.70) below. If, in addition, $(u \log u)^+ \in L^1$, then, again by (4.69), we have that $\widetilde{S}[u] \in (-\infty, \infty)$ and also V is real-valued. The functional

$$\widetilde{S}[u(t)] = -\int_{\mathbb{R}^d} \eta(u(t,x))dx,\ u \in \mathscr{P},\ t \ge 0,$$

is called in the literature (see, e.g., [53, 92]) the entropy of the system, while $F[u(t)]$ is the mean field energy. In fact, according to the general theory of thermostatics (see, e.g., [54]), the functional $\widetilde{S} = \widetilde{S}[u]$ is a generalized entropy because its kernel $-\eta$ is a strictly concave continuous functions on $(0, \infty)$ and $\lim\limits_{r \downarrow 0} \eta'(r) = +\infty$.

In the special case $\beta(s) \equiv s$ and $b(s) \equiv 1$, $\eta(r) \equiv r(\log r - 1)$ and so $\widetilde{S}[u] - 1$ reduces to the classical Boltzmann-Gibbs entropy.

As in [67], one proves that, for $\alpha \in \left[\frac{m}{m+1}, 1\right)$, where m is as in assumption (K4),

$$\int_{\{\Phi \ge R\}} |\min(u \log u, 0)|dx \le C_\alpha \left(\int_{\{\Phi \ge R\}} \Phi^{-k} dx\right)^{1-\alpha} \|u\|^\alpha, \tag{4.70}$$

for all $R > 0$. Indeed, for every $\alpha \in (0, 1)$, there exists $C_\alpha \in (0, \infty)$ such that $(r \log r)^- \le C_\alpha r^\alpha$ for $r \in [0, \infty)$. Hence, the left hand side of (4.70) by Hölder's inequality is dominated by

$$C_\alpha \left(\int_{\{\Phi \ge R\}} u\Phi dx\right)^\alpha \left(\int_{\{\Phi \ge R\}} \Phi^{-\frac{\alpha}{1-\alpha}} dx\right)^{1-\alpha}.$$

Therefore, for $\alpha \in \left[\frac{m}{m+1}, 1\right)$, we obtain (4.70) since $\Phi \ge 1$ and this yields

$$V(u) \ge -C(\|u\| + 1)^\alpha,\ \forall u \in D(V). \tag{4.71}$$

We also consider the function $\Psi : D(\Psi) \subset L^1 \to [0, \infty)$ defined by

$$\Psi(u) = \int_{\mathbb{R}^d} \left| \frac{\beta'(u)\nabla u}{\sqrt{ub(u)}} - D\sqrt{ub(u)} \right|^2 dx, \tag{4.72}$$

$$D(\Psi) = \{u \in L^1 \cap W^{1,1}_{\mathrm{loc}}(\mathbb{R}^d);\ u \ge 0,\ \Psi(u) < \infty\}. \tag{4.73}$$

We extend Ψ to all of L^1 by $\Psi(u) = \infty$ if $u \in L^1 \setminus D(\Psi)$. Since $\nabla u = 0$, a.e. on $\{u = 0\}$, we set here and below

$$\frac{\nabla u}{\sqrt{u}} = 0, \quad \text{a.e. on } \{u = 0\}.$$

We note that every solution to the equation $\Psi(u) = 0$ is a stationary solution to NFPE (4.1) and is an extremal state of the function v.

Theorem 4.1 can be viewed as the weak H-theorem for NFPE (4.1).

Theorem 4.1 *Assume that Hypotheses* (K1)–(K5) *hold. Then, the function* V *defined by* (4.67) *is a Lyapunov function for* $S(t)$*, that is, for* $D_0(V) = D(V) \cap \{V < \infty\}$ $(= \{u \in D(V);\ u \log u \in L^1\})$,

$$\begin{aligned} S(t)u_0 \in D_0(V),\ \forall t \geq 0, u_0 \in D_0(V) \text{ and} \\ V(S(t)u_0) \leq V(S(s)u_0),\ \forall u_0 \in D_0(V), 0 \leq s \leq t < \infty. \end{aligned} \tag{4.74}$$

Moreover, we have, for all $u_0 \in D_0(V)$,

$$V(S(t)u_0) + \int_s^t \Psi(S(\sigma)u_0)d\sigma \leq V(S(s)u_0) \text{ for } 0 \leq s \leq t < \infty. \tag{4.75}$$

In particular, $S(\sigma)u_0 \in D(\Psi)$ *for a.e.* $\sigma \geq 0$*. Furthermore, there exists* $u_\infty \in \omega(u_0)$ *(see* (1.7)*) such that* $u_\infty \in D(\Psi)$, $\Psi(u_\infty) = 0$ *and, for any such a* u_∞*, we have either* $u_\infty = 0$ *or* $u_\infty > 0$ *a.e. In the latter case,*

$$u_\infty = g^{-1}(-\Phi + \mu) \text{ for some } \mu \in \mathbb{R}, \tag{4.76}$$

$$g(r) = \int_1^r \frac{\beta'(s)}{sb(s)}\,ds,\ r > 0. \tag{4.77}$$

Proof We note that (4.74) and (4.75) can be formally derived by Eq. (4.1) if $u(t) = S(t)u_0$ is sufficiently regular. However, since this is not the case, a rigorous proof requires a smooth approximation of the semigroup $S(t)$ (and the best candidate is $S_\varepsilon(t)$) as well as of the function V. To this end, we approximate the function $V : L^1 \to (-\infty, \infty]$ by the functional V_ε defined by

$$\begin{aligned} V_\varepsilon(u) = \int_{\mathbb{R}^d} (\eta_\varepsilon(u(x)) + \Phi_\varepsilon(x)u(x))dx,\ \forall u \in D(V), \\ V_\varepsilon(u) = \infty \ \text{ if } u \in L^1 \setminus D(V), \end{aligned}$$

where $\eta_\varepsilon(r) = -\int_0^r d\tau \int_\tau^1 \frac{\beta'(s)}{b_\varepsilon^*(s)+\varepsilon^{2m}}\,ds,\ r \geq 0,\ \varepsilon > 0$. Clearly, $\eta_\varepsilon \to \eta$ as $\varepsilon \to 0$ locally uniformly. We also note that V_ε is convex, and L^1_{loc}-lower semicontinuous on every ball in $\mathcal{M}$. Furthermore, there exists $C > 0$ such that, for all $\varepsilon \in (0, 1]$, we have $|\eta_\varepsilon(u)| \leq C(1 + |u|^2)$. This implies that $V_\varepsilon < \infty$ on L^2 and $V_\varepsilon(u) \to V(u)$ as $\varepsilon \to 0$ for all $u \in D(V) \cap L^2$ and by the generalized Fatou lemma that V_ε is lower semicontinuous on L^2. We set

$$V'_\varepsilon(u) = \eta'_\varepsilon(u) + \Phi_\varepsilon,\ \forall u \in D(V) \cap L^2.$$

It is easy to check that $V'_\varepsilon(u) \in \partial V_\varepsilon(u)$ for all $u \in D(V) \cap L^2$, where ∂V_ε is the subdifferential of V_ε on L^2. As regards the function Ψ, we have

Lemma 4.4

$$D(\Psi) = \{u \in L^1;\ u \geq 0,\ \sqrt{u} \in W^{1,2}(\mathbb{R}^d)\}, \tag{4.78}$$

$$\|\sqrt{u}\|_{W^{1,2}(\mathbb{R}^d)} \leq C(\Psi(u) + 1),\ \forall u \in D(\Psi), \tag{4.79}$$

where $C \in (0, \infty)$ is independent of u. Furthermore, Ψ is L^1_{loc}-lower semicontinuous on L^1-balls.

Proof By (4.72), taking into account (K1), (K2), we have

$$\begin{aligned}\gamma |b|_\infty^{-1} \int_{\mathbb{R}^d} \frac{|\nabla u|^2}{u} dx &\leq \int_{\mathbb{R}^d} \frac{|\beta'(u)|^2 \cdot |\nabla u|^2}{ub(u)} dx \\ &\leq 2\Psi(u) + 2\int_{\mathbb{R}^d} |D|^2 ub(u) dx < \infty,\ \forall u \in D(\Psi).\end{aligned} \tag{4.80}$$

This yields (4.78) and (4.79) since $\nabla(\sqrt{u}) = \frac{1}{2} \frac{\nabla u}{\sqrt{u}}$ and (K5) holds. To show the lower semicontinuity of Ψ, we rewrite it as

$$\Psi(u) = \int_{\mathbb{R}^d} |\nabla j(u) - D\sqrt{ub(u)}|^2 dx,\ u \in D(\Psi), \tag{4.81}$$

where

$$j(r) = \int_0^r \frac{\beta'(s)}{\sqrt{sb(s)}} ds,\ r \geq 0. \tag{4.82}$$

Clearly,

$$0 \leq j(r) \leq \frac{2\gamma_1}{\sqrt{b_0}} \sqrt{r}. \tag{4.83}$$

Let $\{u_n\} \subset L^1$ and $\nu > 0$ be such that $\sup_n |u_n|_1 < \infty$ and

$$\Psi(u_n) \leq \nu < \infty,\ \forall n, \tag{4.84}$$

$$u_n \longrightarrow u \text{ in } L^1_{loc} \text{ as } n \to \infty. \tag{4.85}$$

This yields $\sqrt{u_n b(u_n)} \longrightarrow \sqrt{ub(u)}$ in L^2_{loc} and so, by Hypothesis (K3), we have

$$D\sqrt{u_n b(u_n)} \longrightarrow D\sqrt{ub(u)} \text{ in } L^2_{loc}(\mathbb{R}^d; \mathbb{R}^d). \tag{4.86}$$

Hence (4.84) implies that (selecting a subsequence if necessary) for all balls B_N of radius $N \in \mathbb{N}$ around zero we have

$$\sup_n \int_{B_N} |\nabla j(u_n)|^2 dx < \infty$$
$$j(u_n) \to j(u) \text{ in } L^2_{\text{loc}} \text{ as } n \to \infty.$$

Therefore (again selecting a subsequence, if necessary), for every $N \in \mathbb{N}$,

$$\nabla j(u_n) \to \nabla j(u) \text{ weakly in } L^2(B_N, dx) \text{ as } n \to \infty.$$

Hence, if we define Ψ_N analogously to Ψ, but with the integral over $\mathbb{R}^d$ replaced by an integral over B_N, we conclude that

$$\liminf_{n\to\infty} \Psi_N(u_n) \geq \liminf_{n\to\infty} \int_{B_N} |\nabla j(u_n)|^2 dx - 2\int_{B_N} \nabla j(u) \cdot D\sqrt{ub(u)}dx + \int_{B_N} |D|^2 ub(u)dx \geq \Psi_N(u).$$

Hence, since $u \in L^1$, we get for $N \to \infty$

$$\liminf_{n\to\infty} \Psi(u_n) \geq \Psi(u).$$

Now, we consider the functional

$$\begin{aligned}
\Psi_\varepsilon(u) = &\int_{\mathbb{R}^d} \left| \frac{\beta'(u)\nabla u}{\sqrt{b^*_\varepsilon(u) + \varepsilon^{2m}}} - D_\varepsilon \sqrt{b^*_\varepsilon(u) + \varepsilon^{2m}} \right|^2 dx \\
&+\varepsilon^{2m} \int_{\mathbb{R}^d} D_\varepsilon \cdot \left(\frac{\beta'(u)\nabla u}{b^*_\varepsilon(u) + \varepsilon^{2m}} - D_\varepsilon \right) dx \\
&+\varepsilon \int_{\mathbb{R}^d} \beta(u)(\eta'_\varepsilon(u) + \Phi_\varepsilon)dx, \ \forall u \in D(\Psi_\varepsilon) = D(V) \cap H^1, \\
\Psi_\varepsilon(u) := &\ \infty \ \text{ if } u \in D(V) \setminus H^1.
\end{aligned} \tag{4.87}$$

Lemma 4.5 *For each $\varepsilon > 0$, Ψ_ε is L^1_{loc}-lower semicontinuous on every ball in $\mathcal{M}$. Moreover, for any sequence $\{v_\varepsilon\} \subset D(V) \cap H^1$ such that*

$$\sup_{\varepsilon \geq 0} \|v_\varepsilon\| < \infty, \ \lim_{\varepsilon\to 0} v_\varepsilon = v \text{ in } L^1_{\text{loc}},$$

we have

$$\liminf_{\varepsilon\to 0} \Psi_\varepsilon(v_\varepsilon) \geq \Psi(v). \tag{4.88}$$

Furthermore, there exists $c \in (0, \infty)$ such that, for all $u \in D(V)$, $\varepsilon \in (0, 1]$,

$$\Psi_\varepsilon(u) \geq -c(|u| + \|u\| + 1). \tag{4.89}$$

Proof First of all we note that by the assusmption on u_ε it follows that $\lim\limits_{\varepsilon\to 0} v_\varepsilon = 0$ in L^1, since $\lim\limits_{|x|\to\infty} \Phi(x) = \infty$. We write $\Psi_\varepsilon(u) \equiv \Psi^*_\varepsilon(u) + G_\varepsilon(u)$, where

$$\begin{aligned}
\Psi^*_\varepsilon(u) =& \int_{\mathbb{R}^d} \left| \frac{\beta'(u)\nabla u}{\sqrt{b^*_\varepsilon(u) + \varepsilon^{2m}}} - D_\varepsilon \sqrt{b^*_\varepsilon(u) + \varepsilon^{2m}} \right|^2 dx \\
&+\varepsilon^{2m} \int_{\mathbb{R}^d} D_\varepsilon \cdot \left(\frac{\beta'(u)\nabla u}{b^*_\varepsilon(u) + \varepsilon^{2m}} - D_\varepsilon \right) dx, \\
G_\varepsilon(u) =& \varepsilon \int_{\mathbb{R}^d} \beta(u)(\eta'_\varepsilon(u) + \Phi_\varepsilon) dx.
\end{aligned}$$

We have, since $\eta'_\varepsilon(\tau) \geq \frac{\gamma_1}{b_0}(\log\tau - \varepsilon(1-\tau))$ for $\tau \in (0, 1]$,

$$\begin{aligned}
G_\varepsilon(v_\varepsilon) \geq& \ \varepsilon\gamma_1 \int_{\{v_\varepsilon \leq 1\}} v_\varepsilon \eta'_\varepsilon(v_\varepsilon) dx \geq \varepsilon \frac{\gamma_1^2}{b_0} \int_{\{v_\varepsilon \leq 1\}} (v_\varepsilon \log v_\varepsilon - \varepsilon v_\varepsilon) dx \\
\geq& -\varepsilon \frac{\gamma_1^2}{b_0} \left[C_\alpha \left(\int_{\mathbb{R}^d} \Phi^{-m} dx \right)^{1-\alpha} \|v_\varepsilon\|^\alpha + \varepsilon \int_{\mathbb{R}^d} v_\varepsilon \, dx \right].
\end{aligned} \tag{4.90}$$

Hence, $\liminf\limits_{\varepsilon\to 0} G_\varepsilon(v_\varepsilon) \geq 0$. Now, arguing as in the proof of Lemma 4.4, we represent Ψ^*_ε as (see (4.80))

$$\Psi^*_\varepsilon(u) = \int_{\mathbb{R}^d} |\nabla j^*_\varepsilon(u) - D_\varepsilon \sqrt{b^*_\varepsilon(u) + \varepsilon^{2m}}|^2 dx + \varepsilon^{2m} \int_{\mathbb{R}^d} D_\varepsilon \cdot \left(\frac{\beta'(u)\nabla u}{b^*_\varepsilon(u) + \varepsilon^{2m}} - D_\varepsilon \right) dx,$$

where $u \in D(V) \cap H^1$ and

$$j^*_\varepsilon(r) = \int_0^r \frac{\beta'(s) ds}{\sqrt{b^*_\varepsilon(s) + \varepsilon^{2m}}}.$$

We may assume that $\Psi^*_\varepsilon(v_\varepsilon) \leq \nu < \infty$, $\forall \varepsilon > 0$. Then, as in (4.80), we see that

$$\begin{aligned}
\int_{\mathbb{R}^d} \frac{|\beta'(v_\varepsilon)|^2 |\nabla v_\varepsilon|^2}{b^*_\varepsilon(v_\varepsilon) + \varepsilon^{2m}} dx \leq& \ 2 \left(\Psi^*_\varepsilon(v_\varepsilon) + \int_{\mathbb{R}^d} |D_\varepsilon|^2 (b^*_\varepsilon(v_\varepsilon) + 2\varepsilon^{2m}) dx \right) \\
&+ 2\varepsilon^{2m} \int_{\mathbb{R}^d} \frac{|D_\varepsilon| |\beta'(v_\varepsilon)| |\nabla v_\varepsilon|}{b^*_\varepsilon(v_\varepsilon) + \varepsilon^{2m}} dx.
\end{aligned} \tag{4.91}$$

Taking into account that

$$\begin{aligned}
&2\varepsilon^{2m}\int_{\mathbb{R}^d}\frac{|D_\varepsilon|\beta'(v_\varepsilon)|\nabla v_\varepsilon|}{b^*_\varepsilon(v_\varepsilon)+\varepsilon^{2m}}\,dx\\
&\le\frac{1}{2}\int_{\mathbb{R}^d}\frac{|\beta'(v_\varepsilon)|^2|\nabla v_\varepsilon|^2}{b^*_\varepsilon(v_\varepsilon)+\varepsilon^{2m}}\,dx+2\varepsilon^{4m}\int_{\mathbb{R}^d}\frac{|D_\varepsilon|^2}{b^*_\varepsilon(v_\varepsilon)+\varepsilon^{2m}}\,dx\\
&\le\frac{1}{2}\int_{\mathbb{R}^d}\frac{|\beta'(v_\varepsilon)|^2|\nabla v_\varepsilon|^2}{b^*_\varepsilon(v_\varepsilon)+\varepsilon^{2m}}\,dx+2\varepsilon^{2m}\int_{\mathbb{R}^d}|D_\varepsilon|^2dx,
\end{aligned}\tag{4.92}$$

and that $\lim\limits_{\varepsilon\to 0}v_\varepsilon=v$ in L^1 by our assumption, it follows by (4.31) and (4.91) that, for some $C>0$ independent of ε,

$$\int_{\mathbb{R}^d}\frac{|\nabla v_\varepsilon|^2}{b^*_\varepsilon(v_\varepsilon)+\varepsilon^{2m}}\,dx\le C,\ \forall\varepsilon>0,$$

and so $\{\nabla j^*_\varepsilon(v_\varepsilon)\}$ is bounded in L^2. Then, arguing as in Lemma 4.4 (see (4.85) and (4.86)), we get for $\varepsilon\to 0$

$$D_\varepsilon\sqrt{b^*_\varepsilon(v_\varepsilon)+\varepsilon^{2m}}\longrightarrow D\sqrt{b(u)u}\ \text{ in } L^2(\mathbb{R}^d;\mathbb{R}^d),$$

and, therefore,

$$\liminf_{\varepsilon\to 0}\Psi_\varepsilon(v_\varepsilon)\ge\liminf_{\varepsilon\to 0}\Psi^*_\varepsilon(v_\varepsilon)\ge\Psi(v),$$

as claimed. By a similar (even easier) argument, one proves that Ψ_ε is $L^1_{\rm loc}$-lower semicontinuous on balls in $\mathscr{M}$. The last part of the assertion is an immediate consequence of (4.90) and (4.92), which hold for all $u\in D(V)\cap H^1$ replacing v_ε. Hence, the lemma is proved.

We denote by $S_\varepsilon(t)$ the semigroup generated on L^1 by the operator A_ε defined by (4.27) and (4.28), that is (see (4.64)),

$$S_\varepsilon(t)u_0=\lim_{n\to\infty}\left(I+\frac{t}{n}A_\varepsilon\right)^{-n}u_0,\ \forall t\ge 0,\ u_0\in L^1.$$

Then, by (4.65) we have

$$\lim_{\varepsilon\to 0}S_\varepsilon(t)u_0=S(t)u_0,\ \forall u_0\in L^1,\ \forall t\ge 0,\tag{4.93}$$

strongly in L^1 and uniformly in t on compact time intervals.

We shall prove first (4.75) for $S_\varepsilon(t)$. Namely, one has

Lemma 4.6 *For each $u_0 \in L^2 \cap D(V)$, we have $S_\varepsilon(\sigma)u_0 \in D(\Psi_\varepsilon)$ for ds-a.e. $\sigma \geq 0$, and*

$$V_\varepsilon(S_\varepsilon(t)u_0) + \int_s^t \Psi_\varepsilon(S_\varepsilon(\sigma)u_0)d\sigma \leq V_\varepsilon(S_\varepsilon(s)u_0),\ 0 \leq s \leq t < \infty, \tag{4.94}$$

and all three terms are finite.

Proof First, we shall prove that, for all $\varepsilon > 0$,

$$V_\varepsilon((I+\lambda A_\varepsilon)^{-1}u_0) + \lambda\Psi_\varepsilon((I+\lambda A_\varepsilon)^{-1}u_0) \leq V_\varepsilon(u_0),\ \lambda \in (0,\lambda_0). \tag{4.95}$$

We set $u_\varepsilon^\lambda = (I+\lambda A_\varepsilon)^{-1}u_0$ and note that, by (4.39) and (4.40), we have

$$u_\varepsilon^\lambda \in H^1(\mathbb{R}^d),\ \beta(u_\varepsilon^\lambda) \in H^1(\mathbb{R}^d),\ \forall \lambda \in (0,\lambda_0),\ \varepsilon > 0, \tag{4.96}$$

$$V_\varepsilon'(u_\varepsilon^\lambda) = \eta_\varepsilon'(u_\varepsilon^\lambda) + \Phi_\varepsilon \in \partial V_\varepsilon(u_\varepsilon^\lambda), \tag{4.97}$$

where $\eta_\varepsilon'(u_\varepsilon^\lambda) \in H^1(\mathbb{R}^d)$. Taking into account that, by Lemma 4.3,

$$\operatorname{div}(\nabla\beta(u_\varepsilon^\lambda) - D_\varepsilon b_\varepsilon^*(u_\varepsilon^\lambda)) = \frac{1}{\lambda}(u_\varepsilon^\lambda - u_0) + \varepsilon\beta(u_\varepsilon^\lambda) \in \mathcal{M} \cap L^2, \tag{4.98}$$

it follows, since $\Phi_\varepsilon \in L^2$ and $\operatorname{div} D_\varepsilon \in L^2 + L^\infty$ by (4.55) and Hypothesis (K3), that

$$\int_{\mathbb{R}^d} (-\Delta\beta(u_\varepsilon^\lambda) + \operatorname{div} D_\varepsilon b_\varepsilon^*(u_\varepsilon^\lambda))\Phi_\varepsilon\, dx = -\int_{\mathbb{R}^d} (\nabla\beta(u_\varepsilon^\lambda) - D_\varepsilon b_\varepsilon^*(u_\varepsilon^\lambda)) \cdot D_\varepsilon\, dx.$$

This yields, by (4.97),

$$\begin{aligned}
&\langle A_\varepsilon(u_\varepsilon^\lambda), V_\varepsilon'(u_\varepsilon^\lambda)\rangle_2 \\
&\quad = \langle -\Delta(\beta(u_\varepsilon^\lambda)) + \varepsilon\beta(u_\varepsilon^\lambda) + \operatorname{div}(D_\varepsilon b_\varepsilon^*(u_\varepsilon^\lambda)), \eta_\varepsilon'(u_\varepsilon^\lambda) + \Phi_\varepsilon\rangle_2 \\
&\quad = \int_{\mathbb{R}^d} (\beta'(u_\varepsilon^\lambda)\nabla u_\varepsilon^\lambda - D_\varepsilon b_\varepsilon^*(u_\varepsilon^\lambda)) \cdot \left(\frac{\beta'(u_\varepsilon^\lambda)}{b_\varepsilon^*(u_\varepsilon^\lambda) + \varepsilon^{2m}}\nabla u_\varepsilon^\lambda - D_\varepsilon\right) dx \\
&\qquad + \varepsilon\langle\beta(u_\varepsilon^\lambda), \eta_\varepsilon'(u_\varepsilon^\lambda) + \Phi_\varepsilon\rangle_2 \\
&\quad = \int_{\mathbb{R}^d} \left|\frac{\beta'(u_\varepsilon^\lambda)\nabla u_\varepsilon^\lambda}{\sqrt{b_\varepsilon^*(u_\varepsilon^\lambda) + \varepsilon^{2m}}} - D_\varepsilon\sqrt{b_\varepsilon^*(u_\varepsilon^\lambda) + \varepsilon^{2m}}\right|^2 dx + \varepsilon\langle\beta(u_\varepsilon^\lambda), \eta_\varepsilon'(u_\varepsilon^\lambda) + \Phi_\varepsilon\rangle_2 \\
&\qquad + \varepsilon^{2m}\int_{\mathbb{R}^d}\left(D_\varepsilon \cdot \frac{\beta'(u_\varepsilon^\lambda)\nabla u_\varepsilon^\lambda}{b_\varepsilon^* + \varepsilon^{2m}} - D_\varepsilon\right) dx = \Psi_\varepsilon(u_\varepsilon^\lambda),\ \forall \varepsilon > 0,\ \lambda \in (0,\lambda_0).
\end{aligned}$$

This yields (4.95) because, by (4.97) and the convexity of V_ε, we have

$$V_\varepsilon(u_\varepsilon^\lambda) \le V_\varepsilon(u_0) + \langle V_\varepsilon'(u_\varepsilon^\lambda), u_\varepsilon^\lambda - u_0\rangle_2, \; u_\varepsilon^\lambda - u_0 = -\lambda A_\varepsilon(u_\varepsilon^\lambda).$$

To prove (4.94), we set

$$\lambda\delta(\lambda, v) = V_\varepsilon((I+\lambda A_\varepsilon)^{-1}v) + \lambda\Psi_\varepsilon((I+\lambda A_\varepsilon)^{-1}v) - V_\varepsilon(v),$$
$$\forall \lambda \in (0,\lambda_0),\; v \in L^2 \cap D(V),$$

and note that, by (4.95), $\delta(\lambda, u_0) \le 0$, $\lambda \in (0, \lambda_0)$. This yields

$$V_\varepsilon((I+\lambda A_\varepsilon)^{-j}u_0) + \lambda\Psi_\varepsilon((I+\lambda A_\varepsilon)^{-j}u_0) - V_\varepsilon((I+\lambda A_\varepsilon)^{-j+1}u_0)$$
$$= \lambda\delta(\lambda, (I+\lambda A_\varepsilon)^{-j+1}u_0),\; \forall j \in \mathbb{N}.$$

Then, summing up from $j=1$ to $j=n$ and taking $\lambda = \frac{t}{n}$, we get

$$\begin{aligned} &V_\varepsilon\left(\left(I+\frac{t}{n}A_\varepsilon\right)^{-n}u_0\right) + \sum_{j=1}^n \frac{t}{n}\Psi_\varepsilon\left(\left(I+\frac{t}{n}A_\varepsilon\right)^{-j}u_0\right) \\ &= V_\varepsilon(u_0) + \sum_{j=1}^n \frac{t}{n}\,\delta\left(\frac{t}{n}, \left(I+\frac{t}{n}A_\varepsilon\right)^{-(j-1)}u_0\right). \end{aligned} \tag{4.99}$$

Note also that, if $n > \frac{t}{\lambda_0}$, then

$$\delta\left(\frac{t}{n}, \left(I+\frac{t}{n}A_\varepsilon\right)^{-j}u_0\right) \le 0,\quad 1 \le j \le n. \tag{4.100}$$

We consider the step function

$$f_n(\sigma) = \Psi_\varepsilon\left(\left(I+\frac{t}{n}A_\varepsilon\right)^{-j}u_0\right) \text{ for } \frac{(j-1)t}{n} < \sigma \le \frac{jt}{n},$$

and note that, for each $t > 0$,

$$\sum_{j=1}^n \frac{t}{n}\Psi_\varepsilon\left(\left(I+\frac{t}{n}A_\varepsilon\right)^{-j}u_0\right) = \int_0^t f_n(\sigma)d\sigma.$$

Then, by (4.50) and (4.60) and the L^1_{loc}-lower semicontinuity of Ψ_ε on balls in $\mathscr{M}$, we conclude, by the Fatou lemma, which is applicable because of (4.89), that

$$-\infty < \int_0^t \Psi_\varepsilon(S(\sigma)u_0)d\sigma \le \liminf_{n\to\infty} \int_0^t f_n(\sigma)d\sigma, \tag{4.101}$$

while, by the L^1_{loc}-lower semicontinuity of V_ε on balls in $\mathscr{M}$, we have

$$\liminf_{n\to\infty} V_\varepsilon\left(\left(I + \frac{t}{n}A_\varepsilon\right)^{-n} u_0\right) \ge V_\varepsilon(S_\varepsilon(t)u_0).$$

Then, by (4.99)–(4.101), we get

$$V_\varepsilon(S_\varepsilon(t)u_0) + \int_0^t \Psi_\varepsilon(S_\varepsilon(\sigma)u_0)d\sigma \le V_\varepsilon(u_0),\ \forall t \ge 0.$$

In particular, $V_\varepsilon(S_\varepsilon(t)u_0) < \infty$ since $V_\varepsilon(u_0) < \infty$. Taking into account the semigroup property $S_\varepsilon(t+s)u_0 = S_\varepsilon(t)S_\varepsilon(s)u_0$, we get (4.94), as claimed.

Proof of Theorem 4.1 (Continued) We shall assume $u_0 \in L^2 \cap D_0(V)$. We want to let $\varepsilon \to 0$ in (4.94), where $s = 0$. We note first that we have

$$\liminf_{\varepsilon\to 0} V_\varepsilon(S_\varepsilon(t)u_0) \ge V(S(t)u_0),\ \forall t \ge 0. \tag{4.102}$$

Here is the argument. We note that, if $v_\varepsilon \to v$ in L^1 as $\varepsilon \to 0$ and $\sup\limits_{\varepsilon>0} \|v_\varepsilon\| < \infty$, then $v_\varepsilon(\log v_\varepsilon)^- \to v(\log v)^-$ in L^1_{loc} as $\varepsilon \to 0$. Furthermore, for $\delta > 0$, and $\alpha \in \left[\frac{m+\delta}{m+\delta+1}, 1\right)$, by (4.70),

$$\int_{\{\Phi\ge R\}} v_\varepsilon(\log v_\varepsilon)^- dx \le C_\alpha \frac{1}{R^{\delta(1-\alpha)}} \left(\int \Phi^{-m} dx\right)^{1-\alpha} \|v_\varepsilon\|^\alpha,$$

hence

$$\lim_{R\to\infty} \sup_{\varepsilon>0} \int_{\{\Phi\ge R\}} v_\varepsilon(\log v_\varepsilon)^- dx = 0,$$

therefore, $v_\varepsilon(\log v_\varepsilon)^- \to v(\log v)^-$ in L^1. Applying this to $v_\varepsilon = S_\varepsilon(t)u_0$, and because $\eta_\varepsilon \to \eta$ as $\varepsilon \to 0$ locally uniformly on $[0,\infty)$ and, because for all $\varepsilon \in (0,1]$, $r \in [0,\infty)$,

$$\eta_\varepsilon(r) \ge -\frac{\gamma_1}{b_0}(r\wedge 1)(\log(r\wedge 1)^- - 2(r\wedge 1)),$$

we can apply the generalized Fatou lemma to conclude that

$$\liminf_{\varepsilon\to\infty}\int_{\mathbb{R}^d}\eta_\varepsilon(S_\varepsilon(t)u_0)dx \geq \int_{\mathbb{R}^d}\eta(S(t)u_0)dx,$$

and we get (4.102), as claimed.

By Lemma 4.6, (4.50) and (4.92), we have that $v_\varepsilon = S_\varepsilon(t)u_0$, $\varepsilon > 0$, satisfy for dt-a.e. $t > 0$ the assumptions of Lemma 4.5, hence

$$\liminf_{\varepsilon\to 0}\Psi_\varepsilon(S_\varepsilon(t)u_0) \geq \Psi(S(t)u_0), \text{ a.e. } t > 0.$$

Moreover, again by Fatou's lemma, which is applicable by (4.89), it follows that

$$\liminf_{\varepsilon\to 0}\int_0^t \Psi_\varepsilon(S_\varepsilon(s)u_0)ds \geq \int_0^t \Psi(S(s)u_0)ds,\ \forall t \geq 0. \tag{4.103}$$

Because, as seen earlier, $V_\varepsilon(u) \to V(u)$ as $\varepsilon \to 0$, if $u \in D(V)\cap L^2$, while (4.102) and (4.94) with $s = 0$ imply

$$V(S(t)u_0) + \int_0^t \Psi(S(\sigma)u_0)d\sigma \leq V(u_0),\ \forall u_0 \in D(V)\cap L^2,\ t \geq 0. \tag{4.104}$$

We note that, by (4.25) and (4.71), we have

$$\begin{aligned} V(S(t)u_0) &\geq -C(\|S(t)u_0\| + 1)^\alpha \\ &\geq C(\|u_0\| + t|u_0|_1)^\alpha,\ \alpha \in \left[\tfrac{m}{m+1}, 1\right). \end{aligned} \tag{4.105}$$

Hence

$$0 \leq \int_0^t \Psi(S(\sigma)u_0)d\sigma < \infty,\ \forall t \geq 0,$$

which implies that

$$S(\sigma)u_0 \in D(\Psi) \text{ a.e. } \sigma > 0. \tag{4.106}$$

Now, to extend (4.104) to all $u_0 \in D_0(V)$, take $u_0^n \in D(V)\cap L^2(\subset D_0(V))$ with $u_0^n \leq u_0$ and $u_0^n \to u_0$ as $n \to \infty$ in L^1. Then, because

$$\eta(r) \geq -\frac{\gamma_0}{b_0}\left[(r\wedge 1)(\log(r\wedge 1)^- + (r\wedge 1))\right],\ \forall r \geq 0,$$

arguing as above, we conclude the monotone convergence applies to get

$$\lim_{n\to\infty} V(u_0^n) = V(u_0)$$

and the generalized Fatou lemma applies to get eventually (4.104) and (4.106) for all $u_0 \in D_0(V)$. Since $S(t)u_0 \in D_0(V)$, if $u_0 \in D_0(V)$, the first part including (4.75) follows.

To prove (4.76), we note that since $\alpha < 1$, by (4.75) and (4.105), we have

$$\begin{aligned} 0 = \lim_{t\to\infty} \frac{1}{t}\int_0^t \Psi(S(\sigma)u_0)d\sigma &\geq \lim_{t\to\infty} \frac{1}{t}\int_n^t \inf_{r\geq n} \Psi(S(r)u_0)d\sigma \\ &= \inf_{r\geq n} \Psi(S(r)u_0) \ \text{ for all } n \in \mathbb{N}. \end{aligned} \tag{4.107}$$

Hence, there exists $t_n \to \infty$ such that

$$\lim_{n\to\infty} \Psi(S(t_n)u_0) = 0. \tag{4.108}$$

Furthermore, we obtain by Lemma 4.4 the first inequality in (4.107), that

$$\sup_{t\geq 0} |S(t)u_0|_1 + \limsup_{t\to\infty} \frac{1}{t}\int_0^t |\nabla(\sqrt{S(s)u_0})|_2 ds < \infty.$$

Hence, similarly as above (selecting a subsequence of (t_n), if necessary),

$$\sup_n \|\sqrt{S(t_n)u_0}\|_{W^{1,2}(\mathbb{R}^d)} < \infty. \tag{4.109}$$

So, by the Rellich-Kondrachov theorem (see, e.g., [30], p. 284), the set $\{S(t_n)u_0 \mid n \in \mathbb{N}\}$ is relatively compact in L^1_{loc}. Hence, along a subsequence $\{t_{n'}\} \to \infty$, we have

$$\lim S(t_{n'})u_0 = u_\infty \text{ in } L^1_{loc} \tag{4.110}$$

for some $u_\infty \in L^1$. Since Ψ is L^1_{loc}-lower semicontinuous on L^1-balls by Lemma 4.2, this together with (4.108) implies that $u_\infty \in D(\Psi)$ and $\Psi(u_\infty) = 0$.

If $u_\infty \in D(\Psi)$, such that $\Psi(u_\infty) = 0$, then

$$\frac{\beta'(u_\infty)\nabla u_\infty}{\sqrt{u_\infty b(u_\infty)}} = D\sqrt{u_\infty b(u_\infty)}, \ \text{a.e. in } \mathbb{R}^d. \tag{4.111}$$

Let us prove now that either $u_\infty \equiv 0$ or $u = u_\infty > 0$, a.e. in $\mathbb{R}^d$. To this end, we consider the solution $y = y(t, x)$ to the system

$$\begin{aligned} y_i'(t) &= \widetilde{D}_i(y_i(t)),\ t \geq 0,\ i = 1, \dots, d, \\ y_i(0) &= x_i, \end{aligned}$$

where $\widetilde{D}_i \in C^1(\mathbb{R})$, $i = 1, \dots, d$, is an arbitrary vector field on $\mathbb{R}$ of at most linear growth, and $y(t) = \{y_i(t)\}_{i=1}^d$, $x = \{x_i\}_{i=1}^d$. If j is defined by (4.82), we have

$$\begin{aligned} \frac{d}{dt} j(u(y(t,x))) &= j_u(u(y(t,x)))\nabla u(y(t,x)) \cdot \frac{d}{dt} y(t,x) \\ &= \frac{\beta'(u(y(t,x)))}{\sqrt{b(u(y(t,x)))u(y(t,x))}} \nabla u(y(t,x)) \cdot \mathscr{D}(y(t,x)),\ \forall t \geq 0, \end{aligned}$$

where $\mathscr{D}(y) = (\widetilde{D}_i(y_i))_{i=1}^d$. Let $D = \{D_i\}_{i=1}^d$. Then, by (4.111),

$$\frac{d}{dt} j(u(y(t,x))) = \sum_{i=1}^d \widetilde{D}_i(y_i(t,x)) D_i(u(y(t,x)))\,(u(y(t,x))b(u(y(t,x))))^{\frac{1}{2}}.$$

We note that

$$C_2 j(r) \leq \sqrt{rb(r)} \leq C_1 j(r),\ \forall r \geq 0,$$

where $C_1, C_2 > 0$. We set $\alpha(t,x) = (u(y(t,x))b(u(y(t,x))))^{\frac{1}{2}}(j(u(y(t,x))))^{-1}$. Then $\alpha \in L^\infty((0,\infty) \times \mathbb{R}^d)$ and

$$\frac{d}{dt} j(u(y(t,x))) = \alpha(t,x) \sum_{i=1}^d \widetilde{D}_i(y_i(t,x)) D_i(u(y(t,x))) j(u(y(t,x))),\ \forall t \geq 0.$$

Hence

$$j(u(y(t,x))) = j(u(x)) \exp\left(\int_0^t \alpha(s,x)\mathscr{D}(e^{\mathscr{D}s}x) \cdot E(u(e^{\mathscr{D}s}x))\right),$$
$$\forall t \geq 0,\ x \in \mathbb{R}^d,$$

and, therefore,

$$j(u(x)) = j(u(e^{\mathscr{D}t}x)) \exp\left(-\int_0^t \alpha(s,x)\mathscr{D}(e^{\mathscr{D}s}x) \cdot E(u(e^{\mathscr{D}s}x))\right),$$

where $e^{\mathscr{D}t}$ is the flow generated by $\mathscr{D}$. Since $\mathscr{D}$ is an arbitrary vector field on $\mathbb{R}^d$, it follows that, for fixed x and t, $\{e^{\mathscr{D}t}x\}$ covers all $\mathbb{R}^d$. We infer that, if $u \not\equiv 0$, then

$j(u(x)) > 0$, $\forall x \in \mathbb{R}^d$, and this implies that $u = u_\infty > 0$, a.e. on $\mathbb{R}^d$. For such a u_∞, this yields, because $\Psi(u_\infty) = 0$,

$$\nabla(g(u_\infty) + \Phi) = 0, \text{ a.e. in } \mathbb{R}^d, \tag{4.112}$$

where

$$g(r) = \int_1^r \frac{\beta'(s)}{sb(s)}\, ds, \ \forall r > 0.$$

By (4.112), we see that $g(u_\infty) + \Phi = \mu$ for some $\mu \in \mathbb{R}$, in $\mathbb{R}^d$ and, since g is strictly monotone, we have

$$u_\infty(x) = g^{-1}(-\Phi(x) + \mu), \ \ x \in \mathbb{R}^d. \tag{4.113}$$

The Strong H-Theorem

We shall prove here that, under the additional Hypothesis (K6), the transient solution $u(t) = S(t)u_0$ is strongly convergent in L^1 to a stationary solution to NFPE (4.1). Namely, we denote by $\widetilde{\omega}(u_0)$ the *omega-limit set* of u_0 in L^1, that is,

$$\widetilde{\omega}(u_0) = \left\{u = \lim_{t_n \to \infty} S(t_n)u_0 \ \text{ in } L^1\right\}.$$

Our aim here is to show that, under additional Hypothesis (K6), $\widetilde{\omega}(u_0) = \omega(u_0)$ consists of a single element u_∞ which is the equilibrium state for NFPE (4.1) and, therefore,

$$\lim_{t \to \infty} S(t)u_0 = u_\infty \ \text{ in } L^1.$$

Namely, we have the following strong form of the H-theorem for NFPE (4.1).

Theorem 4.2 *Assume that Hypotheses* (K1)–(K6) *hold and let* $u_0 \in D_0(V)\backslash\{0\}$. *Then*

$$\omega(u_0) = \widetilde{\omega}(u_0) = \{u_\infty\}, \tag{4.114}$$

and $u_\infty \in \mathcal{P}$, $u_\infty > 0$, *a.e. on* $\mathbb{R}^d$. *Furthermore,* $u_\infty \in D_0(V) \cap D(\Psi)$, $\Psi(u_\infty) = 0$, $S(t)u_\infty = u_\infty$ *for* $t \geq 0$, $|u_\infty|_1 = |u_0|_1$, *and it is given by*

$$u_\infty(x) = g^{-1}(-\Phi(x) + \mu), \ \forall x \in \mathbb{R}^d, \tag{4.115}$$

where μ *is the unique number in* $\mathbb{R}$ *such that*

$$\int_{\mathbb{R}^d} g^{-1}(-\Phi(x) + \mu) dx = \int_{\mathbb{R}^d} u_0\, dx, \tag{4.116}$$

where

$$g(r) = \int_1^r \frac{\beta'(s)}{sb(s)}\,ds,\ r > 0.$$

In particular, for all $u_0 \in D_0(V)$ *with the same* L^1*-norm, the sets in* (4.114) *coincide, and thus* u_∞ *is the only element in* $D_0(V)$ *with given* L^1*-norm such that* $S(t)u_\infty = u_\infty$ *for all* $t \geq 0$.

For proof, we need the following lemma.

Lemma 4.7 *Under Hypotheses* (K1)–(K6), *we have, for all* $u_0 \in \mathscr{M}_+$,

$$\|(I + \lambda A)^{-1}u_0\| \leq \|u_0\|,\ \forall \lambda \in (0, \lambda_0), \tag{4.117}$$

$$\|S(t)u_0\| \leq \|u_0\|,\ \forall t \geq 0. \tag{4.118}$$

Proof We may assume that by approximation $u_0 \in \mathscr{M}_+ \cap L^2$. Arguing as in the proof of Lemma 4.3 and taking into account that $u_\varepsilon \geq 0$, we get by (4.51)–(4.53),

$$\int_{\mathbb{R}^d} u_\varepsilon \varphi_\nu dx \leq -\lambda \int_{\mathbb{R}^d} ((h_\varepsilon^*(u_\varepsilon)|\nabla\Phi_\varepsilon|^2 + \nabla\Phi_\varepsilon \cdot \nabla\beta(u_\varepsilon))(1 - \nu\Phi_\varepsilon)\exp(-\nu\Phi_\varepsilon))dx + \int_{\mathbb{R}^d} u_0\varphi_\nu\,dx. \tag{4.119}$$

Since, by (4.30) and Hypotheses (K3), (K4), we have that $|\nabla\Phi_\varepsilon| \in L^2$ and $\beta(u_\varepsilon) \in H^1$, we may pass to the limit $\nu \to 0$ in (4.119) to find after integrating by parts using Hypothesis (K5) that

$$\int_{\mathbb{R}^d} u_\varepsilon \Phi_\varepsilon\,dx \leq \lambda \int_{\mathbb{R}^d} \left(-b_0 \cdot \frac{u_\varepsilon}{1 + \varepsilon|u_\varepsilon|}|\nabla\Phi_\varepsilon|^2 + \Delta\Phi_\varepsilon\beta(u_\varepsilon)\right)dx + \int_{\mathbb{R}^d} u_0\Phi_\varepsilon\,dx. \tag{4.120}$$

We note that integrating by parts is justified here, since $\beta(u_\varepsilon) \in L^1 \cap L^2$ and $\Delta\Phi_\varepsilon \in L^2 + L^\infty$ because of (4.55) and Hypothesis (K3). Now, we want to let $\varepsilon \to 0$ (along a subsequence) in (4.120). To this end, we note that, since by Hypothesis (K3) $\Delta\Phi = f_2 + f_\infty$ for some $f_2 \in L^2$, $f_\infty \in L^\infty$, it follows by (4.55), (4.56) that

$$\Delta\Phi_\varepsilon = g_\varepsilon(f_2 + f_\infty) + \varepsilon h_\varepsilon|D|^2,$$

where $g_\varepsilon, h_\varepsilon : \mathbb{R}^d \to \mathbb{R}$ such that $g_\varepsilon \to 1$, a.e. as $\varepsilon \to 0$, with $|g_\varepsilon| \leq m + 1$ and $|h_\varepsilon| \leq m(m + 3)$. Since $\beta(u_\varepsilon) \to \beta(u)$ in L^1 by Lemma 4.3 and also weakly in L^2

by (4.42) and since $|\nabla\Phi_\varepsilon|^2 \to |\nabla\Phi|^2$, a.e. as $\varepsilon \to 0$, by (4.30), by virtue of Fatou's lemma we can pass to the limit $\varepsilon \to 0$ (along a subsequence) in (4.120) to obtain

$$\|u\| \leq \lambda \int_{\mathbb{R}^d} (-b_0|\nabla\Phi|^2 u + \Delta\Phi\beta(u))dx + \|u_0\|,$$

where $u = J_\lambda u_0 = (I + \lambda A)^{-1} u_0$ is as in (4.48). By Hypothesis (K6), this implies (4.117), which in turn implies (4.118) by the same argument as in the proof of Proposition 4.2.

By Lemma 4.7, inequality (4.25) holds and, therefore, by (4.71) and (4.118), we have

$$V(S(t)u_0) \geq -C(\|S(t)u_0\| + 1)^\alpha \geq -C(\|u_0\| + 1)^\alpha, \ \forall t \geq 0,$$

and so, by (4.75),

$$\int_0^\infty \Psi(S(\sigma)u_0)d\sigma < \infty. \tag{4.121}$$

This implies that

$$\omega(u_0) \subset \{u \in D(\Psi);\ \Psi(u) = 0\}. \tag{4.122}$$

To prove this, we shall argue as follows.

Let $u_\infty \in \omega(u_0)$ and $\{t_n\} \to \infty$ such that $S(t_n)u_0 \to u_\infty$ in L^1_{loc}. Assume that $\Psi(u_\infty) > \delta > 0$ and argue from this to a contradiction. This implies that there is a bounded open subset $\mathcal{O}$ of $\mathbb{R}^d$ such that

$$\Psi_{\mathcal{O}}(u_\infty) > \frac{\delta}{2}, \tag{4.123}$$

where $\Psi_{\mathcal{O}}$ is the integral for (4.72) restricted to $\mathcal{O}$. Since $\Psi_{\mathcal{O}}$ is lower semicontinuous in L^1, it follows by (4.123) that there is a $\mu = \mu(\delta) > 0$ such that

$$\Psi_{\mathcal{O}}(u) \geq \frac{\delta}{4} \text{ if } |u_\infty - u|_1 \leq \mu. \tag{4.124}$$

Since $S(t)$, $t > 0$, is a semigroup of contractions in L^1, we have

$$|S(t)u_0 - S(s)u_0|_1 \leq \nu(|t - s|), \ \forall s, t \geq 0, \tag{4.125}$$

where $\nu(r) := \sup\{|S(s)u_0 - u_0|_1 : 0 \leq s \leq r\}$, $r > 0$. Clearly, $\nu(r) \to 0$ as $r \to 0$. By (4.125), we have

$$|S(t)u_0 - u_\infty|_1 \leq |S(t)u_0 - S(t_n)u_0|_1 + |S(t_n)u_0 - u_\infty|_1 \leq \mu,$$

for $|t - t_n| \le \nu^{-1}\left(\frac{\mu}{2}\right)$, $n \ge N(\mu)$, where ν^{-1} is the inverse function of ν. By (4.124), this yields

$$\Psi_{\mathcal{O}}(S(t)u_0) \ge \frac{\delta}{4} \text{ for } |t - t_n| \le \nu^{-1}\left(\frac{\mu}{2}\right),$$

and $n \ge N(\mu)$. But this contradicts (4.121).

Then, (4.122) and Theorem 4.1 imply (4.115). By (4.118), we also have

$$\lim_{R\to\infty} \sup_{t\ge 0} \int_{\{\Phi\ge R\}} S(t)u_0\, dx = 0,$$

which implies that the orbit $\{S(t)u_0,\ t \ge 0\}$ is compact in L^1, $\omega(u_0) = \widetilde{\omega}(u_0)$ and that $|u_\infty|_1 = |u_0|_1$ by (4.21). Hence, (4.116) and (4.114) follow. By Fatou's lemma, it follows that $u_\infty \in D(V)$ and, by (4.113), (4.74) and the L^1_{loc}-lower semicontinuity of V on balls in $\mathscr{M}$, we conclude that $u_\infty \in D_0(V)$. Now, let us check that $S(t)u_\infty = u_\infty$, for $t \ge 0$. So, let $t_n \to \infty$, such that $\lim_{n\to\infty} S(t_n)u_0 = u_\infty$. Then, for all $t > 0$, by the semigroup property and the L^1-continuity of $S(t)$,

$$S(t)u_\infty = \lim_{n\to\infty} S(t + t_n)u_0 \in \widetilde{\omega}(u_0) = \{u_\infty\}.$$

The last part of the assertion is obvious by (4.116).

Corollary 4.1 *Let u_∞ be as in Theorem 4.2. Then*

$$|u_\infty|_\infty \le \max\left(1, e^{\frac{|b|_\infty}{\gamma}(\mu-1)}\right),$$

where $\mu \in \mathbb{R}$ is as in (4.115).

Proof For g as above, we have that g is strictly increasing and $g : (0, \infty) \to \mathbb{R}$ is bijective. Furthermore, by (4.68), we have, for $r \in (0, \infty)$,

$$\frac{\gamma_1}{b_0}\mathbf{1}_{(0,1]}(r)\log r + \frac{\gamma}{|b|_\infty}\mathbf{1}_{(1,\infty)}(r)\log r \le g(r).$$

Hence, replacing r by $e^{\frac{b_0}{\gamma_1}r}$, $r \le 0$, we get

$$g^{-1}(r) \le e^{\frac{b_0}{\gamma_1}r},\ r \in (-\infty, 0],$$

and, replacing r by $e^{\frac{|b|_\infty}{\gamma}r}$, $r \in (0, \infty)$, we obtain

$$g^{-1}(r) \le e^{\frac{|b|_\infty}{\gamma}r},\ r \in (0, \infty).$$

This implies, by (4.115), for all $x \in \mathbb{R}^d$,

$$(0 <)u_\infty(x) = g^{-1}(\mu - \Phi(x)) \leq \mathbf{1}_{\{\mu \leq \Phi\}}(x) e^{\frac{b_0}{\gamma_1}(\mu - \Phi(x))} + \mathbf{1}_{\{\mu > \Phi\}}(x) e^{\frac{|b|_\infty}{\gamma}(\mu - \Phi(x))}$$
$$\leq \max\left(1, e^{\frac{|b|_\infty}{\gamma}(\mu - 1)}\right),$$

since $\Phi \geq 1$.
We show now that Theorem 4.2 implies the uniqueness of solutions $u^* \in \mathscr{M} \cap \mathscr{P} \cap \{V < \infty\}$ of the stationary version of Eq. (4.1), that is,

$$-\Delta\beta(u^*) + \operatorname{div}(Db(u^*)u^*) = 0 \text{ in } \mathscr{D}'(\mathbb{R}^d). \tag{4.126}$$

We note that the set of all $u^* \in L^1(\mathbb{R}^d)$ satisfying (4.126) is just $A_0^{-1}(\{0\})$.

Theorem 4.3 *Under Hypotheses* (K1)–(K6), *there is a unique solution* u^* *to* (4.126) *such that* $u^* \in L^1 \cap L^\infty$. *In addition,* $u^* \in \mathscr{M} \cap \mathscr{P} \cap \{V < \infty\}$.

Proof By Theorem 4.2 and Corollary 4.1, it follows that u_∞ is a solution to (4.126), which is in $\mathscr{M} \cap \mathscr{P} \cap \{V < \infty\} \cap L^\infty$. So it only remains to prove the uniqueness. But this follows from Theorem 2.6 taking into account that its hypotheses are satisfied in this case.

As an application to the McKean–Vlasov SDE, we have

Theorem 4.4 *Let* $X^i(t)$, $t \geq 0$, $i = 1, 2$, *be two stationary nonlinear distorted Brownian motions, i.e., both satisfy the McKean–Vlasov SDE*

$$dX(t) = D(X(t))b(u(t, X(t))dt + \sqrt{\frac{2\beta(u(t, X(t)))}{u(t, X(t))}}\, dW(t), \tag{4.127}$$

with $(\mathscr{F}_t^i)$*-Wiener processes* $W^i(t)$, $t \geq 0$, *on probability spaces* $(\Omega^i, \mathscr{F}^i, \mathbb{P}^i)$ *equipped with normal filtrations* $\mathscr{F}_t^i$, $t \geq 0$, *with*

$$\mathbb{P}^i \circ (X^i(t))^{-1} = u_\infty^i\, dx,$$

and $u(t, x)$ *in* (4.127) *replaced by* $u_\infty^i(x)$ *for* $i = 1, 2$, *respectively. Assume that* $u_\infty^i \in \mathscr{M} \cap \{V < \infty\} \cap L^\infty$, $i = 1, 2$. *Then*

$$\mathbb{P}^1 \circ (X^1)^{-1} = \mathbb{P}^2 \circ (X^2)^{-1},$$

i.e., we have uniqueness in law of stationary nonlinear distorted Brownian motions with stationary measures in $\mathscr{M} \cap \{V < \infty\} \cap L^\infty$.

Proof By Itô's formula, both u^1_∞ and u^2_∞ satisfy (4.126). Hence, by Theorem 4.3, we have $u^1_\infty = u^2_\infty = u_\infty$. Fix $T > 0$ and let

$$\Phi(r) := \frac{\beta(r)}{r}, \; r \in \mathbb{R}.$$

Then Theorem 2.6 implies that, for each $s \in [0, T]$ and each $v_0 \in L^1 \cap L^\infty$, there is at most one solution $v = v(t, x)$, $t \in [s, T]$, to

$$\begin{aligned} &v_t - \Delta(\Phi(u_\infty)v) + \operatorname{div}(Db(u_\infty)v) = 0 \text{ in } \mathscr{D}'((0, T) \times \mathbb{R}^d, \\ &v(0, \cdot) = v_0, \end{aligned}$$

such that $v \in L^\infty((s, T) \times \mathbb{R}^d)$ and $t \mapsto v(t, x)dx$, $t \in [s, T]$ is narrowly continuous. But u_∞, the time marginal law of X^i under $\mathbb{P}^i$, $i = 1, 2$, is such a solution with $v_0 = u_\infty$, since $u_\infty \in L^\infty$ by Corollary 4.1. Hence, Lemma 2.12 in [98] implies the assertion, since by Itô's formula $\mathbb{P}^i \circ (X^i)^{-1}$, $i = 1, 2$, both satisfy the martingale problem for the Kolmogorov operator

$$L_{u_\infty} = \Phi(u_\infty)\Delta + b(u_\infty)D \cdot \nabla.$$

4.2 Attractors for Nonlinear Fokker–Planck Flows

As seen earlier, under Hypotheses (K1)–(K6), the transient solutions $u(t) = S(t)u_0$ to Eq. (4.1) converge for $t \to \infty$ to the unique stationary solution of this equation. However, the situation might be different if the stationary solution has multiple solutions which is the case if the diffusion function β is degenerate in origin. We shall analyze this situation on Eq. (4.1), that is,

$$\begin{aligned} &u_t - \Delta\beta(u) + \operatorname{div}(Db(u)u) = 0 \;\text{ in } (0, \infty) \times \mathbb{R}^d, \\ &u(0, x) = u_0(x), \;\; x \in \mathbb{R}^d, \; d \geq 1, \; u_0 \in L^1, \end{aligned}$$

under the following assumptions on $\beta : \mathbb{R} \to \mathbb{R}$ and $D : \mathbb{R}^d \to \mathbb{R}^d$

(L1) $\beta \in C^1(\mathbb{R})$, $\beta'(r) > 0$, $\forall r \in \mathbb{R} \setminus \{0\}$, $\beta(0) = 0$, and

$$\mu_1 \min\{|r|^\nu, |r|\} \leq |\beta(r)| \leq \mu_2|r|, \; \forall r \in \mathbb{R}, \tag{4.128}$$

for $\mu_1, \mu_2 > 0$ and $\nu > \frac{d-1}{d}$.

(L2) $D \in L^\infty(\mathbb{R}^d;\mathbb{R}^d) \cap W^{1,1}_{\rm loc}(\mathbb{R}^d;\mathbb{R}^d)$, $\operatorname{div} D \in (L^1+L^\infty)\cap(L^2+L^\infty)$ and $D = -\nabla\Phi$, where $\Phi \in C(\mathbb{R}^d)\cap W^{1,1}_{\rm loc}(\mathbb{R}^d)$, satisfies the conditions

$$\Phi(x) \ge 1,\ \forall x \in \mathbb{R}^d,\quad \lim_{|x|\to\infty}\Phi(x) = +\infty,\ \Phi^{-m} \in L^1 \text{ for some } m \ge 2, \tag{4.129}$$

$$\mu_2\Delta\Phi(x) - b_0|\nabla\Phi(x)|^2 \le 0,\ \text{a.e. } x \in \mathbb{R}^d. \tag{4.130}$$

(L3) $b \in C^1(\mathbb{R})\cap C_b(\mathbb{R})$, $b(r) \ge b_0 > 0$ for all $r \in [0,\infty)$.

We note that, in particular, (4.130) implies that $(\operatorname{div} D)^- \in L^\infty$. It should also be noted that assumption (L1) does not preclude the degeneracy of the nonlinear diffusion function β in the origin and it covers the power-law case $\beta(r) \equiv |r|^{m-1}r$ discussed earlier, and this is the main difference with respect to the case considered in Sect. 4.1. For instance, any continuous, increasing function $\beta : \mathbb{R} \to \mathbb{R}$ of the form

$$\beta(r) = \begin{cases} \mu_1 r|r|^{d-1} & \text{for } |r| \le r_0,\\ \mu_2 h(r) & \text{for } |r| > r_0,\end{cases}$$

where $r_0 > 0$, $\mu_1,\mu_2 > 0$, $|h(r)| \le L|r|$, $\forall r \in \mathbb{R}$, $L > 0$, satisfies (4.128) for a suitable γ_1. As regards Hypothesis (L2), as mentioned earlier in Sect. 4.1, an example of such a function Φ is

$$\Phi(x) = \begin{cases} |x|^2\log|x| + \mu & \text{for } |x| \le \delta = \exp\left(-\frac{d+2}{2d}\right),\\ \varphi(|x|) + \eta|x| + \mu & \text{for } |x| > \delta.\end{cases} \tag{4.131}$$

Let $A_0 : D(A_0) \subset L^1 \to L^1$,

$$A_0(u) = -\Delta\beta(u) + \operatorname{div}(Db(u)u),\ \forall u \in D(A_0),$$
$$D(A_0) = \{u \in L^1;\ -\Delta\beta(u) + \operatorname{div}(Db(u)u) \in L^1\}.$$

As seen earlier in Theorem 2.10,

$$R(I + \lambda A_0) = L^1,\ \forall \lambda > 0,$$

and, for each $\lambda > 0$, there is $J_\lambda : L^1 \to D(A_0)$ satisfying Proposition 2.3. Then the operator $A : D(A) \subset L^1 \to L^1$, defined by

$$\begin{aligned} A(u) &= A_0(J_\lambda(u)),\ \forall u \in D(A),\\ D(A) &= J_\lambda(L^1),\end{aligned} \tag{4.132}$$

is m-accretive in L^1 and generates a semigroup of contractions on $C = \overline{D(A)}$ the closure of $D(A)$ in L^1. In other words, the Cauchy problem

$$\frac{du}{dt} + A(u) = 0, \quad t \geq 0,$$
$$u(0) = u_0,$$

has a unique mild solution $u \in C([0, \infty); L^1)$ and $S(t)u_0 = u(t)$, $t \geq 0$, is a semigroup of nonlinear contractions in L^1.

As seen earlier in Proposition 2.3, $J_\lambda(f) = \lim\limits_{\varepsilon \to 0} J_\lambda^\varepsilon(f)$ in L^1, where $J_\lambda^\varepsilon(f) = u_\varepsilon$ is the solution to (2.208), that is,

$$u_\varepsilon - \lambda(\varepsilon I - \Delta)(\widetilde{\beta}_\varepsilon(u_\varepsilon)) + \lambda \operatorname{div}(D_\varepsilon b_\varepsilon(u_\varepsilon)u_\varepsilon) = f.$$

We recall that

$$\|\widetilde{\beta}_\varepsilon(u_\varepsilon)\|_{L^q(K)} \leq C_K|f|_1, \ \forall f \in L^1 \cap L^\infty,$$

where $1 < q < \frac{d}{d-1}$ and K is any compact set of $\mathbb{R}^d$. This yields (see (2.242))

$$\|\beta(J_\lambda(f)\|_{L^q(K)} \leq C_K|f|_1, \ \forall f \in L^1 \cap L^\infty. \tag{4.133}$$

Moreover, by Lemma 2.5 it follows also that (see (2.103))

$$|\nabla \beta(J_\lambda(f))|_{L^q(K)} \leq C(|f|_1 + |J_\lambda(f)|_1) \leq C_K|f|_1. \tag{4.134}$$

Now, consider the orbit $\gamma(u_0) = \{S(t)u_0, \ t \geq 0\}$ where $u_0 \in C = \overline{D(A)}$. We associate to u_0 the ω-limit set

$$\omega(u_0) = \left\{u_\infty = \lim S(t_n)u_0 \text{ in } L^1 \text{ for some } \{t_n\} \to \infty\right\}$$
$$= \bigcap_{s \geq 0} \overline{\bigcup_{t \geq s} S(t)u_0}.$$

In particular, if $\omega(u_0) \neq \emptyset$ and consists of one element u_∞ only, then we have

$$\lim_{t \to \infty} S(t)u_0 = u_\infty \text{ in } L^1.$$

In Theorem 4.2, it was proved that this happens if β is not degenerate in the origin, that is,

$$0 < \gamma_0 \leq \beta'(r) \leq \gamma_1, \ \forall r \in \mathbb{R}, \tag{4.135}$$

(which again implies that $C = L^1$) and if

$$u_0 \ln(u_0) \in L^1, \ \|u_0\| = \int_{\mathbb{R}^d} u_0(x)\Phi(x)dx < \infty. \tag{4.136}$$

More precisely, in this case one has $\omega(u_0) = \{u_\infty\}$, where u_∞ is an equilibrium solution to (4.1), and moreover it is the unique solution in $L^1 \cap L^\infty$ to the stationary equation

$$-\Delta\beta(u) + \operatorname{div}(Db(u)u) = 0 \ \text{ in } \mathscr{D}'(\mathbb{R}^d). \tag{4.137}$$

The situation is different in the more general case considered here. As we shall see below, under Hypotheses (L1)–(L3), $\omega(u_0)$ *is a nonempty*, compact subset of L^1 and, for every fix point of $S(t)$, $t > 0$, it is contained in some sphere centered at this fix point. This means that there is a compact set $\omega(u_0)$ of probability densities which attracts the trajectory which starts from a nonequilibrium state u_0 and which is specific to open systems far from thermodynamic equilibrium (see, e.g., [40, 92]). In the following, we shall use the notation

$$\|u\| = \int_{\mathbb{R}^d} \Phi(x)|u(x)|dx, \ \ \forall u \in L^1, \tag{4.138}$$

and we shall denote by $\mathscr{M}$ the subspace of L^1 with the finite norm (4.138). We also set

$$\mathscr{M}_\eta = \{u \in \mathscr{M};\ \|u\| \le \eta\},\ \eta > 0.$$

We have

Theorem 4.5 *Assume that Hypotheses* (L1)–(L3) *hold and let $\eta > 0$ be arbitrary but fixed. Then, there is a continuous semigroup of contractions in L^1, $S(t) : C \to C$ corresponding to NFPE* (4.1) *such that, for $u_0 \in \mathscr{M}_\eta \cap \mathscr{P} \cap C$, $\omega(u_0) \subset \mathscr{M}_\eta \cap \mathscr{P} \cap C$ is nonempty and compact in L^1, and for all $t \ge 0$, $\omega(u_0)$ is invariant under $S(t)$. Moreover, $S(t)$ is, for every $t \ge 0$, an isometry on $\omega(u_0)$ and it is a homeomorphism from $\omega(u_0)$ onto itself for each $t \ge 0$. If $a \in \mathscr{M}_\eta \cap \mathscr{P} \cap C$ is such that*

$$S(t)a = a, \ \ \forall t \ge 0, \tag{4.139}$$

then $\omega(u_0) \subset \{y \in \mathscr{M}_\eta \cap \mathscr{P} \cap C;\ |y - a|_1 = r\}$, for some $0 \le r \le |u_0 - a|_1$.

Here, $C = \overline{D(A)}$ is the closure of the domain $D(A)$ of the operator A in L^1.

In particular, it follows by Theorem 4.5 that $S(t)$, $t \ge 0$, is *a continuous group* on $\omega(u_0)$. Moreover, *the function $t \to S(t)v$ is equi-almost periodic* in L^1 for each

$v \in \omega(u_0)$, i.e., for every $\varepsilon > 0$ there exists $\ell_\varepsilon > 0$ such that for every interval I in $\mathbb{R}$ of length ℓ_ε there exists $\tau \in I$ such that

$$|S(t+\tau)y - S(t)y|_1 \le \varepsilon, \ \forall t \in \mathbb{R}, \ y \in \omega(u_0).$$

To apply this theorem, it is necessary to know that the semigroup $S(t)$ has a stationary point $a \in \mathcal{M}_\eta \cap \mathcal{P} \cap C$, that is, (4.139) holds. In the following, we shall construct such a semigroup. (Of course, if one further assumes that Hypotheses (i)–(iv) of Theorem 2.4 hold, then there is a unique semigroup $S(t)$ with this property.)

We come back to the operator $A_\varepsilon : L^1 \to L^1$ defined by (2.249) and recall that by Lemma 2.10 we have for all $\lambda > 0$, $f \in L^1$ and $\{\varepsilon'\} \to 0$ (independent of f and λ),

$$(I+\lambda A_{\varepsilon'})^{-1} f \to J_\lambda(f) = (I+\lambda A)^{-1} f \ \text{ in } L^1,$$

where A is the m-accretive operator (4.132).

Let $S_\varepsilon(t)$ that generated by A_ε. Then, we have for some $\varepsilon \to 0$

$$S_\varepsilon(t)u_0 \to S(t)u_0 \ \text{ in } L^1, \ \forall u_0 \in C, \ t \ge 0,$$

uniformly in t on compact intervals. Moreover, by Theorem 4.2 we have

$$S_\varepsilon(t)a_\varepsilon = a_\varepsilon, \ \forall t \ge 0,$$

where

$$\begin{aligned} a_\varepsilon(x) &= g_\varepsilon^{-1}(\mu_\varepsilon - \Phi(x)), \ \forall x \in \mathbb{R}^d, \\ g_\varepsilon(r) &= \int_1^r \frac{\widetilde{\beta}'_\varepsilon(s)}{s b_\varepsilon(s)} ds, \ \forall r \in \mathbb{R}, \end{aligned} \tag{4.140}$$

where $\widetilde{\beta}_\varepsilon(r) \equiv \beta(r) + \varepsilon r$, and $\mu_\varepsilon \in \mathbb{R}$ is such that

$$\int_{\mathbb{R}^d} g_\varepsilon^{-1}(\mu_\varepsilon - \Phi(x))dx = 1. \tag{4.141}$$

We also have

$$-\Delta\widetilde{\beta}_\varepsilon(a_\varepsilon) + \mathrm{div}(D_\varepsilon b_\varepsilon(a_\varepsilon)a_\varepsilon) = 0 \text{ in } \mathcal{D}'(\mathbb{R}^d),$$

and note that, if

$$a = \lim_{\varepsilon\to 0} a_\varepsilon \ \text{ exists in } L^1, \tag{4.142}$$

and $J_\lambda = (I + \lambda A)^{-1}$, $J_\lambda^\varepsilon = (I + \lambda A_\varepsilon)^{-1}$, then it follows by the inequality

$$|J_\lambda(a) - a|_1 \leq |J_\lambda(a) - J_\lambda(a)|_1 + |J_\lambda^\varepsilon(a) - J_\lambda^\varepsilon(a_\varepsilon)|_1 + |a - a_\varepsilon|_1$$

that $J_\lambda(a) = a$, $\forall \lambda \in (0, \lambda_0)$. Then, by the exponential formula (4.19) it follows that $S(t)a = a$, $\forall t \geq 0$, so that (4.139) holds.

In Proposition 4.4 below, we shall show that under a mild condition in addition to Hypotheses (L1)–(L3) we indeed have (4.142). To this purpose, consider the function $g : (0, \infty) \to \mathbb{R}$ defined by

$$g(r) = \int_1^r \frac{\beta'(s)}{sb(s)}\,ds, \quad \forall r > 0. \tag{4.143}$$

We have

Proposition 4.4 *Assume that, besides Hypotheses* (L1), (L2), (L3), *the following conditions hold:*

$$\lim_{r \to +\infty} g(r) = +\infty, \; if \; \nu \in \left(1 - \frac{1}{d}, 1\right]; \tag{4.144}$$

$$\lim_{r \to 0} g(r) = -\infty, \; if \; \nu \in (1, \infty), \tag{4.145}$$

where ν is as in Hypothesis (L1)*. Let $a_\varepsilon \in \mathscr{P} \cap \mathscr{M} \cap L^\infty$ be as in* (4.140)*. Then* (4.142) *holds and*

$$a(x) = g^{-1}(\mu - \Phi(x)), \quad x \in \mathbb{R}^d, \tag{4.146}$$

where $\mu \in \mathbb{R}$ is the unique number such that

$$\int_{\mathbb{R}^d} g^{-1}(\mu - \Phi(x))dx = 1.$$

In particular, it follows that a is a stationary solution to Eq. (4.1).

Proof of Proposition 4.4 We first note that by an elementary calculation it follows by (4.144) and (4.145) that $g : (0, \infty) \to \mathbb{R}$ is bijective, since g is strictly increasing. Furthermore, $g((0, 1)) \subset (-\infty, 0)$, $g([1, \infty)) \subset [0, \infty)$, $g \in C^1((0, \infty))$ and for its inverse $g^{-1} : \mathbb{R} \to (0, \infty)$, we have

$$g^{-1} \in C^1(\mathbb{R}), (g^{-1})' > 0, g^{-1}([0, \infty)) \subset [1, \infty), g^{-1}((-\infty, 0)) \subset (0, 1),$$

and, by (4.140),

$$g_\varepsilon(r) := \int_1^r \frac{\beta_\varepsilon'}{sb(s)}\,ds = \int_1^r \frac{\beta'(s) + \varepsilon}{sb(s)}\,ds, \quad r > 0.$$

Then g_ε and its inverse g_ε^{-1} have the same properties as g, g^{-1} above. Clearly, as $\varepsilon \to 0$, $g_\varepsilon \to g$ locally uniformly on $\mathbb{R}$, and, therefore,

$$g_\varepsilon^{-1} \to g^{-1} \text{ locally uniformly on } \mathbb{R}. \tag{4.147}$$

Since $g_\varepsilon^{-1}([0,\infty)) \subset [1,\infty)$, by (4.141) we have

$$1 \geq \int_{\{\Phi \leq \mu_\varepsilon\}} 1\, dx,$$

and, therefore, $\sup\limits_{\varepsilon \in (0,1]} \mu_\varepsilon < \infty$.

Indeed, otherwise there exist $\varepsilon_n \in [0,1]$, $n \in \mathbb{N}$, such that $\mu_{\varepsilon_n} \to -\infty$ as $n \to \infty$, and therefore,

$$1 = \lim_{n\to\infty} \int_{\mathbb{R}^d} g_{\varepsilon_n}^{-1}(\mu_{\varepsilon_n} - \Phi(x))dx.$$

But, as easily seen, the limit on the right hand side is equal to zero. The contradiction we arrived at implies that $\inf\limits_{\varepsilon \in (0,1]} \mu_\varepsilon > -\infty$, and, therefore, $\{\mu_\varepsilon;\ \varepsilon \in (0,1]\}$ is bounded. This implies that there exist $\varepsilon_n \in (0,1]$, $n \in \mathbb{N}$, such that $\lim\limits_{n\to\infty} \varepsilon_n = 0$ and $\mu := \lim\limits_{n\to\infty} \mu_{\varepsilon_n}$ exists in $\mathbb{R}$.

Furthermore, we have

$$\sup_n \int_{\mathbb{R}^d} g_{\varepsilon_n}^{-1}(\mu_{\varepsilon_n} - \Phi(x))\Phi(x)dx < \infty,$$

$$g_{\varepsilon_n}^{-1}(\mu_{\varepsilon_n} - \Phi) \xrightarrow[n\to\infty]{} g^{-1}(\mu - \Phi)\ ,$$

uniformly on compact subsets on $\mathbb{R}^d$. Then, by Fatou's lemma, it follows that

$$\int_{\mathbb{R}^d} g^{-1}(\mu - \Phi)dx < \infty \quad \text{and} \quad g_{\varepsilon_n}^{-1}(\mu_{\varepsilon_n} - \Phi) \to g^{-1}(\mu - \Phi) \text{ in } L^1,$$

as claimed. □

By Proposition 4.4, it follows that there is a semigroup $S(t)$ generated by NFPE (4.1) for which there is a stationary point a to $S(t)$.

Corollary 4.2 *Assume that Hypotheses* (L1)–(L3) *and* (4.144) *and* (4.145) *are satisfied. Then all the conclusions of Theorem* 4.5 *hold. In particular, for each* $u_0 \in \mathcal{M}_\eta \cap \mathcal{P} \cap C$, *the* ω*-limit set* $\omega(u_0)$ *lies on the set*

$$\{y \in \mathcal{M}_\eta \cap \mathcal{P} \cap C;\ |y - a|_1 = r\},$$

where $a \in \mathscr{M}_\eta \cap \mathscr{P} \cap C$ *satisfies* (4.146) *and* $0 \le r \le |u_0 - a_{\mu^*}|_1$, $S(t)a = a$, $\forall t \ge 0$.

Of course, if the set $\omega(y_0) \cap \{a \in C;\ S(t)a = a,\ \forall t > 0\}$ is nonempty, it follows by Corollary 4.2 that $\omega(y_0)$ contains only one element (the equilibrium state a) and so $\lim\limits_{t\to\infty} S(t)y_0 = a$ strongly in L^1. This provides a weak form of the H-theorem for NFPE (4.1) and, roughly speaking, it amounts to saying that $S(t)u_0 \to \omega(u_0)$ in L^1 for $t \to \infty$, that is, $\omega(u_0)$ *attracts* u_0.

One simple example of a function β which satisfies all conditions of Theorem 4.5 is the following:

$$\beta(r) = \int_0^r \theta(s)ds, \quad \forall r \ge 0,$$

$$\theta(s) = \begin{cases} -\dfrac{1}{\log s} & \text{for } 0 < s \le \delta < 1, \\ \zeta(s) & \text{for } s > \delta, \end{cases}$$

and, for $r \in (-\infty, 0)$,

$$\beta(x) = -\beta(-r),$$

where $\delta > 0$, $\zeta \in C^1[\delta, +\infty)$, bounded, $\zeta \ge \zeta_0 \in (0, \infty)$, and ζ is such that $\theta \in C^1(0, \infty)$. Then it is elementary to check that β satisfies Hypothesis (i) with $\nu = 2$. Furthermore, obviously $g(r) = \text{const} - \log|\log r|$ for $r \in (0, 1]$. Hence (4.145) also holds and Theorem 4.5 applies.

Proof of Theorem 4.5 We shall prove Theorem 4.5 in three steps. Let $\eta > 0$ and let $K = \mathscr{M}_\eta \cap \mathscr{P} \cap C$. Clearly, K is a closed and bounded set of L^1. □

Lemma 4.8 *We have*

$$\|(I + \lambda A)^{-1} y\| \le \|y\|, \quad \forall y \in \mathscr{M} \cap \mathscr{P},\ \lambda > 0, \tag{4.148}$$

and

$$(I + \lambda A)^{-1}(K) \subset \mathscr{M}_\eta \cap \mathscr{P} \cap D(A) \subset D(A) \cap K,\ \forall \lambda > 0. \tag{4.149}$$

In particular,

$$\|S(t)u_0\| \le u_0, \quad \forall u_0 \in C,\ t \ge 0, \tag{4.150}$$

and $S(t)(K) \subset K, \quad \forall t \ge 0.$

Proof By Lemma 4.7, we have $\|J_\lambda^\varepsilon(y)\| \le \|y\|$ for all $y \in \mathscr{M} \cap \mathscr{P}$, $\lambda > 0$, which implies (4.148) for $\varepsilon \to 0$. Then, (4.150) follows by the exponential formula

$$S(t)u_0 = \lim_{n\to\infty}\left(I + \frac{t}{n}A\right)^{-n} u_0.$$

In the following, we shall denote by $\widetilde{A}$ the restriction of the operator A to K, that is,

$$\widetilde{A}u = Au, \quad \forall\, u \in D(\widetilde{A}) = D(A) \cap K. \tag{4.151}$$

By (4.149) it follows that, for every $\lambda > 0$,

$$\overline{D(\widetilde{A})} \subset K \subset (I + \lambda\widetilde{A})(D(\widetilde{A})) = R(I + \lambda\widetilde{A}) \tag{4.152}$$

and we have by definition that

$$(I + \lambda\widetilde{A})^{-1} = (I + \lambda A)^{-1} \text{ on } R(I + \lambda\widetilde{A}). \tag{4.153}$$

Furthermore, $(\widetilde{A}, D(\widetilde{A}))$ is accretive on L^1 and, since $\overline{D(\widetilde{A})} \subset K \subset \overline{D(A)}$, we conclude that $\forall u_0 \in \overline{D(\widetilde{A})}$, $t \ge 0$,

$$S(t)u_0 = \lim_{n\to\infty}\left(I + \frac{t}{n}\widetilde{A}\right)^{-1} u_0 \in \overline{D(\widetilde{A})}.$$

Therefore, $\widetilde{S}(t) := S(t)_{\overline{D(\widetilde{A})}}$ (the restriction of $S(t)$ to $\overline{D(A)}$) is the contraction semigroup generated by $\widetilde{A}$ on $\overline{D(\widetilde{A})}$. We are going to apply Theorem 6.7 to this semigroup $\widetilde{S}(t)$, $t \ge 0$, to prove Theorem 4.5. For this we need:

Lemma 4.9 *The operator* $(I + \lambda\widetilde{A})^{-1}$ *restricted to K is compact for* $\lambda \in (0, \lambda_0]$.

Proof Let $f_n \in K$, $n \in \mathbb{N}$, be such that

$$\sup_{n\in\mathbb{N}} |f_n|_1 < \infty. \tag{4.154}$$

Then, since $\sup\limits_{n\in\mathbb{N}} \|f_n\| \le \eta$, by (4.153), it suffices to prove that (selecting a subsequence if necessary) for $\lambda \in (0, \lambda_0]$, $J_\lambda(f_n) = (I + \lambda A)^{-1} f_n$ converges in $L^1_{\rm loc}(K)$ as $n \to \infty$ for every compact set $K \subset \mathbb{R}^d$. Let $q \in \left(1, \frac{d}{d-1}\right)$. By (4.133) and (4.134), and the Sobolev compactness theorem, it follows that (selecting a subsequence if necessary)

$$\beta(J_\lambda(f_n)) \xrightarrow[n\to\infty]{} \eta \text{ in } L^q(K)$$

and (since this holds for every such K)

$$\beta(J_\lambda(f_n)) \underset{n\to\infty}{\longrightarrow} \eta \ , \text{ a.e. on } \mathbb{R}^d,$$

hence, since $\beta \in C^1$ and $\beta' > 0$,

$$J_\lambda(f_n) \underset{n\to\infty}{\longrightarrow} \beta^{-1}(\eta) \quad \text{a.e. on } \mathbb{R}^d.$$

Furthermore, because $\nu > \frac{d-1}{d}$, we may choose q so close to $\frac{d}{d-1}$ that $\nu q > 1$, and hence by (4.128)

$$\mu_1^q \min(|r|^q, |r|^{\nu q}) \le |\beta(r)|^q, \ \forall r \in \mathbb{R}.$$

This implies that $\{J_\lambda(f_n);\ n \in \mathbb{N}\}$ is equi-integrable in $L^1(K)$, hence

$$J_\lambda(f_n) \underset{n\to\infty}{\longrightarrow} \beta^{-1}(\eta) \quad \text{in } L^1(K).$$

□

Now, taking into account that $|S(t)u_0|_1 \le |u_0|_1$, $\forall t \ge 0$, $u_0 \in K$, by Theorem 6.7, we get

Lemma 4.10 *For each $u_0 \in K$, the orbit $\gamma(u_0) = \{\widetilde{S}(t)u_0,\ t \ge 0\}$ is precompact in L^1.*

Proof of Theorem 4.5 (Continued) Since, by Lemma 4.10, $\gamma(u_0)$ is precompact, it follows that $\omega(u_0) \ne \emptyset$ and that $\omega(u_0)$ is compact. Then, the conclusions of the theorem and, in particular, that $\omega(u_0) \subset \{y \in \mathscr{M}_\mu \cap P \cap C;\ |y-a|_1 = r\}$, follow by Theorem 6.7. □

Following a standard terminology in infinite dimensional dynamical systems theory (e.g., [97], p. 21), we say that the set $\mathscr{A} \subset L^1$ is an *attractor* for the semigroup $S(t)$ if there is an open neighbourhood $\mathscr{U}$ of $\mathscr{A}$ such that $\lim\limits_{t\to\infty} (\text{dist}\, S(t)u_0, \mathscr{A}) = 0$, $\forall u_0 \in \mathscr{U}$. The largest open set $\mathscr{U}$ satisfying this is called the *basin of attraction* of $\mathscr{A}$. For each bounded set $\mathscr{B} \subset L^1$, the ω-limit set of $\mathscr{B}$ is defined as

$$\omega(\mathscr{B}) = \left\{ \lim_{t_n\to\infty} S(t_n)u_0;\ u_0 \in \mathscr{B} \right\}.$$

We say that the set $\mathscr{B} \subset \mathscr{U}$ is absorbing if, for each bounded set $\mathscr{B}_0 \subset \mathscr{U}$, $\{S(t)(\mathscr{B}_0);\ t \ge \tau\} \in \mathscr{B}$ for some $\tau > 0$.

By Theorem 1.1 in [97], p. 23, we know that *if $S(\tau)$ is compact for some $\tau > 0$* (which happens in our case) *and there exists an open set $\mathscr{U}$ and a bounded set $\mathscr{B}$ of $\mathscr{U}$ such that $\mathscr{B}$ is absorbing in $\mathscr{U}$, then $\mathscr{A} = \omega(\mathscr{B})$ is a compact attractor which attracts bounded sets.*

Applying this result in our case, we get

Corollary 4.3 *For each r the set ω-limit set $\mathscr{A}_r = \omega(\mathscr{B}_r)$ of $\mathscr{B}_r = \{u \in K; |u|_1 < r\}$ is a compact attractor for the semigroup $S(t)$.*

4.3 The Ergodicity of Nonlinear Fokker–Planck Flows

The asymptotic behaviour of the semigroup $S(t)$ constructed above under Hypotheses (L1)–(L3) on the set $K = \mathscr{M}_\eta \cap \mathscr{P} \cap C$ remains unclear in the absence of a fixed point for $S(t)$. However, also in this case one has an ergodic property related to the classical Hopf's *statistical ergodic theorem.*

Theorem 4.6 *Let $\mathscr{X}$ be a Banach space and let $F : K \to \mathscr{X}$ be a continuous mapping. Then, for each $u_0 \in K$, we have*

$$\lim_{T\to\infty} \frac{1}{T} \int_0^T F(S(t)u_0)dt = \int_{\omega(u_0)} F(\xi)d\xi, \tag{4.155}$$

where $d\xi$ is the normalized Haar measure on the compact set $\omega(u_0)$ endowed with its natural commutative group structure.

Proof One applies Theorem 6.8 and take into account that Lemma 4.10 the orbit $\gamma(u_0) = \{S(t)u_0;\ t \geq 0\}$ is precompact in L^1. □

By Theorem 4.6, it follows for $\mathscr{X} = L^1$ and $F(y) \equiv gy$, $\forall y \in L^1$, where $g \in L^\infty$, the following.

Corollary 4.4 *For each $u_0 \in K$ and all $g \in L^\infty$, we have*

$$\lim_{T\to\infty} \frac{1}{T} \int_0^T dt \int_{\mathbb{R}^d} g(x)(S(t)u_0)(x)dx = \int_{\omega(x_0)} \int_{\mathbb{R}^d} g(x)\xi(x)d\xi\, dx. \tag{4.156}$$

In particular, it follows that the semigroup $S(t)$ is mean-ergodic, that is,

$$\lim_{T\to\infty} \frac{1}{T} \int_0^T S(t)u_0 dt = \int_{\omega(u_0)} \xi\, d\xi \text{ strongly in } L^1. \tag{4.157}$$

By (4.157), which is related to the Birkhoff ergodic theorem (see [79], p. 396), it follows that the nonlinear Fokker–Planck flow $t \to S(t)u_0$ satisfies for $y_0 \in K$ the classical Boltzmann hypothesis ([79], p. 386) with the time average $\int_{\omega(u_0)} \xi\, d\xi$.

Coming back to the corresponding McKean–Vlasov equation, by Corollary 4.4 we get the following ergodic result.

Corollary 4.5 *Let $u_0 \in K$ and $g \in L^\infty$. Then, there is a probabilistically weak solution X to* (4.118), *where $\mathcal{L}_{X_0} = u_0 dx$ such that*

$$\lim_{T\to\infty} \frac{1}{T} \int_0^T \mathbb{E}[g(X(t))]dt = \int_{\omega(u_0)} \int_{\mathbb{R}^d} g(x)\xi \, d\xi(x)dx, \ \forall g \in L^\infty. \tag{4.158}$$

In particular, it follows that

$$\lim_{T\to\infty} \frac{1}{T} \int_0^T \mathcal{L}_{X(t)}(B)dt = \int_{\omega(u_0)} d\xi \int_{\mathbb{B}} \xi(x)dx = \mu(B), \tag{4.159}$$

for any Borelian set $B \subset \mathbb{R}^d$.

By (4.158) it follows that the transition semigroup $P(t)(g) = \mathbb{E}g(X(t))$ is ergodic and the Borelian measure μ defined by (4.159) is an ergodic probability measure.

Proof The existence of a probabilistically weak solution follows by the superposition principle (see Chap. 5). As regards (4.158), it follows by (4.156) taking into account that

$$\mathbb{E}[g(X(t))] = \int_{\mathbb{R}^d} g(x)(S(t)u_0)(x)dx.$$

Comments to Chap. 4

There is a large literature on the H-theorem and convergence to equilibrium for Fokker–Planck equations. We mention in this context the works [2, 28, 38, 55, 74, 92]. The main results of this section, Theorems 4.1, 4.2, and 4.3, are however new in this field and were established in the authors work [19]. In particular, this result was used in [6] to prove the approximating congtrollability of Fokker–Planck equations with controller in the drift term. Theorem 4.5 was previously given in a slightly different form in [16]. Other sharp asymptotic results for nonlinear Fokker–Planck equations with singular kernels were obtained recently by V. Bogachev et al. [29]. As regards the ergodicity theorem, we refer to [20].

Chapter 5
Markov Processes Associated with Nonlinear Fokker–Planck Equations

As mentioned earlier, NFPEs studied in the previous chapters are relevant in the construction of solutions to McKean–Vlasov SDEs, in terms of their probabilistic laws. We shall present this connection herein in some details.

5.1 Fokker–Planck Equations and McKean–Vlasov SDEs

In this section, we shall consider the relation between NFPEs (even of a more general type than above) with distribution dependent stochastic differential equations (DDSDE) of type

$$\begin{aligned} dX(t) &= b(t, X(t), \mathscr{L}_{X(t)})dt + \sigma(t, X(t), \mathscr{L}_{X(t)})dW(t), \\ X(0) &= \xi_0, \end{aligned} \tag{5.1}$$

on $\mathbb{R}^d$, where $W(t)$, $t \geq 0$, is an $(\mathscr{F}_t)$-Brownian motion on a probability space $(\Omega, \mathscr{F}, P)$ with normal filtration $(\mathscr{F}_t)_{t\geq 0}$. The coefficients b, σ defined on $[0,\infty)\times \mathbb{R}^d \times \mathscr{P}(\mathbb{R}^d)$ are $\mathbb{R}^d$ and $d \times d$-matrix valued, respectively (satisfying conditions to be specified below). In (5.1), $\mathscr{L}_{X(t)}$ denotes the law of $X(t)$ under P and ξ_0 is an $\mathscr{F}_0$-measurable $\mathbb{R}^d$-valued map. Equations as in (5.1) are also referred to as McKean–Vlasov SDEs. Here, we refer to the classical works [57, 75, 76, 90, 95, 96] and, for example, the more recent papers [48, 49, 62, 64, 65, 77, 78, 101, 102], in particular, the recent books [3, 36] and the references therein.

By Itô's formula, under quite general conditions on the coefficients, the time marginal laws $\mu_t := \mathscr{L}_{X(t)}$, $t \geq 0$, with $\mu_0 :=$ law of ξ_0, of the solution $X(t)$, $t \geq 0$,

V. Barbu, M. Röckner, *Nonlinear Fokker-Planck Flows and their Probabilistic Counterparts*, Lecture Notes in Mathematics 2353,
https://doi.org/10.1007/978-3-031-61734-8_5

to (5.1) satisfy (1.9) (in the sense of (1.7)), but with the more general Kolmogorov operator

$$(L_\mu\varphi)(t,x) := \frac{1}{2}\sum_{i,j=1}^{D} a_{ij}(t,x,\mu)\,\frac{\partial^2}{\partial x_i\partial x_j}\,\varphi(x) + \sum_{i=1}^{d} b_i(t,x,\mu)\,\frac{\partial}{\partial x_i}\,\varphi(x),$$
$$x\in\mathbb{R}^d,\ t\geq 0, \tag{5.2}$$

where $a_{ij} := (\sigma\sigma^T)_{i,j},\ 1\leq i,\ j\leq d$. Hence, if one can solve (5.1), one obtains a solution to (1.9) this way.

In this section, we want to go in the opposite direction, that is, we first want to solve (1.9) and, using the obtained $\mu_t,\ t\geq 0$, we shall obtain a (probabilistically) weak solution to (5.1) with the time marginal laws of $X(t),\ t\geq 0$, given by these $\mu_t,\ t\geq 0$.

The above framework, in particular, includes the singular case, where the coefficients in (5.1) are of *Nemytskii type*, that is, the following situation, where b_i, a_{ij} depend on μ in the following way:

$$\begin{aligned} b_i(t,x,\mu) &:= \bar{b}_i\left(t,x,\frac{d\mu}{dx}(x)\right),\\ a_{ij}(t,x,\mu) &:= \bar{a}_{ij}\left(t,x,\frac{d\mu}{dx}(x)\right), \end{aligned} \tag{5.3}$$

for $t\geq 0,\ x\in\mathbb{R}^d,\ 1]\leq i,\ j\leq d$, where $\bar{b}_i, \bar{a}_{ij} : [0,\infty)\times\mathbb{R}^d\times\mathbb{R}\to\mathbb{R}$ are measurable functions. Then, (5.1) has the form

$$\begin{aligned} dX(t) &= \bar{b}\left(t,X(t),\frac{d\mathscr{L}_{X(t)}}{dx}(X(t))\right) + \bar{\sigma}\left(t,X(t),\frac{d\mathscr{L}_{X(t)}}{dx}(X(t))\right)\\ X(0) &= \xi_0, \end{aligned} \tag{5.4}$$

with $(\bar{\sigma}\bar{\sigma}^T)_{i,j} = \bar{a}_{ij}$.

Indeed, for the coefficients in (5.3), the SDE (5.4) is a special case of (5.1) if one chooses the Lebesgue version of $\frac{d\mu}{dt}$. We refer to [60, Subsection 4.2] for a detailed proof. Consider the following conditions:

Hypotheses 5.1 *There exists a solution* $(\mu_t)_{t\geq 0}$ *to* (1.9) *with Kolmogorov operator* (5.2) *such that*

(i) $\mu_t\in\mathscr{P}(\mathbb{R}^d)$ *for all* $t\geq 0$.
(ii) *For* $1\leq i,\ j\leq d$, *the maps* $(t,x)\mapsto a_{ij}(t,x,\mu_t)$ *and* $(t,x)\mapsto b_i(t,x,\mu_t)$ *are measurable and*

$$\int_0^T\int_{\mathbb{R}^d}(|a_{ij}(t,x,\mu_t)| + |b_i(t,x,\mu_t)|)\mu_t(dx)dt < \infty, \text{ for all } T\in(0,\infty).$$

(iii) $[0,\infty)\ni t\mapsto\mu_t$ *is narrowly continuous.*

Then, we have the following result (see [11, Section 2]).

Theorem 5.1 *Suppose Hypothesis 5.1 holds. Then, there exists a probabilistically weak solution to* (5.1). *In particular, we have the probabilistic representation*

$$\mu_t(dx) = \mathscr{L}_t(dx),\ t \geq 0. \tag{5.5}$$

Proof Under Hypothesis 5.1, we can apply the superposition principle (see Theorem 2.5 in [98], which in turn is a generalization of a result in [51]) for *linear* FPEs applied to the *linear* Kolmogorov operator

$$L_{\mu_t} := \frac{1}{2}\sum_{i,j=1}^{d} a_{ij}(t,x,\mu_t)\frac{\partial^2}{\partial x_i \partial x_j} + \sum_{i=1}^{d} b_i(t,x,\mu_t)\frac{\partial}{\partial x_i}, \tag{5.6}$$

with $(\mu_t)_{t\geq 0}$ from Hypothesis 5.1 fixed.

More precisely, by Theorem 2.5 in [98], there exists a probability measure P on $C([0,T];\mathbb{R}^d)$ equipped with its Borel σ-algebra and its natural normal filtration obtained by the evaluation maps π_t, $t \in [0,T]$, defined by

$$\pi_t(w) := w(t), \quad w \in C([0,T],\mathbb{R}^d),$$

solving the martingale problem (see [88, Definition 2.4]) for the time-dependent *linear* Kolmogorov operator $\frac{\partial}{\partial t} + L_{\mu_t}$ (with $(\mu_t)_{t\geq 0}$ as above fixed) with time marginal laws

$$P \circ \pi_t^{-1} = \mu_t,\ t \geq 0.$$

Then, [60, Proposition 2.2.3], which in turn is based on [68, Prop. 4.1, Problem 4.14, Corollaries 4.8 and 4.9], implies that there exists a d-dimensional $(\mathscr{F}_t)$-Brownian motion $W(t)$, $t \geq 0$, on a stochastic basis $(\Omega, \mathscr{F}, (\mathscr{F}_t)_{t\geq 0}, \mathscr{Q})$ and a continuous $(\mathscr{F}_t)$-progressively measurable map $X : [0,\infty) \times \Omega \to \mathbb{R}^d$ satisfying the following SDE:

$$dX(t) = b(t, X(t), \mu_t)dt + \sigma(t, X(t), \mu_t)dW(t),\ t \geq 0, \tag{5.7}$$

with the law

$$\mathscr{Q} \circ X^{-1} = P.$$

In particular, we have, for the time marginal laws,

$$\mathscr{L}_{X(t)} := \mathscr{Q} \circ X(t)^{-1} = \mu_t,\ t \geq 0, \tag{5.8}$$

which completes the proof of the theorem. □

Remark 5.1 Because of (5.8), the process $X(t),\ t \geq 0$, is also called a *probabilistic representation* of the solution $(\mu_t)_{t\geq 0}$ for the nonlinear FPE (1.9).

Remark 5.2 There is an analogue of the superposition principle and an analogue of Theorem 5.1 in the case, where the noise in (5.1) is replaced by a jump-type (e.g., Lévy) noise, which was proved in [89].

Conclusion To weakly solve the McKean–Vlasov SDE (5.1), we have to solve the corresponding nonlinear FPE (1.9) and then check Hypothesis 5.1 above.

5.2 Uniqueness of Weak Solutions to McKean–Vlasov SDEs and Corresponding Nonlinear Markov Processes

Uniqueness of solutions to (5.1) is closely related to the fact that the path space measures of solutions to (5.1) form a nonlinear Markov process. In order to explain this more precisely, we need a few definitions and notations. This section is based on [88] to which we refer the reader for more details. We shall concentrate on three cases most relevant for these lecture notes and which are special cases of the much more general results in [88].

First of all we note that (5.1) can, of course, be considedred, on $[s,\infty)$ for any starting time $s \geq 0$, and Theorem 5.1 will still be valid in this case. So, we set $\zeta := \mu_s$, for any solution $(\mu_t)_{t\geq s}$ as in Hypothesis 5.1 with s replacing the initial time zero and adjust the notation by writing $\mu_t^{s,\zeta}$ instead of $\mu_t,\ t \geq s$. Then, by Theorem 5.1 there exists a probabilistically weak solution to (5.1) considered on $[s,\infty)$ with ζ =law of ξ_0. Then, for such $(s,\zeta) \in \mathbb{R}_+ \times \mathscr{P}$ we denote its law by $P_{s,\zeta}$. More precisely,

$$P_{s,\zeta} := P \circ (X_t)_{t\geq s}^{-1},\quad \zeta = P \circ X_s^{-1}.$$

$P_{s,\zeta}$ is a probability measure on Ω_s, where for $s \geq 0$

$$\Omega_s := C([s,\infty),\mathbb{R}^d) = \text{ space of continuous paths in } \mathbb{R}^d \text{ starting at time } s$$

with Borel σ-algebra $\mathscr{B}(\Omega_s)$. Furthermore, for $\tau \geq s$ we define

$$\pi_\tau^s := \Omega_s \to \mathbb{R}^d,\quad \pi_\tau^s(w) := w(\tau),\ w \in \Omega_s,$$

and, for $r \geq s$,

$$\mathscr{F}_{s,r} := \sigma(\pi_\tau^s \mid s \leq \tau \leq r).$$

A nonlinear Markov process in the sense of [75] is defined as follows:

Definition 5.1 Let $\mathscr{P}_0 \subseteq \mathscr{P}$. A nonlinear Markov process is a family $(P_{s,\zeta})_{(s,\zeta)\in\mathbb{R}_+\times\mathscr{P}_0}$ of probability measures $P_{s,\zeta}$ on $\mathscr{B}(\Omega_s)$ such that

(i) The marginals $\mu_t^{s,\zeta} := P_{s,\zeta} \circ (\pi_t^s)^{-1}$ belong to $\mathscr{P}_0$ for all $0 \le s \le t$,
(ii) The nonlinear Markov property holds, i.e., for all $0 \le s \le r \le t$, $\zeta \in \mathscr{P}_0$,

$$P_{s,\zeta}(\pi_t^s \in A \mid \mathscr{F}_{sr})(\cdot) = p_{(s,\zeta),(r,\pi_r^s(\cdot))}(\pi_t^r \in A)\ P_{s,\zeta}\text{-a.s.} \quad \text{for all } A \in \mathscr{B}(\mathbb{R}^d), \tag{MP}$$

where $p_{(s,\zeta),(r,y)}$, $y \in \mathbb{R}^d$, is a regular conditional probability kernel from $\mathbb{R}^d$ to $\mathscr{B}(\Omega_r)$ of $P_{r,\mu_r^{s,\zeta}}[\cdot \mid \pi_r^r = y]$, $y \in \mathbb{R}^d$ (i.e., in particular, $p_{(s,\zeta),(r,y)} \in \mathscr{P}(\Omega_r)$ and $p_{(s,\zeta),(r,y)}(\pi_r^r = y) = 1$).

The term *nonlinear* Markov property originates from the fact that in the situation of the above definition the map $\mathscr{P}_0 \ni \zeta \mapsto \mu_t^{s,\zeta}$ is, in general, not convex.

Remark 5.3 The one-dimensional time marginals $\mu_t^{s,\zeta} = \mathbb{P}_{s,\zeta} \circ (\pi_t^s)^{-1}$ of a nonlinear Markov process satisfy the *flow property*, i.e.,

$$\mu_t^{s,\zeta} = \mu_t^{r,\mu_r^{s,\zeta}}, \quad \forall 0 \le s \le r \le t,\ \zeta \in \mathscr{P}_0. \tag{5.9}$$

For a solution $(\mu_t^{s,\zeta})_{t\ge s}$ of (1.9) in the sense of (1.7) with $\mu_s^{s,\zeta} = \zeta$ and for L_μ as in (5.2), we consider the corresponding linearized Kolmogorov operator $L_{\mu_t^{s,\zeta}}$, i.e., we fix μ in (5.6) to be $\mu_t^{s,\zeta}$ and consider the corresponding linearlized Fokker–Planck equation

$$\frac{\partial \nu_t}{\partial t} = L_{\mu_t^{s,\zeta}} \nu_t, \quad t \ge s,\ \nu_s := \zeta, \tag{5.10}$$

and denote the set of all its narrowly continuous solutions by $M_{\mu^{s,\zeta}}$. Furthermore, we define

$$\mathscr{A}_{\le \mu^{s,\zeta}} := \{\nu \in C([s,\infty),\mathscr{P}) : \exists c > 0 \text{ s.th. } \nu_t \le c\mu_t^{s,\zeta},\ \forall t \ge s\},$$

where $C([s,\infty),\mathscr{P})$ denotes the set of all narrowly continuous paths on $[s,\infty)$ with values in $\mathscr{P}$. Then, we have the following two theorems.

Theorem 5.2 *Let* $\mathscr{P}_0 := \{\zeta \in \mathscr{P} \mid \zeta << dx \text{ and } \frac{d\zeta}{dx} \in L^\infty\}$. *Assume that, for every* $(s,\zeta) \in \mathbb{R}_+ \times \mathscr{P}_0$,

(1) *there exists a solution* $\mu_t^{s,\zeta}$, $t \geq s$, *to* (1.9) *in the sense of* (1.7) *with* $\mu_s^{s,\zeta} = \zeta$ *and for* L_μ *as in* (5.2), *satisfying Hypothesis* 5.1 *such that* $\mu_t^{s,\zeta} \in \mathcal{P}_0$ *for all* $t \geq 0$ *and*

$$\mu_t^{s,\zeta} = \mu_t^{r,\mu_r^{s,\zeta}} \quad \text{for all } 0 \leq s \leq r \leq t; \text{ ("flow property")} \tag{5.11}$$

(2) $\#(\mathcal{A}_{\leq \mu^{s,\zeta}} \cap M_{\mu^{s,\zeta}}) = 1$.

Then, for every $(s, \zeta) \in \mathbb{R}_+ \times \mathcal{P}_0$, *the McKean–Vlasov equation* (5.1) *has a unique probabilistically weak solution and the family* $P_{s,\zeta}$, $(s, \zeta) \in \mathbb{R}^+ \times \mathcal{P}_0$, *of their laws form a nonlinear Markov process in the sense of Definition* 5.1.

Theorem 5.3 *Assume that, for every* $(s, \zeta) \in \mathbb{R}_+ \times \mathcal{P}$,

(0) *there exists a solution* $\mu_t^{s,\zeta}$, $t \geq s$, *to* (1.9) *in the sense of* (1.7) *with* $\mu_s^{s,\zeta} = \zeta$ *and for* L_μ *as in* (5.2), *satisfying Hypothesis* 5.1 *such that the flow property* (5.11) *holds;*
(1) $\mu_t^{s,\zeta} \in \mathcal{P}_0$, $t > s$, *where* $\mathcal{P}_0$ *is as defined in Theorem* 5.2, $t > s$;
(2) $\#(\mathcal{A}_{\leq \mu^{s,\zeta}} \cap M_{\mu^{s,\zeta}}) = 1$, *if* $\zeta \in \mathcal{P}_0$.

Then, for every $(s, \zeta) \in \mathbb{R}_+ \times \mathcal{P}$, *the McKean–Vlasov SDE* (5.1) *has a unique probabilistically weak solution and the corresponding laws* $P_{s,\zeta}$, $(s, \zeta) \in \mathbb{R}_+ \times \mathcal{P}$, *form a nonlinear Markov process in the sense of Definition* 5.1.

Remark 5.4 Clearly, if

$$M_{\mu^{s,\zeta}} \cap \bigcap_{\substack{N>s \\ N\in\mathbb{N}}} L^\infty((s, N) \times \mathbb{R}^d) = 1,$$

then condition (2) in the above theorems holds.

Remark 5.5 In the case of Sect. 2.8, i.e., when we consider the nonlocal NLFP (2.324) of (1.9) and the McKean–Vlasov SDE (2.328) with jump noise instead of (5.1) there is a complete analogue of Theorem 5.2. This immediately follows from the proofs of Corollaries 3.9 and 3.10 in [88]. Of course, also Remark 5.4 remains valid.

Proof of Theorem 5.2 The existence follows by Theorem 5.1 and the remaining part of the assertion follows by Rehmeier and Röckner [88, Corollary 3.9]. □

Proof of Theorem 5.3 The existence follows by Theorem 5.1 and the remaining part of the assertion follows by Rehmeier and Röckner [88, Corollary 3.10]. □

5.3 Strong Solutions to McKean–Vlasov SDEs

It is an interesting and important question whether the probabilistically weak solutions to the McKlean–Vlasov equation (5.1) are actually strong, i.e., can be written as a (measurable) function of the Brownian motion W. This is the topic of the PhD-thesis [60] in which a general approach is developed to prove this under natural (additional) assumptions. We would like to emphasize two main results in [60]:

(1) If we consider the situation of Theorem 2.8, in which we proved that we have weak uniqueness for (5.1) and a corresponding nonlinear Markov process, it is proved in [60, Theorem 4.51] that these weak solutions are in fact strong.
(2) In the situation of Theorem 3.1, where coefficients are allowed to be time dependent, it is proved in [60, Theorem 4.6.1] that under the additional condition (H3) stated on p. 71 in [60] these solutions are in fact strong.

We refer to [60] for more details and further references.

Chapter 6
Appendix

6.1 Nonlinear m-Accretive Operators

Throughout this section, $\mathscr{X}$ is a real Banach space with the norm $\|\cdot\|$ and dual $\mathscr{X}^*$. We denote by ${}_{\mathscr{X}}(\cdot,\cdot)_{\mathscr{X}^*}$ the duality pairing between $\mathscr{X}$, $\mathscr{X}^*$ and by $J:\mathscr{X}\to\mathscr{X}^*$ the duality mapping of $\mathscr{X}$.

Definition 6.1 A subset A of $\mathscr{X}\times\mathscr{X}$ (equivalently, an operator from $\mathscr{X}$ to itself) is called *accretive* if

$$ {}_{\mathscr{X}}(x_1-x_2,y_1-y_2)_{\mathscr{X}^*}\ge 0,\ \ \forall y_i\in Ax_i,\ i=1,2. \tag{6.1}$$

Here, we briefly survey without proof the basic properties of accretive operators and refer to [4, 5] for details.

An accretive operator A is said to be *m-accretive* if $R(I+\nu A)=\mathscr{X}$ for some $\nu>0$. It turns out that, if A is m-accretive, then $R(I+\lambda A)=\mathscr{X}$, $\forall\lambda>0$. Here, we have denoted by I the unity operator in $\mathscr{X}$.

We denote by $D(A)=\{x\in\mathscr{X};\ Ax\ne\emptyset\}$ the domain of A and by $R(A)=\{y\in Ax;\ [x,y]\in A\}$ the range of A. As in the case of operators from $\mathscr{X}$ to $\mathscr{X}^*$, we identify an operator (eventually multivalued) $A:D(A)\subset\mathscr{X}\to\mathscr{X}$ with its graph $\{[x,y];\ y\in Ax\}$, and so view A as a subset of $\mathscr{X}\times\mathscr{X}$.

Given $\omega\in\mathbb{R}$, the operator A is said to be *quasi-accretive* (*quasi-m-accretive*), if $A+\omega I$ is accretive (m-accretive, respectively) for some $\omega\in\mathbb{R}$.

The accretiveness of A can be, equivalently, expressed as

$$\|x_1-x_2\|_{\mathscr{X}}\le\|x_1-x_2+\lambda(y_1-y_2)\|_{\mathscr{X}},\ \ [x_i,y_i]\in A,\ i=1,2, \tag{6.2}$$

V. Barbu, M. Röckner, *Nonlinear Fokker-Planck Flows and their Probabilistic Counterparts*, Lecture Notes in Mathematics 2353,
https://doi.org/10.1007/978-3-031-61734-8_6

for some $\lambda > 0$ (equivalently, for all $\lambda > 0$). Hence, if A is accretive, then the operator $(I + \lambda A)^{-1}$ is single-valued and nonexpansive on $R(I + \lambda A)$, that is,

$$\|(I + \lambda A)^{-1}x - (I + \lambda A)^{-1}y\|_{\mathscr{X}} \leq \|x - y\|_{\mathscr{X}}, \ \forall \lambda > 0, \ x, y \in R(I + \lambda A). \tag{6.3}$$

If A is ω-accretive, then it follows that the operator $(I+\lambda A)^{-1}$ is single-valued and Lipschitzian with Lipschitz constant not great than $\frac{1}{1-\lambda\omega}$ on $R(I+\lambda A), 0 < \lambda < \frac{1}{\omega}$.

Theorem 6.1 *The operator A is accretive if and only if, for each* $\lambda > 0$ *and* $y \in R(I + \lambda A)$*, the equation* $(I + \lambda A)x \ni y$ *has at most one solution and* (3) *holds.*

Let us define for $\lambda > 0$ the operators $J_\lambda, \ A_\lambda : \mathscr{X} \to \mathscr{X}$,

$$J_\lambda(x) = (I + \lambda A)^{-1}x, \ x \in R(I + \lambda A);$$
$$A_\lambda x = \lambda^{-1}(x - J_\lambda x), \ x \in R(I + \lambda A).$$

We have

Theorem 6.2 *Let A be accretive in* $\mathscr{X} \times \mathscr{X}$*. Then:*

(a) $\|J_\lambda(x) - J_\lambda(y)\|_{\mathscr{X}} \leq (1 - \lambda\omega)^{-1}\|x - y\|_{\mathscr{X}}, \ \forall \lambda > 0, \ \forall x, y \in R(I + \lambda A)$.
(b) A_λ *is accretive and Lipschitz continuous for all* $\lambda > 0$.
(c) $A_\lambda x \in AJ_\lambda x, \ \forall x \in R(I + \lambda A), \ 0 < \lambda < \frac{1}{\omega_0}$.
(d) $\|A_\lambda x\|_{\mathscr{X}} \leq |Ax| = \inf\{\|y\|_{\mathscr{X}}; \ y \in Ax\}, \ \forall x \in D(A) \cap R(I + \lambda A)$.
(e) $\lim_{\lambda \to 0} J_\lambda x = x, \ \forall x \in \overline{D(A)} \ \bigcap_{0<\lambda<\infty} R(I + \lambda A)$.

We also have the following surjectivity theorem.

Theorem 6.3 *If* $A : \mathscr{X} \to \mathscr{X}$ *is m-accretive and coercive, that is,*

$$\lim_{\|u\|_{\mathscr{X}} \to \infty} \frac{(Au, u)}{\|u\|_{\mathscr{X}}} = +\infty,$$

then $\mathbb{R}(A) = \mathscr{X}$.

We also note

Theorem 6.4 *Let* $\mathscr{X}$ *be a Banach space,* $A : \mathscr{X} \to \mathscr{X}$ *be m-accretive and let* $B : \mathscr{X} \to \mathscr{X}$ *be continuous and accretive. Then,* $A + B$ *is m-accretive.*

6.2 Semigroups of Contractions in Banach Spaces

Let $\mathscr{X}$ be a real Banach space with the norm $\|\cdot\|$ and the dual $\mathscr{X}^*$ and let $A \subset \mathscr{X} \times \mathscr{X}$ be a quasi-accretive set of $\mathscr{X} \times \mathscr{X}$. Denote by $\overline{D(A)}$ the closure of A in $\mathscr{X}$. Consider the Cauchy problem

$$\begin{cases} \frac{dy}{dt}(t) + Ay(t) \ni 0, & t \in (0,T), \\ y(0) = y_0. \end{cases} \tag{6.4}$$

Definition 6.2 The function $y : [0,\infty) \to \mathscr{X}$ is said to be a *mild solution* to the Cauchy problem (6.4) if $y \in C([0,\infty); \mathscr{X})$ and, for each $T > 0$,

$$y(t) = \lim_{h \to 0} y_h(t) \text{ uniformly on } [0,T],$$

where $y_h : [0,T] \to \mathscr{X}$ is the step function

$$\begin{aligned} &y_h(t) = y_h^j, \ \forall t \in [jh,(j+1)h), \ j = 0,1,\dots,N = \left[\tfrac{T}{n}\right], \\ &y_h^{j+1} + hAy_h^{j+1} \ni y_h^j, \ j = 0,1,\dots,N_h, \\ &y_h^0 = y_0. \end{aligned}$$

Equivalently,

$$\begin{aligned} &\frac{1}{h}(y_h(t) - y_h(t-h)) + Ay_h(t) = 0 && \text{for } t > 0, \\ &y_h(t) = y_0 && \text{for } t \le 0. \end{aligned}$$

Theorem 6.5 (Crandall & Liggett) *Let $A : \mathscr{X} \to \mathscr{X}$ be a quasi-accretive operator such that*

$$R(I + \lambda A) \supset \overline{D(A)} = K, \ \forall \lambda > 0. \tag{6.5}$$

Then, for each $y_0 \in \overline{D(A)}$, the Cauchy problem (6.5) *has a unique mild solution y. Moreover, y is equivalently given by the exponential formula*

$$y(t) = \lim_{n \to \infty} \left(I + \frac{t}{n} A\right)^{-n} y_0 \ \text{in } \mathscr{X}, \tag{6.6}$$

uniformly in t on compact intervals.

If we denote $S(t)y_0 = y(t)$, we have

$$S(t+s)y_0 = S(t)S(s)y_0,\ \forall t, x \geq 0,\ y_0 \in \overline{D(A)}, \tag{6.7}$$

$$\lim_{t\to 0} S(t) = y_0 \text{ in } \mathscr{X}, \tag{6.8}$$

$$\|S(t)y_0 - S(t)\bar{y}_0\|_{\mathscr{X}} \leq \|y_0 - \bar{y}_0\|_{\mathscr{X}},\ \forall t \geq 0,\ y_0, \bar{y}_0 \in \overline{D(A)}. \tag{6.9}$$

A family of operators $\{S(t) : K \to K,\ t \geq 0\}$ satisfying (6.7)–(6.9) is called a *continuous semigroup of contractions* on K. The operator A is called the *generator* of $S(t)$. The semigroup $S(t)$ generated by A is also denoted by e^{-tA} or $\exp(-tA)$. In particular, condition (4) holds if A is m-accretive.

If the space $\mathscr{X}$ is reflexive and $y_0 \in D(A)$, then $t \to S(t)y_0$ is absolutely continuous, a.e. differentiable and satisfies (4), a.e. $t \in (0, T)$.

A *Lyapunov function* for the semigroup $S(t) : [0, \infty) \times K \to K$, where K is a closed subset of $\mathscr{X}$, is a continuous function $V : K \to [0, \infty)$ which is nonincreasing on every trajectory $\gamma(y_0) = \{S(t)y_0,\ t \geq 0\}$, that is,

$$V(S(t)y_0) \leq V(S(\tau)y_0),\ \ \forall\, 0 \leq t \leq \tau,\ y_0 \in K. \tag{6.10}$$

Theorem 6.6 ([44]) *Assume that A is accretive in a Banach space $\mathscr{X}$, $\overline{D(A)} \subset R(I + \lambda A)$, $\forall \lambda > 0$, and that*

$$\|e^{-tA}y_0\|_{\mathscr{X}} \leq C(1 + \|y_0\|_{\mathscr{X}}),\ y_0 \in \overline{D(A)},\ \forall t \geq 0. \tag{6.11}$$

If $(I + \lambda A)^{-1}$ is compact in $\mathscr{X}$ for some $\lambda > 0$, then $\gamma(y_0)$ is precompact in $\mathscr{X}$ for each $y_0 \in K = \overline{D(A)}$.

(We note that, in particular, (6.11) holds if $0 \in \mathbb{R}(A)$.)

For each $y_0 \in \overline{D(A)} = K$, denote by $\omega(y_0)$ the *ω-limit set*

$$\omega(y_0) = \left\{\xi = \lim_{t_n\to\infty} S(t_n)y_0\right\}.$$

We have (see [44])

Theorem 6.7 *If $\omega(y_0) \neq \emptyset$, then $\omega(y_0)$ is invariant under the semigroup $S(t)$ and $S(t)$ is an homeomorphism of $\omega(y_0)$ onto itself. Moreover, for each $t \geq 0$, $S(t)$ is an isometry on $\omega(y_0)$ and, if $\xi \in K$ is a fixed point of S (i.e., $S(t)\xi = \xi$, $\forall t \geq 0$), then $\omega(y_0) \subset \{y \in \mathscr{X};\ \|y - \xi\| = r\}$, where $0 \leq r \leq \|\xi - y_0\|$.*

Moreover, if X is strictly convex and K is convex, then

$$\xi = \lim_{T\to\infty} \frac{1}{T} \int_0^T S(t)\eta\, dt,\ \ \eta \in \overline{\mathrm{Conv}(\omega(y_0))}.$$

In a general Banach space, this ergodic property of the semigroup $S(t)$ holds in a weaker sense. Namely, one has (see [61])

Theorem 6.8 *Let $\mathscr{X}$ and $\widetilde{\mathscr{X}}$ be real Banach spaces, $C \subset \mathscr{X}$ be a closed set and let $S(t) : C \to C$ be a semigroup of contractions on C. If $y_0 \in C$ is such that $\gamma(y_0)$ is precompact, then $\omega(y_0)$ is a compact commutative group and for every mapping $F : C \to \widetilde{\mathscr{X}}$ which is uniformly continuous on bounded subsets of C we have*

$$\lim_{T\to\infty} \frac{1}{T}\int_0^T F(S(t)u_0)dt = \int_{\omega(y_0)} F(\xi)d\xi, \tag{6.12}$$

where $d\xi$ is the normalized Haar's measure on $\omega(u_0)$.

(We recall that, by a classical theorem of A. Weil, there is a unique normalized measure on $\omega(u_0)$.)

The next theorem is the *Trotter–Kato theorem* for nonlinear semigroups of contractions (see [4, 5]).

Theorem 6.9 *Let $\{A_n\}_{n=1}^{\infty}$ and A be m-accretive operators in a real Banach space $\mathscr{X}$. If*

$$\lim_{n\to\infty} (I+\lambda A_n)^{-1}x = (I+\lambda A)^{-1}x,\ \forall x \in \mathscr{X},\ 0 < \lambda < \lambda_0,$$

then $\lim\limits_{n\to\infty} e^{-tA_n}x = e^{-tA}x,\ \forall x \in \mathscr{X}$, uniformly in t on compacts intervals.

Consider now the time dependent Cauchy problem

$$\frac{du}{dt} + A(t)u = 0,\ \ t \in [0,T], u(0) = x, \tag{6.13}$$

where $A(t) : D(A(t)) = D \subset \mathscr{X} \to \mathscr{X}$ is quasi-m-accretive for all $t \in [0,T]$.
Denote by $\overline{D}$ the closure of D and assume that

B.1 There is a continuous function $f : [0,T] \to \mathscr{X}$ which is of bounded variation on $[0,T]$, and a monotone increasing function $L : [0,\infty) \to [0,\infty)$ such that $\|J_\lambda(t,)(x) - J_\lambda(\tau)(x)\|_{\mathscr{X}} \le \lambda\|f(t) - f(\tau)\|_{\mathscr{X}} L(\|x\|_{\mathscr{X}})(1 + \|A(\tau)x\|_{\mathscr{X}})$, for $0 < \lambda < \lambda_0,\ 0 \le t,\ \tau \le T$ and $x \in \overline{D}$, $J_\lambda(t)(x) = (I+\lambda A(t))^{-1}x$.

We also have for the time-dependent system (1.12) (see [43])

Theorem 6.10 *Let $\mathscr{X}$ be a reflexive Banach space and let the above assumption hold. Then, for every $x \in D$ and $0 \le s < T$, the initial value problem* (6.13) *has a unique strong solution $u(t)$ given by*

$$u(t) = U(t,s)x = \lim_{n\to\infty} \prod_{k=1}^{n} \left(I + \frac{t-s}{n} A\left(s + k\,\frac{t-s}{n}\right)\right)^{-1} x.$$

If $A(t)$ is single-valued and

$$\|A(t)x - A(\tau)x\|_{\mathscr{X}} \leq |t-\tau| L(\|x\|_{\mathscr{X}})(1 + \|A(t)x\|_{\mathscr{X}}), \tag{6.14}$$

or $x \in D$ and $0 \leq \tau, t \leq T$, then

$$\|J_\lambda(t)(x) - J_\lambda(\tau)(x)\|_{\mathscr{X}} \leq \lambda |t-\tau| L_1(\|x\|_{\mathscr{X}})(1 + \|A(t)x\|_{\mathscr{X}}). \tag{6.15}$$

By strong solution to (6.13) we mean an absolutely continuous function $u : [0, t)$ which satisfies (6.13), a.e. $t \in (0, T)$.

6.3 Riesz Potentials

Given $0 < m < d$ and $f \in L^1_{loc}(\mathbb{R}^d)$, the *Riesz potential* of order m is defined by

$$I_m(f)(x) = \int_{\mathbb{R}^d} \frac{f(\xi)}{|x-\xi|^{d-m}}\, d\xi, \quad x \in \mathbb{R}^d. \tag{6.16}$$

We note that, if $d > 2$ and $m = 2$, then $I_m(f) = (d-2)\omega_d E_d * f$ and, therefore,

$$\Delta I_m = (d-2)\omega_d f \quad \text{in } \mathscr{D}'(\mathbb{R}^d),$$

where E_d is the fundamental solution of the Laplace operator $-\Delta$ and ω_d is the value of unity ball in $\mathbb{R}^d$.

We have (see, e.g., [25, 93, 94])

Theorem 6.11 *Let $1 < m < d$ and $1 \leq p < \frac{d}{m}$, $q = \frac{dp}{d-mp}$. Let $f \in L^p(\mathbb{R}^d)$. Then,*

(i) *if $p > 1$, we have*

$$|I_m(f)|_q \leq C(m, d, p)|f|_p, \tag{6.17}$$

(ii) *if $p = 1$, we have*

$$\|I_m(f)\|_{M^q(\mathbb{R}^d)} \leq C(m, d)|f|_1. \tag{6.18}$$

Here, $M^q(\mathbb{R}^d)$ is the Marcinkiewicz class of order q, i.e.,

$$M^q(\mathbb{R}^d) = \Big\{ u \in L^1_{loc}(\mathbb{R}^d),\ \textstyle\int_K |u(x)|dx \leq \alpha K (\text{meas } K)^{\frac{1}{q'}} \text{ for all Lebesgue measurable sets } K \subset \mathbb{R}^d \Big\};\ \frac{1}{q'} = 1 - \frac{1}{q}.$$

We note that, for $1 \leq p < q < \infty$, $M^q(\mathbb{R}^d) \subset L^p_{loc}(\mathbb{R}^d)$.

In particular, for $m = 2$ and $d \geq 3$, it follows that

$$|E_d * f|_q \leq C(d, p)|f|_p, \tag{6.19}$$

if $p > 1$ and $q = \frac{dp}{d-2p}$, and

$$\|E_d * f\|_{M^{\frac{d}{d-2}}(\mathbb{R}^d)} \leq C(d)|f|_1, \tag{6.20}$$

if $d \geq 3$. Moreover, since $\nabla E_d * f$ is the Riesz potential of order $m = 1$, we have by Theorem 6.11

$$\|\nabla E_2 * f\|_{M^{\frac{d}{d-1}}(\mathbb{R}^d)} \leq C_1(d)|f|_1, \ \forall f \in L^1, \tag{6.21}$$

$$\|\nabla E_2 * f\|_{M^2(\mathbb{R}^2)} \leq C_2|f|_1, \ \forall f \in L^1. \tag{6.22}$$

By Theorem 6.11, it also follows the Sobolev–Gagliardo–Nirenberg theorem. Namely,

Theorem 6.12 *Let $1 \leq p < d$. Then,*

$$|u|_{p^*} \leq C(d, p)|\nabla u|_p, \ \forall u \in \dot{W}^{1,p}(\mathbb{R}^d),$$

where $p^* = \frac{2p}{2-d}$. Here, $\dot{W}^{1,p}(\mathbb{R}^d)$ is the completion of the space $C_0^\infty(\mathbb{R}^d)$ in the norm $u \to |\nabla u|_p$.

In fact, Theorem 6.12 follows by (6.17) taking into account the inequality

$$|u(x)| \leq C \int_{\mathbb{R}^d} \frac{|\nabla u(\xi)|}{|x - \xi|^{d-1}}\, d\xi, \ \text{ a.e.} x \in \mathbb{R}^d.$$

The kernel $N \in L^1_{\text{loc}}\mathbb{R}^d)$ of the functional

$$K_N(f)(x) = (N * f)(x) = \int_{\mathbb{R}^d} N(x - \xi) f(\xi) d\xi,$$

is said to be of the *Calderon–Zygmund type* if N is homogeneous of order d, that is, $N(\lambda x) = \lambda^{-d} N(x)$, $\forall \lambda > 0$, $x \in \mathbb{R}^d$; N is integrable on $\{x \in \mathbb{R}^d; |x| = 1\}$ and its integral is zero.

Theorem 6.13 (Calderon–Zygmund [35]) *If N is a kernel of the Calderon–Zygmund type, then $|K_N(f)|_p \leq C_p|f|_p$, $\forall p \in (1, \infty)$, $\forall f \in L^p$.*

References

1. Ambrosio, L., Gigli, N., Savaré, G.: Gradient Flows in Metric Spaces and in the Space of Probability Measures. Lecture Notes in Mathematics. ETH Zürich, Birkhäuser Verlag, Bassel (2008)
2. Arnold, A., Markowich, P., Toscani, G., Unterreiter, A.: On convex Sobolev inequalities and the rate of convergence to equilibrium for Fokker-Planck type equations. Commun. Partial Differ. Equ. **26**, 43–100 (2001)
3. Bakry, D., Gentil, I., Ledoux, M.: Analysis and Geometry of Markov Diffusion Operators, xx+552 pp. Springer, Berlin (2014). ISBN:378-3-319-00226-2
4. Barbu, V.: Nonlinear Differential Equations of Monotone Type in Banach Spaces. Springer, New York (2010)
5. Barbu, V.: Semigroup Approach to Nonlinear Diffusion Equations. World Scientific, Singapore (2021)
6. Barbu, V.: Asymptotic controllability of Fokker–Planck equations. Eur. Phys. J. Plus **136**, 896 (2021)
7. Barbu, V.: The controllability of Fokker–Planck equations with reflecting boundary conditions. SIAM J. Control Optim. **59**, 701–725 (2021)
8. Barbu, V.: The Trotter formula for nonlinear Fokker–Planck equations. J. Differ. Equ. **345**, 314–333 (2023)
9. Barbu, V.: Exact controllability of Fokker–Planck and McKean–Vlasov SDEs, generalized solutions to nonlinear Fokker-Planck equations. SIAM J. Control Optim. **61**, 1805–1818 (2023)
10. Barbu, V., Röckner, M.: Probabilistic representation for solutions to nonlinear Fokker-Planck equations. SIAM J. Math. Anal. **50**(4), 4246–4260 (2018)
11. Barbu, V., Röckner, M.: From nonlinear Fokker-Planck equations to solutions of distribution dependent SDE. Ann. Probab. **48**(4), 1902–1920 (2020)
12. Barbu, V., Röckner, M.: Solutions for nonlinear Fokker-Planck equations with measures as initial data and McKean-Vlasov equations. J. Funct. Anal. **280**(7), 1–35 (2021)
13. Barbu, V., Röckner, M.: Uniqueness for nonlinear Fokker-Planck equations and weak uniqueness for McKean-Vlasov SDEs. Stoch. PDEs Anal. Comput. **9**(4), 702–713 (2021)
14. Barbu, V., Röckner, M.: A note on the ergodicity of Fokker–Planck flows in $L^1(\mathbb{R}^2)$. arXiv 2210.13624 [math.PR] (2022)
15. Barbu, V., Röckner, M.: Corrections to: Uniqueness for nonlinear Fokker–Planck equations and weak uniqueness for McKean–Vlasov SDEs. Stoch. PDEs Anal. Comput. **11**, 426–431 (2023)

V. Barbu, M. Röckner, *Nonlinear Fokker-Planck Flows and their Probabilistic Counterparts*, Lecture Notes in Mathematics 2353,
https://doi.org/10.1007/978-3-031-61734-8

16. Barbu, V., Röckner, M.: The invariance principle for nonlinear Fokker–Planck equations. J. Differ. Equ. **315**, 200–221 (2022)
17. Barbu, V., Röckner, M.: Nonlinear Fokker–Planck equations with fractional Laplacian and McKean–Vlasov SDEs with Lévy noise. Probab. Theory Relat. Fields (2024). https://doi.org/10.1007/500440-024-012277-1
18. Barbu, V., Röckner, M.: Nonlinear Fokker–Planck equations with time-dependent coefficient. SIAM J. Math. Anal. **655**(1), 1–18 (2023)
19. Barbu, V., Röckner, M.: The evolution to equilibrium of solutions to nonlinear Fokker-Planck equations. Indiana Univ. Math. J. **72**(1), 89–131 (2023)
20. Barbu, V., Röckner, M.: The ergodicity of nonlinear Fokker–Planck flows in $L^1(\mathbb{R}^d)$. arXiv:2210.13624v2 [math.PR]
21. Barbu, V., Röckner, M.: Uniqueness for nonlinear Fokker–Planck equations for McKean–Vlasov SDEs: the degenerate case. J. Funct. Anal. **285**(4), 109980 (2023)
22. Barbu, V., Röckner, M., Russo, F.: Probabilistic representation for solutions of an irregular porous media equation. The degenerate case. Probab. Theory Relat. Fields **151**(1–2), 1–43 (2011)
23. Barbu, V., Röckner, M., Zhang, D.: Uniqueness of distributional solutions to the $2D$ vorticity Navier–Stokes equation and its associated nonlinear Markov process (2023). https://api.semanticscholar.org/CorpusID:262465383
24. Belaribi, N., Russo, F.: Uniqueness for Fokker–Planck equations with measurable coefficients and applications to the fast diffusion equations. Electron. J. Probab. **17**, 1–28 (2012)
25. Benilan, Ph., Brezis, H., Crandall, M.G.: A semilinear elliptic equation in $L^1(\mathbb{R}^N)$. Ann. Scuola Norm. Sup. Pisa Ser. IV **II**, 523–555 (1975)
26. Blanchard, Ph., Röckner, M., Russo, F.: Probabilistic representation for solutions of an irregular porous media equation. Ann. Probab. **38**, 1870–1900 (2010)
27. Bogachev, V.I., Krylov, N.V., Röckner, M., Shaposhnikov, S.V.: Fokker–Planck–Kolmogorov Equations. Mathematical Surveys and Monographs, vol. 207, xii+479 pp. American Mathematical Society, Providence (2015). ISBN:978-1-4704-2558-6
28. Bogachev, V.I., Röckner, M., Shaposhnikov, S.V.: Convergence in variation of solutions of nonlinear Fokker-Planck-Kolmogorov equations to stationary measures. J. Funct. Anal. **276**(12), 3681–3713 (2019)
29. Bogachev, V.I., Salakhov, D.I., Shaposhnikov, S.V.: The Fokker–Planck–Kolmogorov equation with nonlinear terms of local and nonlocal type. Algebra Anal. **35**, 17–38 (2023)
30. Brezis, H.: Functional Analysis Sobolev Spaces and Partial Differential Equations. Springer, Berlin (2011)
31. Brezis, H., Crandall, M.G.: Uniqueness of solutions of the initial-value problem for $u_t - \Delta\beta(u) = 0$. J. Math. Pures Appl. **58**, 153–163 (1979)
32. Brezis, H., Friedman, A.: Nonlinear parabolic equations involving measures as initial conditions. J. Math. Pures Appl. **IX**(2), 73–97 (1983)
33. Brezis, H., Pazy, A.: Convergence and approximation of semigroups of nonlinear operators in Banach spaces. J. Funct. Anal. **9**, 63–74 (1972)
34. Brezis, H., Strauss, W.: Semilinear elliptic equations in L^1. J. Math. Soc. Jpn. **25**, 565–590 (1973)
35. Calderon, A.P., Zygmund, A.: On singular integrals. Am. J. Math. **78**, 289–309 (1956)
36. Carmona, R., Delarue, F.: Probabilistic Theory of Mean Field Games with Applications, I–II. Springer, Berlin (2017)
37. Carillo, J.A.: Entropy solutions for nonlinear degenerate problems. Arch. Rat. Mech. Anal. **147**, 269–361 (1999)
38. Carillo, J.A., Toscani, G.: Asymptotic L^1-decay of solutions of the porous media equation to self-similarity. Indiana Univ. Math. J. **49**(1), 113–142 (2000)
39. Carillo, J.A., Jüngel, A., Markowich, P.A., Toscani, G., Unterreiter, A.: Entropy dissipation methods for degenerate parabolic problems and generalized Sobolev inequalities. Monatsh. Math. **133**, 1–82 (2001)

40. Chavanis, P.H.: Generalized stochastic Fokker-Planck equations. Entropy **17**, 3205–3252 (2015)
41. Chen, G.Q., Perthame, B.: Well posedness for nonisotropic degenerate parabolic hyperbolic equations. Ann. Inst. H. Poincaré **4**, 645–668 (2003)
42. Crandall, M.G.: The semigroup approach to first order quasilinear equations in several space variables. Israel J. Math. **10**, 108–132 (1972)
43. Crandall, M.G., Pazy, A.: A nonlinear evolution equation in Banach spaces. Israel J. Math. **11**, 1–35 (1972)
44. Dafermos, C., Slemrod, M.: Asymptotic behavior of nonlinear contractions semigroups. J. Funct. Anal. **13**, 97–100 (1973)
45. De Pablo, A., Quirós, F., Rodriguez, A., Vasquez, J.L.: A general fractional porous medium equations. Commun. Pure Appl. Math. **65**, 1242–1281 (2012)
46. Diaz, J.L., Galiano, G., Jürgel, A.: On a quasilinear degenerate system arising in semiconductor theory. Existence and uniqueness of solutions. Nonlinear Anal. Real World Appl. **2**, 305–331 (2001)
47. DiPerna, R.J., Lions, P.L.: Ordinary differential equations, transport theory and Sobolev spaces. Invent. Math. **98**, 511–547 (1089)
48. dos Reis, G., Smith, G., Tankov, P.: Importance sampling for McKean–Vlasov SDEs. Appl. Math. Comput. **453**, 128078 (2023)
49. Eberle, A., Guillin, A., Zimmer, R.: Quantitative Harris-type theorems for diffusions and McKean-Vlasov processes. Trans. Am. Math. Soc. **371**(10), 7135–7173 (2019)
50. Evans, I.C.: Entropy and Partial Differential Equations. Research Notes in Mathematics Series. University of California, Berkeley (1993)
51. Figalli, A.: Existence and uniqueness of martingale solutions for SDEs with rough or degenerate coefficients. J. Funct. Anal. **254**(1), 109–153 (2008)
52. Fokker, A.D.: Die mittlere Energie rotierender elektrischer Dipole im Strahlungsfeld. Ann. Phys. **348**(5), 810–820 (2006)
53. Frank, T.D.: Generalized Fokker-Planck equations derived from generalized linear nonequilibrium thermodynamics. Physica A **310**, 397–412 (2002)
54. Frank, T.D.: Nonlinear Fokker-Planck Equations. Fundamentals and Applications. Springer, Berlin (2005)
55. Frank, T.D., Daffertshofer, A.: *H* theorem for nonlinear Fokker-Planck equations related to generalized thermostatics. Physica A **295**, 455–474 (2001)
56. Friedman, A.: Variational Principles and Free-Boundary Problems. Wiley, New York (1982)
57. Funaki, T.: A certain class of diffusion processes associated with nonlinear parabolic equations. Z. Wahrsch. Verw. Gebiete **67**(3), 331–348 (1984)
58. Gess, B., Hofmanova, M.: Well-posedness and regularity for quasilinear degenerate parabolic-hyperbolic SPDE. Ann. Probab. **46**(5), 2495–2544 (2018)
59. Gess, B., Souganidis, P.E.: Stochastic nonisotropic degenerate parbolic-hyperbolic equations. Stoch. Process. Appl. **127**, 2961–3004 (2017)
60. Grube, S.: Strong solutions to McKean–Vlasov stochastic differential equations with coefficients of Nemytskii-type. Doktor–Degree thesis, Bielefeld, p. 123 (2022)
61. Gutman, S., Pazy, A.: An ergodic theorem for semigroups of contractions. Proc. Am. Math. Soc. **88**, 254–256 (1983)
62. Hammersley, W., Šiška, D., Szpruch, L.: McKean-Vlasov SDEs under measure dependent Lyapunov conditions. Ann. Inst. Henri Poincaré Probab. Stat. **57**(2) (2018). https://doi.org/10.1214/20-AIHP1106
63. Hao, Z., Röckner, M., Zhang, X.: Second order fractional meanfield SDEs with singular kernels and measure initial data. arXiv:2302.04392v2
64. Huang, X., Wang, F.-Y.: Distribution dependent SDEs with singular coefficients. Stoch. Process. Appl. **129**, 4747–4770 (2019)
65. Huang, X., Röckner, M., Wang, F.-Y.: Nonlinear Fokker-Planck equations for probability measures on path space and path-distribution dependent SDEs. Discrete Contin. Dyn. Syst. **39**, 3017–3035 (2019)

66. Jacod, J., Shiryaev, A.N.: Limit Theorems for Stochastic Processes. Springer, Berlin (1987)
67. Jordan, R., Kinderlehrer, D., Otto, F.: The variational formulation of the Fokker-Planck equation. SIAM J. Math. Anal. **29**, 1–17 (1998)
68. Karatzas, I.l, Shreve, E.S.: Brownian Motion and Stochastic Calculus. Graduate Texts in Mathematics, vol. 113, 2nd edn. Springer, New York (1991)
69. Kruzkov, S.: First order quasilinear equations with several independent variables. Sbornic: Mathematics **10**(2), 217–243 (1970)
70. Lasry, J.M., Lions, P.L.: Mean field games. Jpn. J. Math. **2**, 229–260 (2007)
71. Lions, P.L., Perthame, B., Tadmor, E.: A kinetic formulation of multidimensional scalar conservation laws and related equations. J. Am. Math. Soc. **7**, 169–191 (1994)
72. Manita, O.A., Shaposhnikov, S.V.: Nonlinear parabolic equations for measures. St. Petersburg Math. J. **25**(1), 43–62 (2014)
73. Manita, O.A., Romanov, M.S., Shaposhnikov, S.V.: On uniqueness of solutions to nonlinear Fokker-Planck-Kolmogorov equations. Nonlinear Anal. **128**, 199–226 (2015)
74. Markowich, P.A., Villani, C.: On the trend to equilibrium for the Fokker-Planck equations: an interplay between physics and functional analysis. Math. Contemp. **19**, 1–29 (2000)
75. McKean, Jr., H.P.: A class of Markov processes associated with nonlinear parabolic equations. Proc. Natl. Acad. Sci. U.S.A. **56**, 1907–1911 (1966)
76. McKean, Jr., H.P.: Propagation of chaos for a class of non-linear parabolic equations. In: Stochastic Differential Equations (Lecture Series in Differential Equations, Session 7, Catholic Univ., 1967), pp. 41–57. Air Force Office Sci. Res., Arlington (1967)
77. Mehri, S., Scheutzow, M., Stannat, W., Zangeneh, B.Z.: Propagation of chaos for stochastic spatially structured neuronal networks with fully path dependent delays and monotone coefficients driven by jump diffusion noise. Ann. Appl. Probab. **30**(1), 175–207 (2020). https://doi.org/10.1214/19-AAP1499
78. Mishura, Yu.S., Veretennikov, A.Yu.: Existence and uniqueness theorems for solutions of McKean–Vlasov stochastic equations. Theory Probab. Math. Stat. **103** (2016). https://doi.org/10.1090/tpms/1135
79. Nemytskii, V.V., Stepanov, V.V.: Qualitative Theory of Differential Equations (Russian). OGIZ, Moskow, Leningrad (1947)
80. Olivera, C., Richard, A., Tomasevic, M.: Quantitative particle approximation of nonlinear Fokker–Planck equations with singular kernel. arXiv:2011.00537
81. Otto, F.: The geometry of dissipative evolution equation: the porous medium equation. Commun. Partial Differ. Equ. **26**(1–2), 101–174 (2001)
82. Otto, F., Villani, C.: Generalization of an inequality by Talagrand and links with the logarithmic Sobolev inequality. J. Funct. Anal. **173**, 361–400 (2000)
83. Pazy, A.: The Lyapunov method for semigroups of nonlinear contractions in Banach spaces. J. Anal. Math. **40**, 239–262 (1981)
84. Pierre, M.: Uniqueness of the solutions of $u_t - \Delta\varphi(u) = 0$ with initial data measure. Nonlinear Anal. Theory Methods Appl. **6**(2), 175–187 (1982)
85. Planck, M.: Sitzungsber, Preuss, Akad. Wiessens, 324–326 (1917)
86. Porretta, A.: Weak solutions to Fokker–Planck equations and mean field games. Arch. Rat. Mech. Anal. **216**, 1–62 (2015)
87. Rehmeier, M.: Flow selections for (nonlinear) Fokker–Planck–Kolmogorov equations. J. Differ. Equ. **328**, 105–133 (2022)
88. Rehmeier, M., Röckner, M.: On nonlinear Markov processes in the sense of McKean. arXiv:2212.12424v2
89. Röckner, M., Xie, L., Zhang, X.: Superposition principle to non-local Fokker–Planck–Kolmogorov operators. Probab. Theory Relat. Fields **178**(3–4), 699–733 (2020)
90. Scheutzow, M.: Uniqueness and nonuniqueness of solutions of Vlasov–McKean equations. J. Aust. Math. Soc. **43**(2), 246–256 (1987)
91. Schilling, R.L., Song, R., Vondracek, Z.: Bernstein Functions, Theory and Applications, vol. 37. De Gruyter, Berlin (2012)

92. Schwämmle, V., Nobre, F.D., Curado, E.M.F.: Consequences of the H-theorem from nonlinear Fokker-Planck equations. Phys. Rev. E **76**, 041123 (2007)
93. Stein, E.: Singular Integrals and Differentiability Properties of Functions. Princeton Mathematical Series, vol. 30. Princeton University Press, Princeton (1970)
94. Stein, E., Weiss, G.: Introduction to Fourier Analysis and Euclidean Spaces. Princeton University Press, Princeton (1971)
95. Stroock D.W., Srinivasa Varadhan, S.R.: Multidimensional Diffusion Processes. Springer, Berlin (1997)
96. Sznitman, A.-S.: Nonlinear reflecting diffusion process, and the propagation of chaos and fluctuations associated. J. Funct. Anal. **566**, 311–336 (1984)
97. Temam, R.: Infinite Dimensional Dynamical System in Mechanics and Physics. Springer, New York (1988)
98. Trevisan, D.: Well-posedness of multidimensional diffusion processes with weakly differentiable coefficients. Electron. J. Probab. **21**(22), 1–41 (2016)
99. Vasquez, J.L.: The Porous Media Equation. Oxford University Press, Oxford (2006)
100. Veron, L.: Effets régularisants de semi-groupes non linéaires dans des espaces de Banach. Ann. Faculté Sci. Toulouse **1**(2), 171–200 (1979)
101. Wang, F.-Y.: Functional Inequalities, Markov Semigroups and Spectral Theory, xx+379 pp. Science Press, New York (2005). ISBN:7-03-014415-5
102. Wang, F.-Y.: Distribution dependent SDEs for Landau type equations. Stoch. Process. Appl. **128**(2), 595–621 (2018)
103. Yosida, K.: Functional Analysis. Springer, Berlin (1980)

Index

V. Barbu, M. Röckner, *Nonlinear Fokker-Planck Flows and their Probabilistic Counterparts*, Lecture Notes in Mathematics 2353,
https://doi.org/10.1007/978-3-031-61734-8

LECTURE NOTES IN MATHEMATICS 

Editorial Policy

1. Lecture Notes aim to report new developments in all areas of mathematics and their applications – quickly, informally and at a high level. Mathematical texts analysing new developments in modelling and numerical simulation are welcome.

   Manuscripts should be reasonably self-contained and rounded off. Thus they may, and often will, present not only results of the author but also related work by other people. They may be based on specialised lecture courses. Furthermore, the manuscripts should provide sufficient motivation, examples and applications. This clearly distinguishes Lecture Notes from journal articles or technical reports which normally are very concise. Articles intended for a journal but too long to be accepted by most journals, usually do not have this "lecture notes" character. For similar reasons it is unusual for doctoral theses to be accepted for the Lecture Notes series, though habilitation theses may be appropriate.

2. Besides monographs, multi-author manuscripts resulting from SUMMER SCHOOLS or similar INTENSIVE COURSES are welcome, provided their objective was held to present an active mathematical topic to an audience at the beginning or intermediate graduate level (a list of participants should be provided).

   The resulting manuscript should not be just a collection of course notes, but should require advance planning and coordination among the main lecturers. The subject matter should dictate the structure of the book. This structure should be motivated and explained in a scientific introduction, and the notation, references, index and formulation of results should be, if possible, unified by the editors. Each contribution should have an abstract and an introduction referring to the other contributions. In other words, more preparatory work must go into a multi-authored volume than simply assembling a disparate collection of papers, communicated at the event.

3. Manuscripts should be submitted either online at www.editorialmanager.com/lnm to Springer's mathematics editorial in Heidelberg, or electronically to one of the series editors. Authors should be aware that incomplete or insufficiently close-to-final manuscripts almost always result in longer refereeing times and nevertheless unclear referees' recommendations, making further refereeing of a final draft necessary. The strict minimum amount of material that will be considered should include a detailed outline describing the planned contents of each chapter, a bibliography and several sample chapters. Parallel submission of a manuscript to another publisher while under consideration for LNM is not acceptable and can lead to rejection.

4. In general, **monographs** will be sent out to at least 2 external referees for evaluation.

   A final decision to publish can be made only on the basis of the complete manuscript, however a refereeing process leading to a preliminary decision can be based on a pre-final or incomplete manuscript.

   Volume Editors of **multi-author works** are expected to arrange for the refereeing, to the usual scientific standards, of the individual contributions. If the resulting reports can be forwarded to the LNM Editorial Board, this is very helpful. If no reports are forwarded or if other questions remain unclear in respect of homogeneity etc, the series editors may wish to consult external referees for an overall evaluation of the volume.

5. Manuscripts should in general be submitted in English. Final manuscripts should contain at least 100 pages of mathematical text and should always include

   – a table of contents;
   – an informative introduction, with adequate motivation and perhaps some historical remarks: it should be accessible to a reader not intimately familiar with the topic treated;
   – a subject index: as a rule this is genuinely helpful for the reader.
   – For evaluation purposes, manuscripts should be submitted as pdf files.

6. Careful preparation of the manuscripts will help keep production time short besides ensuring satisfactory appearance of the finished book in print and online. After acceptance of the manuscript authors will be asked to prepare the final LaTeX source files (see LaTeX templates online: https://www.springer.com/gb/authors-editors/book-authors-editors/manuscriptpreparation/5636) plus the corresponding pdf- or zipped ps-file. The LaTeX source files are essential for producing the full-text online version of the book, see http://link.springer.com/bookseries/304 for the existing online volumes of LNM). The technical production of a Lecture Notes volume takes approximately 12 weeks. Additional instructions, if necessary, are available on request from lnm@springer.com.

7. Authors receive a total of 30 free copies of their volume and free access to their book on SpringerLink, but no royalties. They are entitled to a discount of 33.3 % on the price of Springer books purchased for their personal use, if ordering directly from Springer.

8. Commitment to publish is made by a *Publishing Agreement*; contributing authors of multiauthor books are requested to sign a *Consent to Publish form*. Springer-Verlag registers the copyright for each volume. Authors are free to reuse material contained in their LNM volumes in later publications: a brief written (or e-mail) request for formal permission is sufficient.

Addresses:
Professor Jean-Michel Morel, CMLA, École Normale Supérieure de Cachan, France
E-mail: moreljeanmichel@gmail.com

Professor Bernard Teissier, Equipe Géométrie et Dynamique,
Institut de Mathématiques de Jussieu – Paris Rive Gauche, Paris, France
E-mail: bernard.teissier@imj-prg.fr

Springer: Ute McCrory, Mathematics, Heidelberg, Germany,
E-mail: lnm@springer.com

Printed in the United States
by Baker & Taylor Publisher Services

Printed in the United States
by Baker & Taylor Publisher Services